Rocks and Relief

By the same author

Geomorphology in the Geographies for Advanced Study Series

Rocks and Relief

B. W. Sparks

University Lecturer in Geography
Fellow and Senior Tutor, Jesus College, Cambridge

Longman

LONGMAN GROUP LIMITED
London

Associated companies, branches and representatives
throughout the world

© Longman Group Ltd 1971

First published 1971
Second impression (and first appearance in paperback) 1972

ISBN 0 582 48140 6   cased
ISBN 0 582 48162 7   paper

Set in Monophoto Baskerville
and printed in Great Britain by William Clowes and Sons, Limited
London, Beccles and Colchester

76 004447

Contents

Contents

6 Dating and correlating rocks   259

7 The Lower Palaeozoic systems   267

Contents

List of text figures

List of plates

Acknowledgements

We are indebted to the following for permission to base diagrams on existing material:

F. H. Hatch, A. K. Wells and M. K. Wells, 1949, *Petrology of the Igneous rocks*, 10th edn., London (Murby), Figs. 109 and 114; *Geological Survey* 1″ sheet No. 270; I. J. Smalley, 1966, *Geological Magazine*, vol. 103, Fig. 2; V. E. Shainin, 1950, *Bulletin Geological Society America*, vol. 61, Figs. 1 and 8; L. U. de Sitter, 1964, *Structural geology*, 2nd edn., New York (McGraw-Hill), Fig. 56; D. L. Linton, 1955, *Geographical Journal*, vol. 121, Figs. 2a, 2b and 2c; G. K. Gilbert, 1880, Report on the geology of the Henry Mts., 2nd edn., *U.S. Dept. of the Interior, Washington*, Fig. 18; J. E. Richey *et al.*, 1930, The geology of Ardnamurchan . . ., *Geological Survey (H.M.S.O.)*, Fig. 4 and Plate 11; J. E. Richey, 1964, Scotland: the Tertiary volcanic districts, 3rd edn., Fig. 50. *Geological Survey (H.M.S.O.)*; W. D. West and V. D. Choubey, 1964, *Journal Geological Society India*, vol. 5, Fig. 5; B. Smith and T. N. George, 1961, North Wales, 3rd edn., *Geological Survey (H.M.S.O.)*, Fig. 13; Emm. de Martonne, 1947, *Géographie Universelle, vol. 6(i), France physique*, Paris (Armand Colin), Fig. 53c; F. H. Hatch and R. H. Rastall, 1965, *Petrology of the sedimentary rocks*, 4th edn., London (Murby), Figs. 18, 13 and 15; F. Hjulström, 1935, *Bulletin Geological Institute Uppsala*, vol. 25; F. Trombe, 1952, *Traité de spéléologie*, Paris, Fig. 58; A. Bögli, 1960, *Zeitschrift für Geomorphologie*, Supplementband 2, Figs. 1, 2, 3, 4 and 5; P. W. Williams, 1966, *Institute of British Geographers, Transactions and Papers*, No. 40, Figs. 3 and 8; T. M. Thomas, 1954, *Geographical Journal*, vol. 120, Figs. 1 and 5; M. M. Sweeting, 1968, *Proceedings of the 4th International Congress of Speleology in Yugoslavia*, Ljubljana, vol. III, Figs. 2a and 2b; A. Bögli, 1964, *Erdkunde*, vol. 18, diagram 2; A. K. Wells and J. F. Kirkaldy, 1966, *An outline of historical geology*, 6th edn., London (Murby), Figs. 19 and 35 and L. J. Wills, 1951, *Palaeogeographical Atlas*, London (Blackie), Plate 15.

R. Coe for Plates 4(a) and 4(b), R. J. Sparks for Plates 7 and 14. The remaining photographs were taken by the author.

Preface

I believe firmly that rocks are fundamentally important in relief. So have most geomorphologists with field experience. Yet the case has often been overstated. Process can dominate rock. A glaciated granite resembles a glaciated sandstone more than it resembles an unglaciated granite. Yet, where the present and past processes are and have been roughly comparable, the rocks show in varying degrees in the relief. Even here the case must not be overstated. I could not look at all British landscapes and guess what the rock is. Even less could I guess the rock type from topographical maps, as many examination candidates have been invited to do in the past.

The study of rock resistance and its effects on relief has two stages, one easy and general and the other detailed and extremely laborious. It is with the first of these—it might almost be called coarse geomorphology—that this book is primarily concerned. Because of the attachment of landform study to geography in Britain, the geologist has tended to ignore lithology and relief and it is not easy to extract information on this subject from the geological literature. Rocks affect relief in two ways. Through their intrinsic properties they govern denudation: hence the first five chapters of this book, which might be described as applied petrography. In their depositional arrangements, or stratigraphy, they affect regional relief variations: hence the rest of the book, which illustrates the theme from Britain and might be described, if not as stratigraphy without tears, at least as stratigraphy without fossils but with landforms.

In dealing with any topic which bridges two of the conventional divisions of knowledge it is always difficult to know how far to delve into the subject from which knowledge is being extracted as the basis for the other subject. I find it completely unsatisfactory, and indeed academically dangerous, to have no appreciation of the validity of the source of the material. On the other hand, as someone once said, if one seeks explanations of explanations in endless progression, one probably ends in either nuclear physics or the Old Testament. Hence, I have included what, in my opinion, is a sufficient amount of geology for a proper appreciation of the origins and properties of rocks, the effects of which in the relief are the prime objects of study. Beyond this the infinite variety of geological structures must also affect the relief in varying degrees, but their proper study would extend this book far beyond its intended scope.

The reader is urged to get as much field experience as possible, and, if practicable, experience of handling rocks and minerals in a laboratory as hand specimens and, ideally, as thin sections studied under a microscope as well.

I hope that this book will help those in the final stages of school and the earlier years of university education, who wish to understand a little more about the dominance of rocks in relief. I say a little more advisedly, because our knowledge is limited, and future advances beyond the stage of coarse geomorphology must reside in the proliferation of detailed experimental, statistical and field studies of the type outlined in the last chapter.

My thanks are due to Mr Michael Young for most of the line diagrams included here; to Mr Robert Coe for the photographs used in Plate 4; and to my son, Mr Robert Sparks, for the photographs used in Plates 7 and 14. The rest of the photographs are my own.

1
Introduction

Although few would care to deny the essential truth of Davis's dictum that landscape was a function of structure, process and time, the proper balance between the three elements has rarely been correctly kept either by individual geomorphologists or by schools of geomorphological thought. Even Davis is said by his critics to have treated the origin of landforms in a partisan way: in general it might be said that he exaggerated the importance of time, especially in his concept of stages in a cycle of erosion, simplified structure, although this was only for the readier comprehension of the essence of his cycle scheme, and to a considerable extent ignored process. Typical of the last charge levelled against Davis are the words of Kirk Bryan (1940, p. 254) concerning Davis's treatment of slopes: 'Slightly bemused by long, though mild intoxication on the limpid prose of Davis's remarkable essays, he [the reader] wakes with a gasp to realize that in considering the important question of slope he has always substituted words for knowledge, phrases for critical observation.'

Although it may be true that Davis neglected processes, the deficit has been made up since by the great amount of research essentially directed towards the evaluation of processes. It seems, indeed, that, if any major aspect of geomorphology may be said to be in temporary neglect at the present time, it is structure which is receiving too little attention. There are reasons for the emphasis on the importance of process and to a lesser extent on stage. They are particularly apparent in the case of process.

The fundamental importance of processes cannot be denied. To geomorphologists brought up in or near the glaciated highlands of Europe and North America the overwhelming dominance of process seems so clear. Feature after feature of glaciated highlands is obviously to be attributed to the processes of glaciation: the corries, the hanging valleys, the glaciated troughs with over-deepened basins, the truncated spurs and the roches moutonnées, all these and many others point to the dominance of glacial processes of erosion even though our knowledge of such processes may not be as precise as might be wished. On a broader and even more elementary scale the differences between the landforms caused by marine erosion, river erosion and wind abrasion are clear and well known.

Secondly, research into the nature of geomorphological processes represents a seeking after natural laws, which with few or no exceptions are applicable everywhere. It is mentally satisfying to work through detailed example after detailed example to emerge at the end with a short though universally applicable generalisation. Such natural laws are unlikely to be revealed by detailed studies of the effects of structure. It is true that some generalisations may be made, but few rocks and few structures are precisely similar, so that the study is to a considerable extent a study of unique examples.

Thirdly, the study of some geomorphological processes has attracted investigators from other sciences, often with great profit to geomorphology. This is especially true of river and marine processes. Problems of irrigation, of reservoirs for irrigation and power generation, and of the provision of coastal defence measures have all involved research and thus added to our knowledge of the physical processes involved in geomorphological work.

Allied with the research into physical processes has been the experimental work undertaken both in indoor and outdoor laboratories. Experimental work has been done, for example, by the Mississippi River Board in the United States and at Wallingford in this country: some of these experiments have been on a large scale and have necessitated the provision of large outdoor models. Smaller scale experiments have also been undertaken in laboratories, for example by Bagnold on the transport of sand and by Lewis on river processes.

Lastly, the emphasis on process may probably be referred to the demands of teaching. Whether this is being done as class instruction or by means of a text-book, it is obviously more gratifying to teach and more pleasurable to learn laws of general applicability than a series of unique examples, such as tend to arise from a study of the effects of rocks.

Closely connected with the study of process is the emphasis placed by some geomorphologists today on the study of the effects of climate. Climate is generally held to govern the nature of the processes operating, so that this really amounts to another aspect of the study of process. Landscapes in arid and semi-arid regions are usually thought to possess features which distinguish them from landscapes developed under more humid conditions, though there are some who would regard the major landforming processes as independent of climate (e.g. King 1953) and seek to apply, for example, arid and semi-arid processes to humid landscapes. The emphasis on the climatic control of process is not confined to the study of such variation at the present time over the earth's surface, but is also extended to a study of the various processes which have operated in the Quaternary at given points on the earth. The recognition at an early date of the role of past glacial processes in the present relief is an example of this. At the present time the emphasis being placed by many Continental geomorphologists, notably in France, on periglacial processes in the past to explain some of the features of the present relief of Europe is yet another example of pre-occupation with the climatic control of process.

The importance of the stage of development reached in the landscape has probably been less emphasised than the importance of process. In the Davisian cycle of erosion youth is a period of developing relief during which lithological differences are etched out until they reach their maximum effect in maturity. Old age is a stage of the increasing dominance of stage over structure, so that structure is only really regarded as being of very great significance in the middle stages of the cycle of erosion. There is one very important exception to this, namely in the development of the drainage pattern, where the adoption by the streams of the major lines of geological weakness proceeds slowly and may not be complete even in the old age stage of the cycle. A second cycle, initiated by rejuvenation, is necessary for the full exploitation of the structure by the drainage.

Davis himself probably stressed the stage of the cycle, because of his concern with the overall view of the landscape. The details, which might be more readily attributed to either structure or process, concerned him less than the total landscape, which is more readily described in terms of stage.

The emphasis of stage in the landscape, though in rather a different fashion, is characteristic of many geomorphologists at the present day. They regard the landscape as a succession of remnants of end-stages of cycles of erosion, whether they be humid, arid or marine cycles. Typical of this view is King's approach to the development of the old stable blocks, which are presumed once to have formed Gondwanaland and which have drifted apart to form the plateaus of South America, Africa, peninsular India and Australia. These are viewed as a succession of pediplains formed under semi-arid conditions and still present as a series of dissected plateaus in the landscape.

Comparable with this is Brown's recent treatment of Welsh relief as being fundamentally due to a series of incomplete cycles of subaerial and marine planation. The higher and more dissected plateaus of Wales are regarded as the remnants of peneplains, while most of the low coastal plateaus are attributed to marine erosion.

In a sense the view that stage is dominant in the landscape is a geomorphologist's point of view rather than a geographer's point of view. The former, concerned in these examples with the evolution of landscape, puts himself in such a position that his eye can glance from one undissected ridge to another and ignore the valleys between. Sometimes the view is so broad that it can only be obtained from the study of maps, for no viewpoint in the field offers the necessary extensive panorama. From such viewpoints, whether in the field or on the map, the deliberate suppression of the dissection of the landscape leads to the emphasis of stage as the dominant factor in the landscape. But, in the same landscape, the observer in the valley probably sees little of stage and very much more of structure.

3

Another reason for the comparative neglect of rocks in the landscape may be found in the fact that landforms are mainly studied by geographers, at least in Great Britain. Rocks, after all, are the occupation of the geologist, so that it is perhaps not surprising that process and stage have received more emphasis. Further, the geologist is interested primarily in features of rocks other than their effects on the landscape. Yet, in the United States, where geomorphology is certainly carried out by the geology departments and not by the geography departments, the emphasis at present is largely on process, especially on attempts to understand process from landscape patterns treated statistically.

It is, of course, impossible to answer the question whether structure is more important than process. All that can be done is to try to ensure that each is given due weight. In a sense process may be regarded as dominant, because it is active, whereas structure in all its aspects is passive. Yet, even in regions where process appears to be utterly dominant, the significance of structure may be greater than is at first realised.

Turning back to the question of glaciated highlands, which were cited above as the sphere of predominating process, it can be seen that the effects of rocks may be considerably greater than had at first been realised. Corries and many other glacial features are best preserved in Britain in rocks which might be termed hard and splintery, for example the metamorphic rocks of the highlands of Scotland and the volcanic rocks of the Snowdon region of North Wales. Comparable corrie forms are far less frequent in the Southern Uplands of Scotland and in the monotonous sedimentary rocks of central Wales. This may be due either to corries never having formed in these regions to the same extent or to the poorer preservation of corries in these regions: both possibilities involve the significance of rocks, either as governing the process or as preserving the landform. Almost all known corries are in hard rocks and the appearance of corries in less resistant rocks is largely a matter of speculation.

Again, the Lewisian Gneiss of the far north-west coast of Scotland and of parts of the Outer Hebrides seems to possess the imprint of very heavy glaciation in its low-lying landscape of hummocky knolls, bare rock, small lakes and indeterminate drainage (Plate 1). Yet in the same area other rocks behave differently, although they must have been subjected to approximately the same intensity of glaciation as the Lewisian Gneiss. The Torridonian rocks usually stand above the Lewisian as bold hills, but, even where they are found at a low elevation, they have none of the hummocky, Laurentian Shield appearance of the Lewisian. Generally too the relief on the Moine metamorphic rocks is different, though something of the hummocky effect of Lewisian relief is achieved on the Moine rocks at the southern end of Loch Shiel and around Arisaig.

PLATE 1. Torridonian Sandstone, Stac Polly, Ross and Cromarty. Weathered, jointed Torridonian Sandstone overlooking glaciated Lewisian Gneiss. Torridonian mountains, Suilven and Quinag, in the background

The overwhelming importance of lithology may be inferred also from a paper once written to suggest that the South Downs were glaciated (Martin 1920). The author saw many of the normal features of South Downs relief as the local equivalents of glacial landforms. Thus, the dry valleys, which are certainly U-shaped, were regarded as glacial troughs, the whale-backed downs were treated as roches moutonnées, the Coombe Rock south of the Downs was regarded as moraine and the sarsen stones, conglomerate relics of Tertiary beds, were treated as erratics. The author concluded that, as the general form of the South Downs did not differ from that of chalk regions known to have been glaciated, it might be inferred that the South Downs too were glaciated. It is known, of course, that the Pleistocene ice sheets did not approach within fifty miles of the South Downs, so that the general thesis of their glaciation cannot be maintained, but the similarity of glaciated and unglaciated chalk relief clearly indicates the overwhelming importance of lithology rather than process in the relief.

In fact, it is always difficult to exclude the possibility of lithological effects, because many lithological differences are only exposed by the operation of the processes of erosion. To detect significant lithological differences in advance is extraordinarily difficult and often the presence of such differences may only be inferred from the way in which the processes of erosion have etched them out. Faced with, for example, inselbergs on a pediplain, an investigator might conclude that they must represent the triumph of process over structure because he was unable to detect any lithological difference responsible for them, or he might conclude that the delicate processes of erosion had discovered a lithological difference, such as a lower frequency of joints, which he was unable to detect. Much would depend on the preconceptions of the investigator.

Although process might be considered to be more important than rocks in the immediate relief, it might be argued that the effects of lithology permeate the landscape more deeply than those of process. Thus, the type of rock not only affects the landform, but also has a large influence on the type of soil and consequently on the natural vegetation and, at a later stage, on the agriculture. When one is dealing with a landscape it is always difficult to disentangle in one's mind the form from the vegetation that clothes that form. Chalk landscapes are sometimes said to be unique in their convexo–concave slope forms, but much of this impression may be attributed to the bareness of many chalk landscapes, for chalk lands, which are forested for some reason or another, appear far less distinct, though the form beneath the vegetation is probably the same. In more primitive societies than ours the nature of the rocks has a pronounced effect on the regional building materials, so that the effects of structure on the landscape are far-reaching.

Both process and time may have comparable effects but they are probably not as profound as that of structure. For example, a process such as glaciation may scrape the soil and weathered rock away from one region and deposit it in another, thus affecting both natural vegetation and land use. Stage exerts an influence, too, especially in the development of great thicknesses of weathered material and very mature soil profiles on peneplains, pediplains and comparable features. However, the time taken for a soil to reach a full development is many times less than that required for a land surface to be peneplained or pediplained, so that all such surfaces, except perhaps the very youngest, may be expected to show similar soils. The association of soil types and vegetation with geomorphological surfaces has been observed in many different regions. Wooldridge and Linton, for example, stress the association of true clay-with-flints with their mid-Tertiary peneplain in south-eastern England, while Beard, in an exhaustive study of American savannas, reached the conclusion that the impeded drainage developed on tropical pediplains was largely responsible for the distribution of American savannas.

Whether one decides that rocks are the most important elements in the formation of the physical landscape or not, their very great importance cannot be denied. For this reason anyone studying landscape must be prepared to consider fully the possible effects of structure: to do this one needs to appreciate certain geological aspects as fully as possible. Structure affects the relief in three main ways:

1. Through the lithology of the rocks. The processes of weathering and erosion are governed in their attack on rocks by the chemical composition of the rocks, by the ways in which the individual particles forming the rocks are bonded together, and by such structural features as joints, bedding planes and cleavage.

2. Through the succession or order of the rocks. A succession of rocks of equal resistance to denudation will result in a uniformity of relief but alternations of resistant and non-resistant beds will form a varied relief. Thus, the juxtaposition of rocks is very important in relief study. At the same time the changes in lithology that occur from area to area in the same succession of rocks needs to be considered very carefully in appreciating relief.

3. Through the structural arrangements of rock successions, i.e. through structure in its narrower and more modern sense as contrasted with the wide meaning given to the word by Davis. In the simplest example it will obviously make a tremendous difference in relief if a 1000-ft thick bed of rock is vertical or horizontal: a comparison of the chalk relief of the Isle of Wight with that of Salisbury Plain affords a convincing example. Varied dips plus folding will obviously introduce a greater complexity into the relief pattern. Structures of extreme complexity, because of the tremendous amount of alternation and repetition of beds, may lead, however, to essentially uniform relief, for example, parts of the highlands of Scotland.

These three aspects of the study of rocks are covered by the geological subjects of petrology, or the study of rocks as rocks, stratigraphy, the study of the succession of rocks, and tectonics, or structural geology. But these sciences have more far-reaching aims than the study of the effects of rocks on relief, for they strive to understand fully all aspects of their particular fields of research. The attempt in the following chapters to select the information which has a bearing on relief development is an attempt to provide a geological background for geomorphology, especially for those without geological knowledge. Most of the illustrations will be drawn from the British Isles, because this is the area with which most people are familiar. The British Isles provide a wealth of examples, though few of them are as spectacular as, for example, the structural features of the western United States or the glacial features of Switzerland, a fact which may be considered a disadvantage by some but an advantage by others.

2

Rock characteristics and denudation

The resistance of rocks to the processes of weathering and erosion, which together constitute denudation, depends on a number of characteristics possessed by the rocks themselves, on the nature of the processes of denudation and on the time during which those processes have acted. It will be assumed that the reader is familiar with the main processes of weathering and erosion, accounts of which are readily available (e.g. Thornbury 1969, Dury 1959, Wooldridge and Morgan 1959 or Sparks 1960), and also with the effects of the operation of these processes through varying lengths of time. Attention will, therefore, be concentrated on those characteristics of rocks which appear to be vital in a consideration of denudation. They are:

(a) The mineral composition of the rock.
(b) The texture of the rock.
(c) The strength of the rock.
(d) The minor structures of the rock, such as joints, cleavage, schistosity and bedding planes.

These various properties, which will be considered and defined more fully below, affect different aspects of the process of denudation. The mineral composition of the rock is of fundamental importance in controlling the action of chemical weathering and may be of some relevance in a consideration of mechanical weathering. The texture of the rocks has a bearing on both the main categories of weathering, and, combined with the mineral composition, exerts a great influence on the strength of the rock. The strength of a rock has probably little to do with weathering, but has a considerable effect on resistance to mechanical abrasion and battering of all types, in fact upon the more violent manifestations of physical erosion. The minor structures are of extreme importance in practically every aspect of weathering and erosion.

The effects of mineral composition

The mineral composition of a rock may have possible effects on both mechanical and chemical weathering, though considerable doubt has been cast on the efficacy of the former, largely because laboratory experiments

9

have seemed to point to the ineffectiveness of mechanical weathering alone. Under natural conditions the two main types of weathering must act together, so that a discussion of the two as separate processes is largely artificial.

Mineral composition and mechanical weathering

The basis for the suggestion that alternate heating and cooling will set up strains within the rock is that different minerals have different coefficients of expansion. Even without different coefficients of expansion the different colours and degrees of surface polish of different minerals should ensure that each mineral reflected and absorbed different proportions of any radiation received at the surface. In addition, the same mineral has different rates of expansion in different directions within its crystal structure. These facts, coupled with the common occurrence of granular debris approximating in size to the crystals in the rock, have provided the support for the hypothesis of mechanical disintegration of rocks on a small scale. Shattering on a larger scale is a different question and will be discussed below.

Yet every laboratory experiment seems to suggest the inability of temperature change alone to shatter rocks as a result of different expansions of different minerals. A well known experiment was performed by Griggs (1936), who subjected the polished face of a block of coarse-grained granite to a series of violent temperature changes. By means of an electric heater and a stream of cool, dry air he produced the equivalent of a diurnal temperature range of 110 degrees C acting over a period of 244 years. This temperature fluctuation, far greater than anything that could be expected regularly in any hot desert, should have set up strains due to differential expansion of different minerals, as well as strains due to the steep temperature gradient in the uppermost layers of the rock, which might have been expected to produce the spalling-off of layers parallel to the surface. But neither the one nor the other took place.

When, however, the cooling was effected by a spray of tap water for the equivalent of a mere 2½ years, microscope examination showed both widening of cracks within crystals and a tendency for the rock to spall off parallel to its surface. Although mechanical disintegration caused by the different expansions of different minerals is, thus, seen to be unlikely to happen without important chemical action, differential expansion undoubtedly facilitates chemical disintegration, a point considered more fully below.

Another possible cause of enormous temperature differences is the natural fire, whether in forest or in bush. Of course, for such a fire to have a full effect, rock must be near the surface or otherwise the effects will be minimised by the insulating effect of the soil and waste mantle. Such fires are not everyday occurrences, like large temperature changes in deserts, but

Blackwelder (1926) compared the effects of one 'good' forest fire with the effects of sun and frost over a thousand years. The effects of such fires seem to be the formation of shells parallel to the surface of the rocks rather than the production of granular debris: such spalling is not fundamentally due to the mineral composition.

The laboratory experiments seem to indicate that, with sound rock, the expansions and contractions due to heating and cooling are insufficient to exceed the strength of the rock and to cause rupture either within or between the individual mineral constituents. It may be, however, that there is one important way in which the laboratory experiments do not precisely simulate natural conditions. A loose block taken into the laboratory presumably has all the stresses caused by the pressure of the environment in which it has been formed already relieved (the effects of these stresses and their relief upon the jointing of rocks are discussed below). In this way it may not be exactly comparable with an unfractured natural surface of rock, where considerable stresses may still be acting in the plane of the surface. If such a condition is possible, the addition of a small extra stress due to the expansion and contraction of different minerals may cause the strength of the rock to be exceeded with consequent fracture possibly along the interfaces between mineral grains. This is virtually the application of the theory that it is the last straw that breaks the camel's back. However, the possibility of such conditions is probably remote, for the chances are that the stresses are already relieved in rocks before they are brought quite to the surface by denudation. At the surface the majority of rocks are characterised by open joints, so that the individual joint blocks are as likely to have had their stresses relieved as the block of rock used in the laboratory.

Mineral composition and chemical weathering

When one turns to the effects of mineral composition on chemical weathering, there seems to be no doubt whatever of its extreme importance. The basis of chemical disintegration is usually the attack of chemical agents at different rates on different constituents of the rock. Almost all rocks, whether they are igneous, metamorphic or sedimentary, are composed of more than one mineral so that in most rocks differential chemical weathering of the minerals takes place.

The common rock-forming minerals are subject to weathering at different rates. In a study of the weathering of igneous rocks Goldich (1938) found it possible to arrange their common minerals in order of resistance to chemical weathering. This order is reproduced below, the two columns containing dark minerals to the left and light minerals to the right.

On the whole the dark, ferromagnesian minerals are more susceptible to chemical attack than the light-coloured minerals, which are predominantly felspars, the exceptions being the white mica, muscovite, and quartz. The

MOST SUSCEPTIBLE	Olivine ———————————————————
	————————————— Lime plagioclase
	Augite ———————————————————
	————————————— Lime–soda plagioclase
	Hornblende ————————— Soda–lime plagioclase
	————————————— Soda plagioclase
	Biotite ———————————————————
	————————————— Orthoclase
	————————————— Muscovite
LEAST SUSCEPTIBLE	————————————— Quartz

effects of this order on the weathering of igneous rocks will be considered more fully in the next chapter. At the moment it is sufficient to note that minerals from the top of the table on the left are usually combined in rocks with minerals from the top of the table on the right. Combinations involving most susceptible minerals from one side and least susceptible from the other are very rare. This is significant because, if such combinations were available, all rocks would seem to be of approximately equal resistance, for their resistance would be that of their most susceptible mineral. In nature, the rocks, consisting mainly of least or most susceptible combinations of minerals, vary widely in resistance to chemical weathering.

The less common constituents of igneous rocks, the accessory minerals, also differ greatly in their resistance to chemical attack. Goldich considered this question by comparing the relative percentages of the various heavy minerals in both the original rock and in the clay derived from it as a weathering product. In the weathering of a granite–gneiss, the Morton Gneiss of Minnesota, heavy minerals were rarer in the residual clay than in the sound gneiss, and their relative proportions had changed. Hornblende, epidote, titanite and apatite were the least stable and were on a par with plagioclase felspar in their resistance, i.e. in the range from lime to soda plagioclase in the preceding table. Ilmenite and magnetite were more stable and on a par with biotite and orthoclase felspar in their resistance to chemical weathering, while zircon seemed to be as resistant as quartz. Comparable analyses of diabase, both from Medford in Massachusetts and from Lake Superior, and of amphibolite from the Black Hills of South Dakota suggested that apatite was a fairly resistant accessory mineral, while the garnet in the last rock was very resistant.

It can be seen from this that there is a considerable variation in the resistance of the accessory minerals in crystalline rocks to chemical weathering. No attempt can be made however to put the whole range of accessory minerals into an order of resistance to chemical attack. Details of the various minerals mentioned in these last two paragraphs may be found in Rutley's *Elements of Mineralogy* (Read 1962).

Metamorphic rocks consist largely of the same minerals as igneous rocks, so that the order of resistance to weathering is also useful in discussing the denudation of metamorphic rocks. Sedimentary rocks may be somewhat different. Many of the constituents of sedimentary rocks (see Chapter 5) are those parts of igneous rocks which resist weathering, notably the quartz and the muscovite, together with the products of previous chemical weathering, essentially the clays. These elements of the rocks cannot be expected to be susceptible to further chemical weathering, but they are often bound together by cements deposited by percolating solutions. Such cements may be liable to chemical attack, especially calcium carbonate. Similar calcium carbonate rocks, limestones, which are formed by the precipitation of the mineral from solution, either by chemical means or through the action of invertebrate organisms in building their shells of calcium carbonate, are more susceptible to chemical attack than any other rocks. They will be discussed more fully in Chapter 5.

By far the most important processes of chemical weathering operating on these rock-forming minerals, which are nearly all silicates, are hydrolysis and hydration. Hydrolysis is the reaction between water, which dissociates into H^+ and OH^- ions, and the elements in minerals. By a complex series of chemical changes the silicates are converted to clay minerals, silica and hydroxides, the last being often removed in solution as carbonates after further reaction with weak natural solutions of carbonic acid. These chemical changes are possible in the presence of pure water, as has been shown by the fact that the agitation of finely ground felspar in water ultimately produces a milkiness, due to the presence of clay minerals. The changes are greatly accelerated by the presence of greater concentrations of H^+ ions, i.e. in the presence of acids, several of which may occur in nature.

The one usually spoken of is carbonic acid, formed by the solution of carbon dioxide in rain water. The sources of carbon dioxide include the atmosphere and the soil atmosphere, and the strength of the acid depends partly on the concentration of the gas and partly on the ambient temperature.

The normal concentration of carbon dioxide in the atmosphere is about 0·03 per cent, but this is increased many times in the soil atmosphere, where large supplies are made available by the metabolism of plants. In such situations concentrations are usually of the order of 1 per cent and may in extreme conditions reach 25 per cent. Thus the presence of soil carbon dioxide is probably of greater importance than atmospheric carbon dioxide.

Like all gases carbon dioxide is more soluble at low temperatures than at high temperatures, its solubility at 20 degrees C being one half that at 0 degrees C. This fact has been used to support the possibility of chemical weathering at low temperatures, for example by Tamm (1924) and Williams (1949). It should not be injudiciously concluded that, because of the greater solubility of carbon dioxide, chemical weathering is more effective at low temperatures than at high (see Chapter 5). Apart from the possibility of other weathering agents, it is probably true to say that the available carbon dioxide in glacial and periglacial climates is very much less than that in warmer climates, for the vegetation and soil are the main sources of the gas and these are favoured by warm conditions.

In addition to carbonic acid, it has been suggested that certain much stronger acids assist in the chemical decay of rocks. Nitric acid is produced during the discharge of lightning and it is possible that, in those areas which have a high incidence of thunderstorms, the natural acidity of rain might be slightly augmented. It is also possible to have the natural formation of sulphuric acid by the oxidation of pyrite and marcasite, which are the naturally occurring minerals composed of iron sulphide. Although such minerals occur in igneous rocks, they are more common in certain sediments, notably clays and shales. Another source of sulphuric acid is in the oxidation of sulphur dioxide, a waste product of industrial combustion. The effect of this second source is to increase the rate of decay of building stones in industrial cities. Presumably, the last century saw the peak of such weathering, for it should be greatly reduced by the increase in the use of smokeless fuels and electricity.

A second possibility is chemical weathering caused by acid clays. The clay fractions of very acid soils are capable of ion exchange with rock-forming minerals, as was shown in a series of experiments by Graham (1941a). Graham prepared a suspension of acid clay from a very acid American soil, the Putnam silt loam. It had a pH value of 3·3. The pH value is a logarithmic index of acidity, with neutrality equal to 7 and lower indices indicating acidity. A value of 3·3 is the order of acidity one would expect on a very acid podsol soil developed in north-western Europe. Into this acid clay Graham inserted samples of various rock-forming minerals. These had been ground down until the particles were silt-sized, so that a very large surface area was available for chemical reactions to take place. The degree of attack, as measured by the rise in the pH value, i.e. by the decrease in the acidity of the acid clay suspension, varied with the mineral and confirmed the order of susceptibility indicated on the table on page 12. The most rapid attack was that on anorthite felspar, which is lime plagioclase in the terminology of that table. In 107 days 3·4 per cent of the calcium had been leached from the felspar in spite of the progressive neutralisation of the acid which was taking place due to the conditions of the experiment. In nature there would be a continuous supply of acid so that a constant rate of attack might be maintained. A felspar very near to orthoclase in its composition was attacked at about one tenth the rate of the attack on anorthite. These are very rapid rates of chemical weathering and clearly indicate the possibility of weathering by acid clays. Two important differences between the experiment and nature should be remembered. The first, the progressive neutralisation of the acid under laboratory but not under natural conditions, has already been mentioned: in fact chemical attack is one of the few geomorphological processes not subject to a very great diminution in activity with time and hence presumably increasing in relative importance late in the cycle of erosion. The second is that nature would not supply the felspar ground to such an ideal size for weathering, but as crystals of varying size. However, though this might slow down the attack, the slow permeation of the rock by acid solution along already corroded crystal margins and along the cleavage planes within crystals would ensure a large surface available to attack.

Thirdly, the effects of organic acids need to be taken into consideration. These are complex in character and occur in natural acid organic deposits, for example upland peats in the west of Britain. The possible effects of such a form of weathering have been considered by a number of investigators over a long period. One of the early references to such an effect was made by Clement Reid in 1899 in writing about the solution holes in the Chalk beneath Reading Beds at Puddletown Heath in Dorset. He suggested that organic acids might be derived from the acid vegetation growing on these heaths and that these acids might be expected to have some weathering effect on the underlying Chalk. A similar sort of effect was inferred by Smith (1941) to explain small closed circular depressions in the granites of South Carolina. Even after dilution by recent heavy rainfall the water in such pits proved to be markedly acid and Smith suggested that the acidity was due to decaying organic matter.

Experimental work on organic acids has again been done by Graham (1941b), who compared the effectiveness of two organic acids with that of acetic acid and also with that of an acid clay. The sample of anorthite felspar was again ground into silt-size particles and care was taken that the total number of H^+ ions available in the reaction should be the same in each case. The organic acids proved to be definitely more effective than the acetic acid and on a par with the acid clay. Graham concluded that it was possible for colloidal humus soils unsaturated with bases to become effective agents in transferring bases from primary rock minerals. Such a condition probably obtains where acid peats overlie igneous rocks containing unweathered silicates.

The chemical process of hydrolysis is thus aided by a number of acids which occur naturally. Full discussions of the chemistry of hydrolysis will be found in Reiche (1950) and Keller (1957).

Mechanical and chemical weathering in combination

At this stage a mechanical effect becomes probably of significance again. Many of the alteration products formed by the process of hydrolysis are capable of hydration, i.e. of taking up water. The simplest example is calcium sulphate, which occurs naturally in two forms, a hydrated form known as gypsum and an unhydrated form known as anhydrite. During the process of hydration there is an accompanying swelling of the minerals concerned and so forces are exerted which tend to disintegrate the rock mechanically. If the rock is so altered as to have weathering products capable of being hydrated, it is also likely that the rock has been so weathered along the crystal interfaces and along any cleavage or fractures within crystals that its strength has been already greatly reduced. Thus smaller mechanical stresses are needed to reduce it to granular debris. This facilitation of mechanical weathering by chemical and vice versa is yet another example of the impossibility of considering these two main classes of weathering in isolation.

Clear examples of the hydration and expansion of alteration products are provided by the weathered olivine crystals commonly found in basic igneous rocks. Reference to the table on p. 12 will show that olivine is the most susceptible to weathering of all the common rock-forming minerals. It usually occurs in rock as large, fractured crystals and is readily altered along the fractures, the usual product being serpentine. An example is shown in Fig. 2.1 of altered olivine crystals in a troctolite. Troctolite is

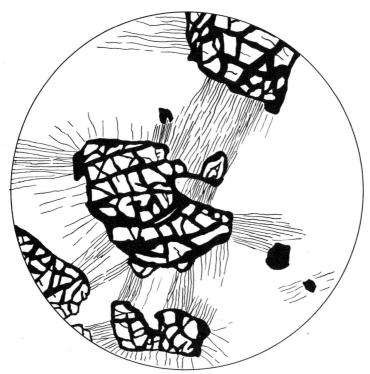

FIG. 2.1. Troctolite in thin section. Large olivine crystals altered along cracks into serpentine and magnetite, with the development of systems of minute fractures in the enclosing plagioclase *(after Hatch, Wells & Wells)*

PLATE 2. Weathered granite, near Mt Aigoual, Central Plateau, France. Weathering to granitic sand with formation of roughly spherical core stones

a coarse-grained basic rock, which differs from gabbro in that its essential constituents are plagioclase and olivine, whereas gabbro is essentially plagioclase and augite (or a closely related pyroxene). In this particular example magnetite, shown as separate small black crystals, is present as an accessory mineral. The olivine has been altered along the fractures to serpentine and magnetite, the whole being shown solidly in black for the sake of simplicity in the figure. The pressures set up by the hydration of these alteration products have been responsible for the systems of minute fractures running through the plagioclase, left entirely in white on the figure.

Mineral composition, weathering and climate

In different climates the breakdown of one or more constituents of rocks composed of several minerals gives rise to somewhat different results. Most rocks of this type are coarse granites and gneisses, which are of widespread occurrence in the wet tropical zones. With the predominating high rates of chemical weathering characteristic of the hot, wet regions of the earth's surface the alteration of coarse crystalline rocks is effected to great depths. The whole of the rock is not reduced to an incoherent pile of coarse sand, for at depths the rock may appear sound in that the texture is preserved, the individual crystals being still visible and the fabric of the rock traceable. But the fact that one of the mineral constituents has been weathered means that the strength of the rock has gone. Weathering is known down to depths of 75 m (250 ft) in some tropical regions, though this must be regarded as exceptional. Presumably, the depth of the weathered layer preserved will depend on the relative rates of weathering and erosion. Where the underlying weathered mantle is protected by vegetation or by being on gentle slopes, it may remain and reach great thicknesses, but where slopes are steep and the protection of vegetation much less, erosive processes acting on this thick mass of altered rock will tend to remove it and thus hinder the preservation of great thicknesses of altered rock.

In both arid and humid temperate regions, the coarse crystalline rocks, principally granites and gneisses, tend to break down in two stages: firstly, joints are weathered out and a pile of boulders formed; secondly, the rock is attacked chemically in one of its main minerals so that it loses its strength and falls to a mass of granular debris, composed largely of the unweatherable minerals. This phenomenon has been observed in deserts, where granites tend to occur as steep hills and escarpments, bounded at their feet by gently sloping pediments. A possible contributory cause is the fact that the weathering of granites and gneisses tends to give rise to the two grades of debris mentioned above. The escarpment or the steep side of the hill has

PLATE 3. Spheroidally weathered granite block, near Chaves, north Portugal. Similar block in background broken up for use as stone

a slope controlled by the joint-bounded boulders, which, when they fall to the foot of the slope, break down into granitic sand due to the fact that the rock has already been weathered in one of its main mineral constituents. This difference in the size of the detritus is possibly partly responsible for the observed breaks of slope at the feet of the hills, for the slope required by the agents of erosion to transport the debris changes abruptly with the change in size from boulders to gravel. It must be emphasised, however, that there are other factors probably involved in the formation of pediments and steep-sided hills in deserts. Rocks, which do not have mineral constituents of such varying resistances to weathering, do not show such a curious double maximum in the size of the debris formed from them. Instead, there is a progressive series of smaller and smaller fragments, as are yielded by a rock such as quartzite, for example. In Britain the change from jointed boulders to gravelly debris can be observed very well on parts of the Arran granite, for example on the north ridge of Goat Fell.

SPHEROIDAL WEATHERING

A more widespread phenomenon is that of spheroidal weathering, that is the formation of approximately spherical boulders from joint-bounded blocks (Plate 3). There are two terms used for the rounding of rock masses, namely exfoliation and spheroidal weathering. By some authors they are used almost interchangeably, but they will be used with different senses here. Exfoliation will be reserved as the name for a large-scale sheeting of rocks, due principally to the formation of joint-bounded concentric shells of rock. It is a feature which affects whole hills and not individual boulders. Spheroidal weathering is the rounding of jointed blocks of rock largely by chemical processes of weathering. The differences between the two may not be as great as suggested by these definitions, for some authorities have attributed exfoliation to chemical weathering. If this hypothesis is true it becomes little more than large scale spheroidal weathering.

Spheroidal weathering (Plate 2) of boulders underground in the waste mantle is a process likely to lead ultimately with other processes to the formation of tors in granitic and gneissic country (see Chapter 3). But spheroidal weathering certainly affects rocks other than coarsely crystalline types, for basalts are known to produce features of this type. Many rocks susceptible to chemical weathering—but not all, for limestones seem to form one exception—may be spheroidally weathered to boulders.

An interesting example of spheroidal weathering has been quoted by Dodge (1947). In this example most of the other possible hypotheses may be excluded. The rock concerned is a mountain of igneous rock, 'a porphyritic and pyroclastic volcanic of andestic composition' (p. 39), in the arid south-western part of the United States. Into this in 1925 an adit was driven for a distance of 70 m (228 ft), so that the weathering, described by Dodge as exfoliation, took place in a period of approximately twenty years. In the adit when examined by Dodge there was no weathering in the outer

8·5 m (28 ft), but spheroidal weathering was best developed at a distance of 14–20 m (45–65 ft) from the mouth of the adit. Further in was a broad zone in which no 'exfoliation' was observable. Dodge suggested that there was no weathering in the outermost part because this part was always dry and there were no 'exfoliation' features in the same rock outside the adit. The lack of weathering features in the more interior part of the adit might have been due to some pre-existing alteration of the rock, but was more probably due to the rock being continually wet. In the intervening zone, where the rock was alternately wet and dry the processes of weathering had reached their maximum effect. In this zone the walls of the adit were converted into a series of rounded knobs from 0·3–0·9 m (1–3 ft) in dia- meter. Mechanical weathering due to temperature changes could be ruled out in this particular example, for the diurnal temperature changes, which might have been adduced as a cause in this hot arid region, did not affect the interior of the adit, and, in any case, their maximum effect should have been exerted near the entrance and decreased along the adit. Similarly, any possible effects of natural forest or bush fires could be excluded for the same reason. The other possibility, that these rounded surfaces were due to the relief of pressure (see p. 46) causing a spalling off of rock layers, could be dismissed because the form of the surface produced was so complex that the idea that it was due to a relief of pressure was incredible. The rounded projections of the wall were, therefore, regarded as the original projections of the adit wall formed during its construction modified by the attack of chemical weathering.

Mineral weathering and minor surface relief

Differential weathering of different minerals also leads to the formation of surfaces irregular in detail. A number of different examples of this type of effect may be quoted. In areas of garnetiferous schists very weathered sections of the rocks may show the garnets, which are more resistant than most minerals, projecting as miniature relief on the surface. Very often in limestones the fossils, which are mostly composed of crystalline calcium carbonate, are much more resistant to weathering than the bulk of the rock. In shelly limestones the full detail of the shell fragments can usually only be appreciated from a weathered surface, for freshly fractured surfaces run through the matrix and the fossils, which do not stand out. This may even be appreciated in towns by studying any freshly prepared limestone used in building and comparing it with limestone which has been weathered for a few tens of years. In many igneous rocks the degree of roughness of the surface is a measure of the grain size of the rocks, coarse-grained rocks presenting much rougher surfaces than fine-grained rocks. Similarly, thin veins of quartz injected into rocks may be weathered out as projections from the general surface of the rock. Porphyritic igneous rocks (see p. 66) may have the phenocrysts projecting or depressed, depending on the relative resistance to weathering of the phenocrysts and the ground mass

PLATE 4a. Weathered glacial erratic of rhomb porphyry. Phenocrysts of felspar form pitted areas. *(Photo by R. Coe)*

PLATE 4b. Igneous glacial erratic pitted by weathering. Deep pits formed by destruction of one constituent mineral. *(Photo by R. Coe)*

of the rock (Plate 4). An example of this is shown in Plate 4a, a photograph of a glacial erratic of rhomb-porphyry, which is a Norwegian rock of complex mineral composition. The rhomb-shaped phenocrysts are of a special type of felspar, about which there is argument but which, in a general way, may be regarded as belonging to the soda end of the plagioclase series. The ground mass consists of dark minerals, special varieties of augite and olivine, with more felspar. The fact that the felspar phenocrysts weather more readily than the bulk of the rock suggests that some process other than chemical weathering is operating for the dark minerals of the ground mass should have proved less resistant than the felspar.

The effects of rock texture

Rock texture might be described as the way in which the different component particles forming the rock are arranged. In the case of igneous rocks it is important to know whether the grains are interlocked in any way and, in sedimentary rocks, the degree of openness of the rock may well be of considerable significance during denudation.

Weathering and igneous rock texture

It was noted long ago by Merrill (1906) that coarsely crystalline rocks yield on the whole more readily than finely crystalline rocks. He cited Lone Mountain on the west side of the Madison valley in Montana as an example. This is an ancient volcano, consisting of a more coarsely crystalline central section and a finer-grained outer part, the differences in grain size being due to the more rapid cooling of the outer part where it was in contact with the country rock. The processes of denudation have attacked and removed the coarse central part and left the outer finer parts intact. Such variations in texture may contribute to the phenomenon of margins of a dyke, which have been more rapidly chilled than the central part, projecting above the centre. Such variations are more likely to occur in thick dykes than in thin ones, for the differences between central and marginal rates of cooling will be greater.

Several types of rock texture lead to an interlocking of crystals, for example poikilitic texture, a specimen of which is shown in Fig. 2.2. It is easier to describe the nature of this poikilitic texture than the reasons why it occurs (Tyrrell 1930). In the example shown in the figure the bulk of the rock section consists of a large crystal of orthoclase felspar (shown in black) in which are enclosed smaller crystals of plagioclase and biotite together with smaller amounts of accessory minerals, which have been omitted from the figure for the sake of simplicity. The rock in question, a quartz monzonite is intermediate in character between a syenite and a diorite (see Chapter 3). It is thought that the texture may be produced by there being one constituent present in large amounts, this being the last to

23

crystallise and so enclosing the earlier formed crystals of other minerals. Usually, however, the last mineral to crystallise fills in the irregular gaps between those previously crystallised, for example the quartz between the felspars and the micas in granites. Whatever the precise origin of the

FIG. 2.2. Quartz monzonite in thin section. Plagioclase (P), biotite (B) and augite (A) enclosed poikilitically in orthoclase (black); quartz (Q) also present *(after Hatch, Wells & Wells)*

texture, which is mainly characteristic of intermediate rocks, such as syenites and monzonites, the more readily weathered minerals, the plagioclase and the biotite, are contained within the orthoclase and thus the rock must really await the weathering of its least susceptible constituent, orthoclase, until it can disintegrate. A similar interlocking texture, known as ophitic texture, is found in dolerites. It is the characteristic texture of such rocks and consists of large crystals of augite enclosing many thin lath-shaped crystals of plagioclase felspar. In this example the augite should prove less resistant to weathering than the felspar so that the interlocking texture may not increase resistance to chemical weathering, though it may still cause the rock to hold together longer than if it had not been present.

Few comments appear to have been made concerning the effects of such rock textures in actual examples, an exception being the statement about the effects of ophitic texture on spheroidal weathering made by Chapman and Greenfield (1949). This investigation was primarily directed towards the relative efficacy of chemical weathering and mechanical expansion and contraction in the production of such spheroids. Some rocks which weather spheroidally do so by shedding granular debris, while others shed sections of shells. According to the authors granular debris was provided by coarse-grained rocks while fineness of grain and ophitic texture favoured the production of shells.

Rock texture, permeability and denudation

As indicated at the beginning of this section, the degree of openness of sedimentary rocks is of importance in its possible effects on the porosity and permeability of sedimentary rocks. The porosity of the rock is merely the total proportion of the rock occupied by spaces or voids and is a measure of the maximum amount of water that may be contained within the rock. The permeability is the ability of the rock to allow water to pass through it. The major factor in rock permeability is the degree and openness of the jointing of the rock, a feature to be described later in this chapter, but texture may play a smaller part in producing the overall permeability of the rock.

Sedimentary rocks with open textures are most likely to be recent and unconsolidated ones, for, in older rocks, many of the spaces between the grains will be occupied by the cement binding the grains together. In an ideal gravel consisting of spherical particles the volume of the voids has a value of 25·9 per cent of the total volume at the minimum, that is with the closest possible arrangement of the particles. It may be more than this if this condition is not fulfilled. On the other hand, with more angular or partially rounded particles the percentage of the total volume occupied by voids may be less. Nevertheless, in all natural gravels the voids occupy a considerable proportion of the total volume. The same is true for sands, though, for the same arrangement of similarly shaped particles, the percolation of water through the sand will be significantly slower. This is due to the resistance caused by friction between the water and the grains and by surface tension effects both increasing to a marked degree.

As a result sands and gravels are all to high degrees permeable and porous rocks. Any rock property which inhibits surface run-off of water effectively protects the rock from erosion, so that these completely un-consolidated rocks may be very resistant to denudation. As always in a discussion it is not sufficient merely to consider the permeability of the rock, for the vegetation cover and the soil may effectively reduce the permeability of a highly permeable rock. Many sands are liable to have podsolised soils developed on them and such soils, in their more extreme forms, are characterised by the development of iron and humus pans. These are impermeable and are responsible for the waterlogging which one sometimes observes on very permeable rocks, for example the Bagshot Beds of the western part of the London Basin.

Interesting examples of the resistance of unconsolidated gravels were given half a century ago by Rich (1911). Rich mentioned factors other than the permeability, notably the fact that gravels are usually composed of the indestructible remnants of earlier beds and hence cannot be expected to be eroded by agents which had been powerless to destroy them before. The resistance of the gravel resides in two facts: its permeability and the large grain size, the latter implying that large volumes of water, which are prevented from forming on the surface by the permeability, are needed to move them. The resistance of gravel is at a maximum in low-lying situations, because at high levels the often heavier rainfall and the generally steeper slopes ensure that run-off is greater and hence gravel will be more readily eroded. Rich quotes examples from south-west Wyoming and north-east Utah of spreads of piedmont gravel, the Bishop Conglomerate, which appear to be more resistant than the mountains from which the gravels were derived. At the eastern end of the Uinta Mountains the mountains are lower than the remnants of the plateaus capped by the Bishop Conglomerate.

Comparable though probably less spectacular examples may be quoted from Britain, where, for example, the gravelly base of the Tertiary locally forms an escarpment on the dip-slope of the Chalk presumably because the Chalk is less permeable than the Tertiary Beds. The area immediately south of London in the neighbourhood of Croydon provides some good examples (Fig. 2.3). The preservation of some glacial landforms composed of gravel, such as terminal moraines, kames and eskers, is probably due to the permeability of their constituent gravel, examples being provided by the complex of constructional glacial features preserved behind Sheringham and Cromer in north Norfolk (Sparks and West 1964).

The effects of rock strength

Although a discussion of the forces acting upon the earth's crust is properly a part of structural geology or geophysics and is usually discussed in works on these subjects (Jeffreys 1929, Hills 1953 and de Sitter 1964), an elementary appreciation of these forces and of the reactions of rocks to them is necessary for a fuller understanding of the origin of features such as jointing and cleavage, which profoundly affect denudation.

The deformation of rocks

The effects of the forces concerned are principally to change the volume and to change the shape of the rocks they act upon. Changes of volume without changes of shape are difficult to envisage happening near the surface of the earth. They require uniform pressure or tension over the

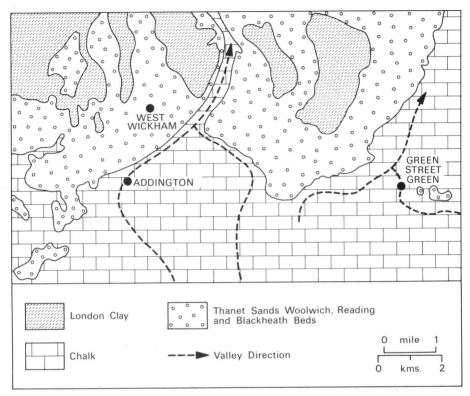

FIG. 2.3. Valley direction and Tertiary escarpment south of London *(after H.M. Geol. Survey)*

whole surface of the mass concerned. Even then there will be a mere change in volume only when the material concerned is isotropic. This state is defined as one in which the dimensions in all directions change in the same ratio and is characteristic of substances which are uniform throughout. Rocks are not uniform, for they contain planes of weakness and are usually composed of several minerals. However, an equigranular rock such as a granite, with a random orientation of its mineral components, may be regarded approximately as isotropic. In addition, uniform pressure in all directions is very rare at the earth's surface, the usual state being one of greater pressures in the horizontal plane than in the vertical plane, in which only atmospheric pressure together with that due to any overlying rocks is operating. Compression under uniform pressure is much more possible at great depths and is, in fact, an important factor in metamorphism (Chapter 4).

A stress which tends to alter angles and, therefore, shape, whether accompanied by changes of volume or not, is known as a shearing stress. The simplest example to envisage is the alteration of a rectangle to a parallelogram on the same base (Fig. 2.4a), an example which involves no

27

change of volume, or, correctly, area in this two-dimensional example. But the alteration shown in Fig. 2.4b is also accompanied by shearing as can be seen by drawing a diagonal in the figure, for the angles between the diagonals and the sides change after the pressure has been applied. There

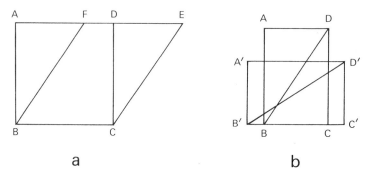

FIG. 2.4. Examples of shear: a. Rectangle ABCD changed to parallelogram BCEF, indicating shear. b. Rectangle ABCD changed to A'B'C'D' by stresses on AD and BC exceeding those on AB and DC. Angle change from DBC to D'B'C' indicates shear

is no change of area in the figure as drawn, but the application of forces at the sides of the figure of smaller magnitude than the vertical force necessary to effect the squashing could have resulted in a volume change as well. Deformations in rocks at the earth's surface, whether caused by tectonic or denudational forces, are usually caused by shearing stresses such as these. Another term for the change in size and shape caused by the stress is strain.

On any plane within a body acted upon by a stress it is possible to resolve that stress into a normal stress, perpendicular to the plane, and a shearing stress along the plane (Fig. 2.5). In any body, acted upon by a system of forces, it is possible to choose three mutually perpendicular planes intersecting at a point and such that the resultant stresses are entirely normal to each. At any point the stresses acting across these planes are known as the principal stresses, and, depending on their relative magnitudes, they are referred to as the greatest, mean and least principal stresses. The difference between the greatest and least stresses is known as the stress difference. It is the stress difference which tends to deform the body.

If a surface of rock at the earth's surface is imagined, it will be realised that the principal stresses are unequal as a normal state of affairs. Two of the principal stresses will be parallel to the earth's surface and at right angles to each other. They are due to the forces exerted by the rock containing the section under consideration and may be considerably augmented by any tectonic forces. The third principal stress direction is vertical and along it, as stated above, there is little stress except that due to atmospheric pressure.

If the stress difference increases the rock will tend to deform. It may

28

resist deformation unless the stress difference exceeds a certain value and, if the stress is removed, it will recover from any deformation imparted to it by the stress. The tendency to resist deformation is known as rigidity, the property of returning to the original shape is elasticity, and any time-lag

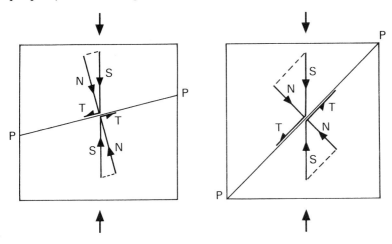

FIG. 2.5. Examples of stress (S) resolved into components, normal stress (N) perpendicular to plane (PP) and shear or tangential stress (T) along that plane

between the removal of the stress and the recovery of the initial shape is known as hysteresis. But, with a stress difference exceeding a certain value, a point is passed beyond which elastic recovery does not occur. The body is then subject to plastic deformation, or plastic flow, a state in which with a given stress difference the change of shape is proportional to the time. If the stress difference is removed, the body does not recover its shape, though it may still recover the extent of deformation covered by its elastic property (de Sitter, 1964). The point beyond which plastic deformation occurs may be called the strength of the material, or the elastic limit, or the set point. Plastic deformation does not, however, continue indefinitely. It is succeeded, either as the result of increasing stress difference or as the result of the same stress difference acting for a longer period, by fracture. The point at which this occurs may be termed the rupture limit or the ultimate strength. Although from a macroscopic point of view there is a great difference between plastic deformation and fracture, that difference is not absolute for plastic deformation may well occur by shearing along a very large number of small plane surfaces, either between crystals or within crystals.

It is possible near the earth's surface for the smallest principal stress to be negative, that is to have tension in one direction, though this state is not possible at depths in the earth's crust. In this case the material may be torn apart before it can be sheared along planes within the rock by the stress difference. This is important with rocks most of which have very low tensile strengths when compared with their strength under compression or even with their strength under shear. With brittle rocks such as granites, sandstones and limestones the compression strength is usually about fifteen times the shear strength and about thirty times the tensile strength according to figures given by de Sitter (1964). In simple terms this means that rocks resist crushing much more than they resist shearing, while their resistance to tension, provided that this can be induced, is very small.

Generally speaking a substance composed of a number of constituents has a smaller ultimate strength than an isotropic substance. With plastic deformation acting on an anisotropic or heterogeneous material a point is reached at which the weakest constituent fractures. The failure of the weakest component puts an increased stress on the other components and they in turn may be suddenly brought to the rupture limit, so that the whole material fractures.

Rock strength and erosion

These physical properties of rocks need to be considered in discussing what happens when rocks are pounded against each other by the forces of erosion. Soft rocks, such as imperfectly cemented sandstones and clays, do not resist such attacks, but hard rocks, such as most igneous rocks, crystalline limestones and well-cemented sandstones, are so tough that it is difficult to imagine how they are fractured into smaller and smaller pieces.

As an example, a sea cliff composed of crystalline rock, without jointing, schistosity, cleavage or any other lines of weakness, may be imagined. Against this the sea will, in times of storm, hurl boulders of varying size. The stress exerted by the boulder on the cliff will depend on the mass of the boulder, on its velocity and on the area of contact. It is not sufficiently accurate to liken the pounding of the boulders against such a cliff to the blows of a stonebreaker's sledgehammer against an isolated slab of rock. Stonebreakers usually take advantage of directions within the rock along which experience has taught them that the rock will yield most readily. Even if it is imagined that such directions are absent, a stonebreaker will ensure that his blows all fall along one line along which he intends fracture to occur. This can be observed by watching a bricklayer splitting brick. It is unlikely that nature would produce a similar pattern of blows, for the impacts of boulders against the cliff would have a random distribution. Further, the stonebreaker's slab is not under high containing pressures as is a comparable section of rock in an unweakened cliff face. These containing pressures tend to increase the resistance of the rock face to splitting as a result of the pounding it receives from boulders thrown against it.

The pounding of boulders against unjointed rocks probably results not in large fractures but in localised 'bruising' as illustrated in Fig. 2.6. Even with moderate sized boulders the stress at the moment of contact, provided that the contact is more or less a point (Fig. 2.6a), must be well in excess of the ultimate strength of the rock so that local shearing takes place. But as the stress decreases with the increasing area of contact extremely rapidly (Fig. 2.6b), very high stress is applied only momentarily. The result is a bruising of the rock surface and the erosion of a small quantity of comminuted material, as indicated by the dotted area of Fig. 2.6b. This is the

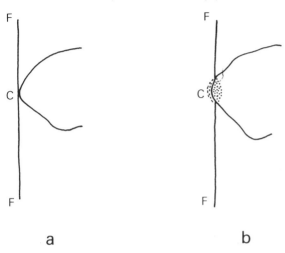

a   b

FIG. 2.6. Impact of boulder on cliff face (FF) showing probable area of local shear (C). For explanation see text

process of attrition and by it boulders may be reduced in size and the asperities rounded off cliffs, but it cannot often result in major fractures. Any flint beach will show that the individual pebbles are covered with small bruises and abrasions due to the hammering of the material against itself, but the number of fresh fractures, even in beach material not under confining pressures, is very small.

Some interesting calculations of the effectiveness of the load in the bed of a glacier in planing off irregularities in the rock floor and in striating that floor have been made by McCall (1960). He starts with the assumption that the yield stresses in ice are of the order of 2 kg/cm² in compression and 1 kg/cm² in shear and that an ideal block, a cube with sides 1 m in length, rests on a horizontal bed as shown in Fig. 2.7. With the direction of ice movement as shown a number of forces are tending to move the block along the bed: they are the 'push' of the ice on face a and the drag of the ice along the sides b and d and along the top c. As the yield stress of ice in compression is 2 kg/cm² the maximum horizontal force that the ice can exert on face a is 2 kg/cm² × 100 cm × 100 cm = 20,000 kg. Any force greater than this would cause the ice to yield and would not be transmitted to the block. On each side of the block the shear force of the ice, provided that the ice adheres firmly to the block, would be 1 kg/cm² × 100 cm ×

100 cm = 10,000 kg. Any force greater than this would cause the ice to shear and would not be transmitted to the block. Thus shear forces of a total of 30,000 kg would be exerted on the sides and top of the block, and coupled with the 20,000 kg force on the back, they would provide a total

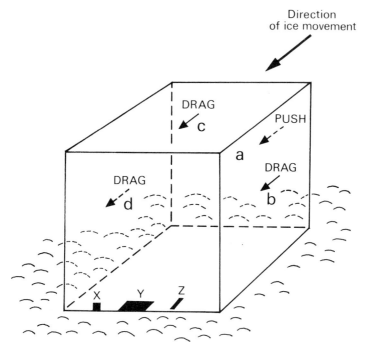

FIG. 2.7. Abrasion likely to be caused by a glacially-moved block. For explanation see text

force of 50,000 kg, provided that there was no ice in contact with the front or downstream side of the block thereby exerting a force opposed to that operating downstream. McCall assumes that the actual force present is never likely to exceed 50 per cent of the calculated force, but even so there is sufficient to cause considerable erosion on the bed.

If the compression strength of granite is taken as 1400 kg/cm² any projection on the floor will resist crushing if the bearing area between the projection and the block, both assumed to be granite, exceeds

$$\frac{25,000 \text{ kg}}{1400 \text{ kg/cm}^2} \text{ or } 18 \text{ cm}^2 \text{ approximately.}$$

On Fig. 2.7 the small black area, X, represents 18 cm² on the scale on which the front face of the cube is drawn. If the bearing area between the block and the projection is less than this, either the block or the projection will be crushed.

The area of the base of the projection required for that projection to resist being sheared off may also be calculated, the shear strength of granite being taken as 160 kg/cm². It is

$$\frac{25{,}000 \text{ kg}}{160 \text{ kg/cm}^2} \text{ or 160 cm}^2 \text{ approximately.}$$

This is represented on the drawing (Fig. 2.7) as the black area, Y. This size as well as that necessary to resist crushing give some idea of the size of the sledgehammer needed to crack the nut in glacial abrasion.

The maximum area of contact between the block and the bed above which striation cannot take place may also be calculated. In this example it is assumed that the top face is under compression and so has a force of 20,000 kg exerted upon it. To this must be added the force due to the weight of the block. If the density is assumed to be 3, which is a little high for granite but very easy for calculation, the block will weigh 3000 kg, thus giving a total downward acting force of 23,000 kg. For the bed to be striated or gouged, i.e. crushed, the bearing area must not exceed

$$\frac{23{,}000 \text{ kg}}{1400 \text{ kg/cm}^2} \text{ or 16 cm}^2 \text{ approximately.}$$

This value is shown by area Z on Fig. 2.7. Again the great size of the block needed to produce a striation in hard rock will be noted. The maximum compression force of 2 kg/cm² on the top face will only be realised when the thickness of the ice reaches 22 m, but above that any increase in thickness will merely result in the compression strength of ice being exceeded. Thus the amount of abrasion is proportional to the area of a fragment provided that the glacier is at least 22 m deep. It is a fallacy to assume that the thicker the ice the greater will be its abrasive power, except in the thickness range of 0 to 22 m.

In practice all sorts of additional and complicating effects may occur, such as the actions of gravity, of imperfect rock to ice adhesion and of rolling movements of blocks, so that these calculations really only indicate a likely order of magnitude for abrasion and must not be taken as precise. Nevertheless, one cannot get away from the fact that large forces are needed to effect abrasion.

It seems, therefore, that crystalline and hard sedimentary rocks have such ultimate strengths that it is difficult to visualise the forces of erosion as capable of more than abrasion. The strength is such that, unless weakness already exists in the rocks, it is difficult to understand how any real erosion can take place.

The effects of minor rock structures

Fortunately, it is the exception rather than the rule to find natural rocks devoid of planes of weakness. Most rocks are already fractured before erosion starts and many contain planes along which they are far less coherent than is normal for the rock as a whole. Actual fractures are represented by joints, the most common and most important class of minor rock structure, while the bedding of sedimentary rocks and the cleavage and schistosity of metamorphic rocks ensure that there are many planes of weakness both in unjointed rocks and within the blocks bounded by the joint pattern.

Bedding

A fundamental characteristic of sedimentary rocks is that they are bedded (Plates 5 and 6). All the possible causes and types of bedding need not be considered here, for it matters little whether the rocks are normally or current bedded (Chapter 5). Bedding represents some variation in deposition: the easiest example to imagine is that quoted by Tyrrell (1930) in which a sample of unsorted material is dropped into water. The coarse fragments settle first and the others more slowly, for the finer the fragment the lower is its settling velocity. This gives rise to a progression from coarse to fine in the sediment deposited on the floor. Repeated injections of such samples of sediment, such as might be supplied annually to a semi-desert or arctic lake by inflowing streams, give rise to bedding. The strength of the rock varies with the lithology and, once cemented, it is likely that a rock of this type will split more easily along its more clayey levels. Variations of current direction may cause alternations of different strata, such as the intercalation of sandstone with clay or silt seams. These again provide planes along which the rock will tend to be more readily eroded. Not only do the processes of erosion find such weaknesses, but the latter may already have been exploited by the development of jointing within the rock by tectonic forces. Thus the bedding may assist the development of jointing, which in turn facilitates erosion.

Any minerals occurring in platy fragments will tend to be deposited parallel to the bed of the stream or sea. The best example is mica, for fragments of the light mica, muscovite, are very resistant to weathering and are often available for incorporation in sedimentary rocks. Mica in a rock, especially when it is confined to definite narrow horizons as in flagstones, ensures that the rock is far from isotropic and that it is very weak in one plane.

Jointing

Although bedding needs consideration it cannot compare in importance with jointing, which ensures that the rock is already broken. Curiously, a joint is very difficult to define with absolute precision. It is, according to

PLATE 5. Coastal erosion of greywackes and mudstones, Aberystwyth Grits, Clarach, near Aberystwyth. Typical of much of the Lower Palaeozoic of central Wales. Note minor control of cliff profile by greywacke beds

Hills (1953), 'a fracture in a rock mass, along which there has been extremely little or no displacement' (p. 99). Many joints probably have the same origin as faults (de Sitter 1964), so that a fault basically is a joint along which appreciable movement has occurred. The limit drawn between a joint and a fault is a commonsense one and probably varies with the scale and nature of the example under consideration. In a Pleistocene sand, fractures with throws of the order of a centimetre or two (one inch) would almost invariably be classed as faults and not as joints, but in a jointed mass of granite broken into blocks of the order of 0·5 to 1 m (2 by 3 ft) a similar displacement would probably not result in such blocks being classed as faulted. It is easy, in thinking about joints, to picture those common illustrations of the mural jointing of granite with three sets of joints planes at right angles to each other and to imagine that the fractures must have such a pattern before they are classed as joints. But in Hills's definition quoted above there is nothing about magnitude nor about the direction of the joints, and such a definition seems to be more common than those which specify magnitude and direction. For example, according to Geikie (1908, p. 144), 'joints are superinduced divisional planes which traverse rocks in different directions and at various angles, so as to allow of their ready separation into larger and smaller blocks and fragments of regular or irregular shape'. Again, Lake and Rastall (1927, p. 20) in describing jointing in chalk say that the 'rock is divided into irregular blocks by peculiar curved joints'. Any chalk face shows itself to be shattered by these small, irregular and closely spaced joints. Indeed any definition of joints which attempts to include regularity and frequency qualifications would seem to be attempting to put a division where none occurred naturally.

There are a number of ways in which rocks may become jointed and the prevailing frequency of joints in practically every rock should be the cause of no surprise at all. The methods of origin of joints are outlined below before a broad survey of their overwhelming importance in erosion is undertaken.

JOINTING CAUSED BY SHRINKAGE

In sedimentary rocks joints may be formed merely by the contraction of the sediment as it dries. The shrinkage caused by drying is greatest in fine-grained sediments especially in sediments containing much clay, a colloidal substance which shrinks greatly as it dries. Joints in clays tend to be irregular in pattern, but the usual form of a joint system in sedimentary rocks is for one set of joints to be parallel to the bedding planes with two sets at right angles to it.

In igneous rocks jointing may be produced during the cooling of the

PLATE 6. Monocline truncated by coastal plateau, Lower Lias limestone, Glamorgan. Fairly thin-bedded limestones interbedded with thin shales

rock. The tendency is for jointing to occur at right angles to isothermal surfaces. This is why sills are vertically jointed and why dykes are horizontally jointed. It is one possible reason for the tendency for granite intrusions to be jointed at right angles to their surfaces. In all examples the isothermal surfaces formed during cooling are parallel to the sides or margins of the intrusion. Such joints may be very well displayed in those structures known as tholoids or cumulo-domes. These are masses of viscous lava squeezed out of a volcano and merely spreading out slightly at the top. Thus they differ slightly from the normal volcanic plug which remains within the vent and from the plug-dome, which represents a plug squeezed out but not changed in shape. As a tholoid cools the isotherms will obviously be parallel to the surface so that a joint pattern as shown in Fig. 2.8 should

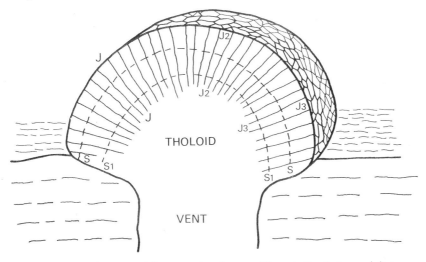

FIG. 2.8. Tholoid squeezed from a volcanic vent. The relation between joint directions (JJ, J2J2 and J3J3) and isothermal surfaces (SS and S1S1) is shown

be produced in an ideal case. Cotton (1944, Fig. 71) has an excellent photograph of one side of a tholoid at Whakaipo Bay, Lake Taupo, New Zealand, showing very clearly the relation between the jointing and the surface.

The polygonal columnar jointing (Plate 7) found in many basalts but also in other dyke rocks, for example quartz porphyry in Arran, is a type of jointing due to contraction during cooling. With the isothermal surfaces parallel to the surface of the basalt flow there will tend to develop horizontal contraction uniformly across those surfaces, provided that the rock is homogeneous. Contraction equal in all directions is mechanically the same as contraction towards centres equally spaced over the sheet (Holmes 1965).

PLATE 7. Rough columnar jointing, Whin Sill dolerite, Crag Lough, near Haltwhistle. The scarp in the background is also developed on the Whin Sill. *(Photo by R. J. Sparks)*

An equal spacing of centres is achieved when they form the corners of equilateral triangles, as shown by the pecked lines on the right hand side of Fig. 2.9. Along any side of any of these equilateral triangles there will be tension as the forces caused by contraction will be directed towards the

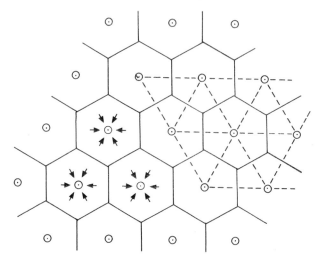

FIG. 2.9. Hexagonal joint pattern, developed as a result of contraction towards equally-spaced centres of cooling

centres of cooling. When these forces exceed the tensile strength of the rock it will fracture midway between the centres of cooling. This repeated will result in a series of hexagonal cracks or joints, because at each centre of cooling there are six directions of greatest tension, as shown by the small arrows on the left hand side of Fig. 2.9. Such cracks extend downwards from the surfaces of cooling and so produce the well-known hexagonal pattern. Many of the hexagonal columns are cross jointed as well, such joints very often being curved, thus producing a cup and ball effect. In practice the cooling surfaces will be more irregular than postulated here and the rock less homogeneous. The result is that the ideal hexagonal pattern will be impaired and the resulting joint pattern is usually polygonal, the number of sides to the jointed columns ranging from three to eight though the majority will probably remain hexagonal.

It might possibly be better to consider not an equally spaced series of cooling centres, as a random distribution of centres may give a more life-like pattern. A model of such a system has been made by Smalley (1966), who used random numbers to provide coordinates for the centres of stress. The ensuing pattern of cracks produces a very irregular series of six-sided figures. Some of these have sides so short, for example at a, b, c, d, and e in Fig. 2.10, that in the field their equivalents might well be passed off as five-sided figures. If they are counted as five-sided the average number of sides in these random patterns is 5·6–5·7, which corresponds very closely with that found in the Giant's Causeway basalt of Northern Ireland but is appreciably higher than that found in three other basalts investigated.

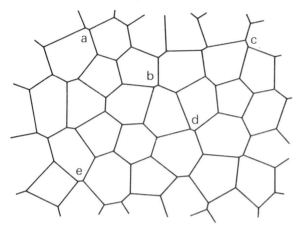

FIG. 2.10. Crack network produced by random centres of cooling *(after Smalley)*

JOINTING CAUSED BY SHEARING

In addition to the joints caused by cooling a much more important class is caused by earth movements. There are very few rocks at the surface of the earth completely unaffected by earth movements. The stresses involved, even in slight movements, are responsible for the formation of complicated joint patterns.

Given homogeneous (isotropic) material and a marked stress difference (deformative stress) at the surface of the earth, slip planes will tend to be set up at an angle of 45° to the direction of the deformative stress (Fig. 2.11a). Although the shearing stress reaches its maximum value at this angle and although shearing has been produced at approximately this angle experimentally, in practice the angle of inclination between the deformative stress and the slip planes usually becomes less than 45°, as shown by the angle α in Fig. 2.11b. The reduction in angle is partly due to the value of the normal stress across the slip plane and partly to the angle of internal friction of the material under consideration (de Sitter, 1964). The latter is a specific property of the material and is larger, for example, in sandstones than in clays. It increases with confining pressures, but the rate of increase varies from rock to rock: the increase is much more rapid in sandstones than in clays. Thus more acute angles will be found between the direction of deformative stress in sandstones than in clays and, in addition, the difference will be accentuated by differences in confining pressure. These conjugate shear planes at an angle to each other are considered to be the basis of much of the rectangular jointing commonly observed in rocks.

Although a repetition of the intersecting pattern of fractures considered above should, under ideal conditions, produce two sets of joints with the

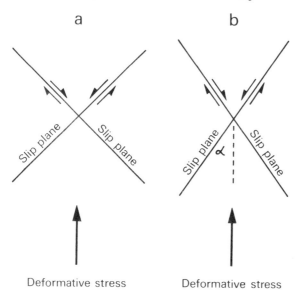

FIG. 2.11. Joint pattern produced by deformative stress. For explanation see text

acute angle between the sets bisected by the direction of the deformative stress, the position in nature is complicated by the anisotropic character of the rocks. Rarely if ever are rocks found with equal resistance to shearing in all directions. Planes of weakness, however ill-defined, are almost invariably present and such planes tend to dominate in the determination of joint directions. The strength of the rock may be so reduced along these planes that it gives along them first regardless of the angle between the weaknesses and the deformative stress direction.

It must not be imagined that great earth movements and very hard rocks are needed for the production of these joint patterns. In a description of the joint patterns affecting Upper Ordovician to Upper Devonian rocks in the Allegheny plateau of southern New York and northern Pennsylvania, Parker (1942) commented upon the lack of systematic relations between the sets of joints on the one hand and the folds, faults and dip on the other. He regarded the formation of the joint pattern as an early phase in the tectonic history of the area for the vertical position of the joints has been disturbed by the folds, which are themselves not very severe in this district.

Even completely uncemented sand is liable to shear when subjected to deformative stress. Mead (1925) showed long ago that when a pile of loose dry sand was put across a joint between two sheets of paper and those sheets of paper moved relative to each other, faulting and rifting was readily reproducible in the sand. The question of shearing in sand has also been considered by de Sitter (1964). If the sand grains are so packed that the volume of voids is at a minimum, the rock cannot be compressed and any distortion is bound to lead to an increase in volume. The increase in volume means that the hydrostatic pressure due to the water between the grains is decreased and, at the same time, the grain pressure is increased: this leads to an increase in resistance to shear. On the other hand, when the sand is not packed as tightly as possible, any compression leads to a diminution of the volume of voids, and hence to an increase in the hydrostatic pressure of the water in the voids provided that it cannot escape immediately. The

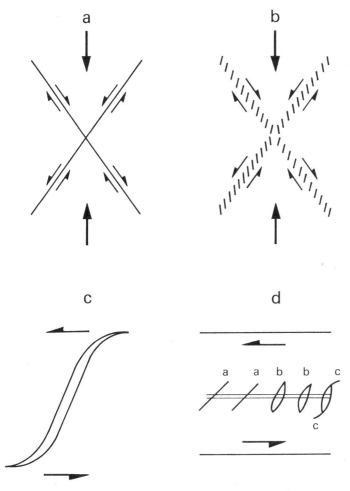

Fig. 2.12. Tension fractures in the Athens Limestone *(after Shainin)*. For explanation see text

43

increase in hydrostatic pressure is accompanied by a decrease in the grain pressure and the sand shears suddenly. This is the mechanism responsible for the sudden collapse of sand faces. Clean breaks in unconsolidated Pleistocene sands and gravels, which seem so odd, are according to Mead the expected state of affairs.

A curious variation on the pattern of two sets of shear joints at an angle to each other has been described by Shainin (1950) from the Lower Ordovician Athens Limestone at Riverton, Virginia. Instead of the normal pattern, with the acute angle bisected by the deformative stress (Fig. 2.12a), the joints have been replaced by rows of tension fractures in echelon, as shown in Fig. 2.12b. These fractures vary between 1 and 18 cm ($\frac{1}{2}$ and 7 in.) in length and they are up to 5 cm (2 in.) wide. In detail their form is curious (Fig. 2.12c) because it is the opposite of what might have been expected from the general frictional drag, in that the ends of the fractures face the direction of the stress. The explanation of this is not clear but Bucher made certain suggestions to the author of the paper. It had been observed that in a sheet of clay covering two boards differential movement of the boards resulted in the formation of fractures at 45° in the clay, as indicated by the two fractures marked 'a' on Fig. 2.12d. Continued movement of the boards would tend to cause these fractures to rotate and gape as shown in the two fractures marked 'b' on the same figure. At this stage one might expect a new set of fractures at 45°, which might coincide with the original fractures to produce the form 'c', which is that displayed by the tension fractures themselves. On this explanation they are regarded as compound forms.

But conjugate sets of joints or fractures in echelon are not the only fractures produced by earth movements, for several other sets begin to appear as soon as flexuring of the strata begins. A very interesting example of a pattern of very small faults, comparable with joints, has been cited by de Sitter (1964) from Algeria. The local rocks are gently folded Cretaceous and Tertiary sediments with a Cretaceous limestone exposed in the core of the anticline and a Tertiary limestone exposed in the axis of the syncline (Fig. 2.13). The pattern of joints on the anticline is different from that in the syncline because the stress conditions are different. Although the general stress (the greatest principal stress) is from left to right in Fig. 2.13, it is offset on the crest of the anticline by local stresses directed outwards from the axis and caused by the 'stretching' occurring there. Thus, locally, the general stress direction becomes one of tension, with the result that the greatest principal stress along the anticline is along the crest of the fold. The shear joints, therefore, form a pattern with their acute included angle bisected by this direction. At the same time tension joints or fissures appear and these are aligned parallel to the direction of the greatest principal stress. In the syncline, on the other hand, the local stresses, caused by the folding, act in the same direction as the general stress. Thus the sets of shear joints, with their acute included angle bisected by this direction, are orientated very differently from those on the anticline. Any tension joints

44

will be orientated differently as well, for they will be parallel to the greatest principal stress, and so across the axis of the fold and not along it. Thus a new set of fractures, due to tension, is introduced here in addition to the shear joints.

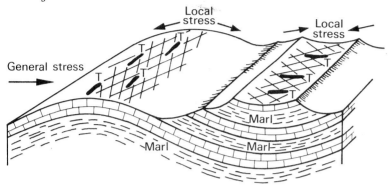

FIG. 2.13. Shear joints and tension joints (T) on folds *(after de Sitter)*

These considerations apply to the top of the beds under discussion. When the bases of the beds are considered the local stresses are different, because in the anticline they will be directed towards the axis of the fold and away from the axis in the syncline, i.e. the reverse of the conditions at the top of the beds. Theoretically this should lead to the patterns of shear and tension joints in folds being different on the upper and lower surfaces of the beds. In fact the patterns characteristic of anticlines and synclines at the top will be reversed at the bottom. In practice this might be expected to result in a very chaotic joint pattern, especially as these are not the only shear joints likely to be formed.

Further sets of joints are liable to be produced, as can be appreciated by considering a section of a bed subjected to folding (Fig. 2.14). Because the direction of local stress is different on the upper and lower surfaces of the bed, there tends to occur shearing parallel to the bedding planes but within the bed, as at X and Y in Fig. 2.14. Again, if one considers a block, such as that shown on the flank of the anticline, the forces on the faces, A and B,

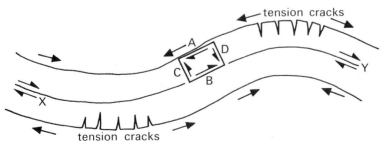

FIG. 2.14. Joints within a competent folded bed. For explanation see text

are in opposite directions and tend to cause a rotational movement to be imparted to the block. This rotational movement is offset, because the block cannot move against the pressure of the confining rocks, by forces such as those shown along the faces, C and D. These effects tend to cause jointing parallel to the fold axes.

It must not be imagined that joint patterns in the field may always, or even usually, be fitted closely into the patterns forecast by considering the effects of stresses on isotropic materials. They are complicated not only by the assumption by jointing of predetermined lines of weakness within the rock, but also by the fact that few rocks have suffered only one series of earth movements. If a rock has experienced several series of movements with the principal stress directions even slightly different, the resulting joint pattern may well be extremely complicated. For the geomorphologist it is important to realise and appreciate that shearing in rocks is likely to lead to every rock being already shattered by joints before erosion starts.

JOINTING CAUSED BY PRESSURE RELEASE

The idea that jointing in rocks might also be caused by expansion of the rocks consequent upon their being relieved of the pressure of the overlying rocks by denudation has also been used as a frequent explanation of certain types of joint patterns. One of the earliest supporters of this idea was Gilbert (1904) and his paper provides an excellent introduction to the subject.

It had been observed that curved sheet structures are often approximately parallel to the ground surface (Plate 8a), so that it is not surprising that a close connection between sheet jointing and the ground surface had been envisaged. As clearly stated by Gilbert, it is not easy to distinguish cause from effect. The present ground surface might owe its shape to the jointing already existing in the rock or, alternatively, the jointing parallel to the surface might be the result of the relief of pressure and hence virtually determined by the shape of the surface. Gilbert was one of the few investigators who sought evidence to enable him to choose between these two ideas. He found it in the fact that sheet structures in the area studied by him extended down only to depths of about 30 m (100 ft). If the surface had been determined by the jointing, one might have expected the joint pattern to extend indefinitely downwards, so that this limitation to the uppermost 30 m (100 ft) suggested that the jointing was due to the surface.

Curved sheets are not characteristic, according to Gilbert, of well-jointed rocks, presumably because any expansion due to pressure relief would be taken up by the existing joint pattern. Rocks were known, however, in which jointing already existed and was not crossed by the curved sheet jointing. This suggested that the sheet jointing was later than the other jointing, another point consistent with the idea that the curved sheets were caused by relief of pressure.

Gilbert considered the possibility of temperature changes and chemical

PLATE 8a. Sheet structure in granite, near Chaves, north Portugal

PLATE 8b. Carboniferous laccolith, Traprain Law, East Lothian. The rock is phonolite

changes and dismissed them as unsuitable as explanations of sheet structures. As shown above, mechanical breakage due to temperature changes has not been adopted by later investigators, but chemical change as a cause of sheet structures has not been excluded. This second point is discussed below.

Other problems which have arisen since Gilbert originally wrote about sheet structure have concerned the nature of the rocks affected by such structures, the recognition of such joints in the field and their differentiation from other joint patterns, and the dating of the sheet structures.

Most investigators seem to have believed in the efficacy of the process. Indeed, as Harland (1957) observed, the difficulty lies not in the explanation of the joints but in their recognition in the field and differentiation from other joints. Provided that rocks have been sufficiently deeply buried and then exposed at the surface by erosion, there seems little reason why such joints should not be widespread. Nor need the burial apparently be very deep for the pressure exerted by about 180 m (600 ft) of granite is equal approximately to the tensile strength of the rock at the surface, so that the denudation of approximately this thickness of rock might be expected to produce sheet structures parallel to the surface. There is plenty of evidence of the expansion of rocks when relieved of the pressure of overlying strata. The sides of boreholes close in on the boring apparatus and rock bursts are common in deep mines, for example in South Africa. A well-known account of such phenomena is that of Bain (1931) concerning marble quarries in Tennessee and Vermont. Apart from the rock closing in on the drills the quarries in this marble were the scenes of spectacular and dangerous bursting of the rocks. These were of such power as to throw a machine weighing one ton from its tracks and to make the quarrying so dangerous that it was abandoned in places. No bursts occurred in quarrying the uppermost 6 m (20 ft) nor in working back a similar distance from a face. In that thickness the strains seem to have been already relieved, but at greater depths there remained the possibility of sudden rock bursts. Bain measured very carefully the expansion in the marble and found it to be appreciable, one example quoted being of an expansion of 0·05 mm in 1 m (one fiftieth of an inch in three feet) i.e. an expansion of the order of 1 in 2000.

Further evidence for pressure release is provided by the arching up of rock shells along joint planes, as shown diagrammatically in Fig. 2.15. The shells to do this must be under some lateral confining pressure. The arching of shells in this way has been commented upon by Bain (1931), Matthes (1930) and Lewis (1954). The last two authors have excellent photographs of the feature. Such arching must be developed in sound rock as solution along joints might produce a comparable form, which would need to be inspected carefully before its origin was decided.

Further confirmation that the sheet structures are most closely related to the form of the ground and not to the intrusion surface has been provided by Jahns (1943). He described sheet structures in New England granites, which appeared to be independent of the granite boundary in that they ran through minor post-granite intrusive bodies, through xenoliths (masses of some other rock incorporated in the intrusion), through roof pendants

Ground surface

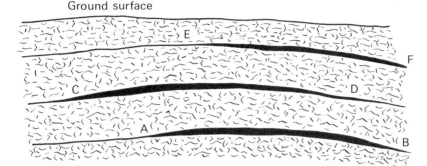

FIG. 2.15. Pressure relief joints, AB, CD and EF, resulting from arching of rock shells

(fragments of the original country rocks incorporated in the upper parts of an intrusion), and across the banding in places where the granite was somewhat foliated. These relationships are shown diagrammatically in Fig. 2.16.

Opinions regarding the types of rocks prone to pressure relief jointing have varied considerably. Jahns, who has just been quoted, took an intermediate position on this question. His general description of granite included a number of comparable rocks such as granitic gneiss, quartz monzonite, acid and intermediate intrusives of other types, and anorthosite, a basic rock composed of calcic plagioclase. All these rocks showed sheet structures, although they were best developed in unjointed granite and, where the sheets ran through rocks such as gneiss, they tended to be less well developed.

Balk (1939), in describing parts of New England and adjacent parts of

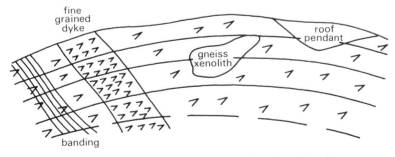

FIG. 2.16. Sheet structures in an intrusion, passing through and so independent of various features of the intrusion

Canada, mentioned sheet structures in a wider variety of rocks. He thought that they were found in every type of intrusive rock except massive gabbroic rocks, in most injection gneisses, and also in some of the metamorphosed sediments of the area, though he specifically excluded marbles and quartzites. The omission of marbles is curious in view of Bain's description of rock bursts from marble quarries. Balk agreed, however, that sheet structures were most likely in poorly jointed intrusives of deep-seated origin.

A more limited view was adopted by Matthes (1937), who considered that sheet structures were confined to siliceous granitic rocks, a term which would probably include some of the intermediate intrusives as well as certain coarse acid gneisses. This view as well as that of Balk in excepting gabbros is at variance with the experiences of glaciologists in Norway. The occurrence of pressure relief jointing in the Jotunheim has been very strongly supported by Lewis (1954) and by Battey (1960) in metamorphic rocks, which are probably best described loosely as basic gneisses: they approximate to gabbro in general, but not in detailed composition. Even here, exfoliation according to Battey is less well developed in the most basic rocks, massive pyroxenites and peridotites, than in the basic gneiss.

Although quartzites were excluded by Balk from the list of rocks liable to show sheet structures, exfoliation on a small scale in such rocks has been described by Farmin (1937) and attributed to pressure relief. This exfoliation affects pebbles of pure close-grained Cambrian quartzite which occur in pebble dykes near Salt Lake City. He rejects all weathering hypotheses for, on any of these, surface gravels should possess comparable features, while chemical weathering at depth can probably be excluded for quartzite is one of the rocks least likely to be attacked. As rock bursts are known from depths of 750 m (2500 ft) in mines and as 3000 m (10,000 ft) of rock is believed to have been removed from the Sierra Nevada in the Tertiary era, pressure relief is advanced as the cause of the exfoliation.

The age of sheet structures and the variations of pressure necessary for their formation are still debatable questions. Several authors have described structures parallel to surfaces laid bare by glacial erosion. Matthes (1937), for example, states categorically that glacial erosion of a monolith is followed by the development of a new set of sheet structures unconformable to the older sets. On the other hand Jahns (1943) attributes all the sheet structures in the New England granites, which he studied, to preglacial erosion and, from this, deduced the amount of glacial plucking by reconstructing the form of the sheets. Even Matthes stressed what he believed to be the slowness of the formation of sheets when he observed that there seemed to be an absence of postglacial shells on surfaces laid bare by the Wisconsin, or last, glaciation.

These views are at variance with those expressed by Battey and Lewis concerning the formation of pressure release joints in the Jotunheim region of Norway. The joints described by Battey are related to the shape of a corrie still occupied by a small glacier and presumably still developing.

Further, Lewis (1954) was convinced that jointing may develop as a result of variations in pressure at the bottom of a glacier, and he used such a suggestion to explain some of the jointing and plucking characteristic of roches moutonnées. It is not impossible to reconcile these apparently divergent views, because the rocks in one glaciated area may already have had great thicknesses removed by fluviatile erosion before they were glaciated and so have been in a state when little further erosion was necessary to produce sheet structures. On the other hand, other glaciated areas may have required the removal of considerable thicknesses of rock before such structures were developed. In other words, the glacial erosion of the former region may be regarded as the last straw that broke the camel's back.

Among alternative suggestions put forward to account for exfoliation the most persistent has probably been the hypothesis of chemical weathering. Usually this has been applied to spheroidal weathering, which has been called exfoliation by some authors. Most writers on the subject have regarded the role of chemical weathering as contributory but a few have regarded it as dominant. White (1945), for example, considered that the apparent absence of sheet structures, the presence of a weathered layer several feet in thickness, and the presence of granitic debris at the base of granitic domes in the south-eastern Piedmont region of the United States was a sign of the dominance of chemical weathering. These facts could, of course, also be regarded as indications of a strong contributory action of chemical weathering rather than its fundamental role.

Harland (1957) had an interesting alternative hypothesis to explain the formation of joints parallel to glaciated surfaces. Upon the retreat of a glacier meltwater would penetrate into normal joints and freeze in them. This would cause a considerable compressive force to act on the blocks between the joints, so that further joints parallel to the surface might be formed.

It seems, then, that there are variations in opinion concerning many aspects of the formation of sheet structures. Practically all investigators agree that such joints do occur and that they are most frequently and best developed in massive, poorly jointed, plutonic igneous rocks, especially granitic rocks. The diversity of opinion about their occurrence in other rocks may reflect genuine regional differences, while the lack of agreement about their relation to glaciated surfaces may also be a measure of the amount of glacial erosion required to produce pressure release joints in different areas.

Once sheets are developed their movement downhill will tend to produce minor jointing within the sheets. Two sets of joints might be expected: one set at right angles to the direction of creep, i.e. approximately along the contours, and another set at right angles to this. These two sets are termed annular and radial patterns by Chapman (1958) in a study of such features in the Acadia National Park in Maine.

51

Cleavage and schistosity

Before considering in general terms the significance of joints in denudation, cleavage and schistosity in rocks need to be discussed for they also affect the resistance of rocks to denudation. They are approximately parallel planes developed as a result of some degree of metamorphism of a rock and along which the rock may be split into thin sheets. There are various types of cleavage, full accounts of which may be found in Wilson (1946), Hills (1953) and de Sitter (1964).

The most widespread type of cleavage is known as slaty cleavage and anyone familiar with the slates more commonly used on roofs in the past than at present will know how these rocks split into very thin sheets indeed. Slaty cleavage is developed by stresses in rocks and is usually set up at right angles to the greatest principal stress, or, as Wilson puts it, at right angles to the direction of greatest shortening of the rock mass in question. The cleavage is, therefore, parallel to the axial planes of folds set up by the stresses and may have practically any sort of relation with the original bedding of the rock, as shown in Fig. 2.17. Cleavage is due to the presence

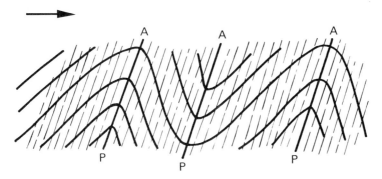

FIG. 2.17. Relations between bedding, slaty cleavage and axial planes of folds (AP). Solid lines are bedding planes; pecked lines show cleavage

within the rock of orientated flakes of minerals with platy characteristics, especially micas. The re-orientation of the mineral flakes is the result of a low grade of metamorphism and is probably partly a mechanical re-orientation and partly a chemical extension of the flakes. On the first view, flakes already within the rock are rotated by the pressures involved until they adopt the position normal to the principal stress direction. Such a view which was accepted early in the history of thought about cleavage would be supplemented today by the hypothesis that, as well as being re-orientated, the mineral flakes were added to chemically. This second view arose as a result of failure to be able to account for all the detrital grains in the parent material from which the slates were developed. With an increasingly high grade of metamorphism, involving probably considerable chemical change, there is a much greater development of micas and the resulting rocks grade through phyllites, in which considerable mica is developed along the cleavage, into schists, rocks dominated by orientated platy minerals, especially micas, but no longer cleavable.

A different type of cleavage is known as fracture cleavage. It consists of very closely spaced planes of fracture along which the rock will cleave, but the rock between the fracture planes does not have its mineral constituents orientated parallel to them. As the fracture planes become more widely spaced they grade into joints (Wilson 1946). Cleavage of this type often affects slate beds in tight folds, whether these are on a large scale or merely minor features. In a fold consisting of an alternation of sandstones and slates (Fig. 2.18) the sandstones are the competent beds and the incompetent slates are forced to accommodate themselves to the form adopted by the sandstones. This results in a thinning of the slates on the flanks of the fold and a thickening at the crest, as indicated in Fig. 2.18. The changes in

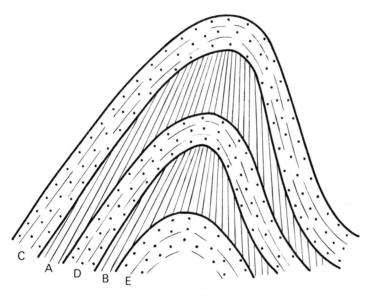

FIG. 2.18. Cleavage in slate in a fold. Fracture cleavage in slate beds (A and B) but not in sandstone (C, D and E)

thickness are effected by very close shearing, or fracture cleavage, within the slates. In the competent sandstones concentric shearing parallel to the bedding takes place and this shearing truncates the fracture cleavage of the intervening slates. Thus one set of forces acting upon different beds in a sequence produces in the one concentric shear planes and in the other a multitude of shear planes along which it yields in adapting itself to the form imposed upon it by the competent bed.

53

Minor structures and denudation

The above discussion of the many forces combining to produce jointing and cleavage seems to make it inevitable that every rock at the surface of the earth is already fractured and so prepared for denudation. The rocks are already weakened in their resistance to erosion and the lines of weakness allow the ingress of agents of weathering which themselves effect further great weakening.

JOINTS AND MECHANICAL WEATHERING

A well-jointed rock may not be more prone to mechanical weathering, because the expansions and contractions caused by temperature changes may be absorbed by the joints rather in the way in which expansion and contraction in railway lines or concrete roads are absorbed respectively by gaps in the lines and tar-filled joints in the roads. Another force, that exerted by the roots of plants, is undoubtedly allowed fuller play by the presence of joints. In completely unjointed rock plant roots, even if plants could grow, could not split the rock, and all that might be expected to result would be the sort of surface corrosive effect which seems to be effected by mosses and lichens. But the action of plant roots in prising apart jointed rocks is well known. An example that comes readily to mind is that of the beech tree overhanging the face of a chalk quarry and penetrating deeply into the chalk with its roots. The destructive action of the roots of woody plants on brick walls and building foundations, some-times the subject of legal action, is another well known effect of what amounts to the denudation of a jointed structure.

The mechanical weathering process of freeze–thaw action is another which demands that the rock should be fractured in advance. In a completely unjointed rock there seems to be no possibility of the process having any appreciable destructive effects, for the water needs to enter into confined spaces and freeze there to effect any disruption of the rock. It cannot create the crevices itself. Even within the rock certain conditions have to be fulfilled (Grawe 1936). The water needs to be as confined as possible so that it cannot migrate when freezing starts. Water in a cavity on the surface of a rock is far less likely to cause damage to the rock on freezing because the expansion can be readily taken up. On the other hand water confined in a narrow joint may completely fill the joint and be incapable of migrating when the pressure caused by freezing is produced, so that the pressures are exerted against the containing rock. The water subject to freezing also needs to be devoid of contained air or gases because these are capable of compression when the water is frozen and so likely to absorb the pressure exerted by the expansion on freezing. Thus, for freeze–thaw

PLATE 9. Coarse jointing, Barmouth Grits, Foel Penolau, near Trawsfynydd. Massive joint blocks with joints hardly opened up. Compare with plates 10 and 11

to be an effective agent of disintegration of rock pre-existing fractures, such as joints, are required.

JOINTS AND CHEMICAL WEATHERING

It is with respect to chemical weathering, however, that the significance of joints becomes paramount. The greater the degree of jointing the greater the surface area available for the agents to operate on and hence the faster the destruction of the rock. In fact variations in the spacing of joints have been held to be the basic cause why in some igneous rocks certain parts develop into well formed inselbergs (King 1948) or tors (Linton 1955); in both examples it is the most jointed parts of the rocks which are denuded to form the lower lying areas between the inselbergs or tors. Most chemical weathering is caused or assisted by the carbon dioxide dissolved in rain water and this solution, even where it can only percolate with difficulty into the joints, will slowly widen them until they are so open that the access of weathering agents is facilitated. Such a process can be most clearly and easily seen on a small scale in the weathering of clay faces exposed in road cuttings and similar artificial excavations. Most clays, including many varieties of boulder clay, are broken by a series of cracks into lumps. The latter are oxidised brown along their faces by the penetration of the atmosphere, while the lumps remain blue or grey in their centres where they are not affected by oxidation. The effects of this oxidation can also be seen on fossils contained in the clay. Fragments of plants, which are readily destroyed by oxidation, are found to be badly altered, while shells of Mollusca are sometimes found to be weathered on the sides which have been exposed in the cracks and unweathered on the other side protected in the clay. Similar penetration of oxidation along the joints of basic igneous rocks results in a rusty staining along the joints. Pebbles of dolerite, which are common in the glacial drifts of East Anglia, have the superficial appearance of coarse sandstones until they are broken open, when they reveal centres of dark, blackish, unweathered rock.

JOINTS AND EROSION

When one turns to the erosive as distinct from the weathering processes, the fundamental requirement of a rock already broken is again apparent. This can be readily appreciated with reference to marine erosion. Earlier in this chapter the probable effects of boulders flung against unjointed cliffs of hard rocks were discussed and it was concluded that these effects would be limited to bruising and abrasion and as such would not be a very powerful process of erosion. Most accounts of marine erosion of hard rocks stress the importance of the compression of air in cracks in the rock, and to this might be added the effects of marine scouring in such cracks, the 'suction'

PLATE 10. Progressive opening of joints, Barmouth Grits, Foel Penolau, near Trawsfynydd. Joints wide open at edge of outcrop but nearly closed lower down in picture

exerted by the waves, and possibly the effects of marine animals in forcing such cracks wider apart. All these processes demand that the rocks should already be jointed.

An example of the progressive opening of joints in Cambrian Barmouth Grits on Foel Penolau is shown by Plates 9, 10 and 11. Whether the process was freeze–thaw or glacial drag or both is uncertain.

Again many glacial processes of erosion are conditioned by the jointing of the rocks. The action of glacial plucking, which is important on the back wall of a corrie, on the downsteam side of a roche moutonnée, in fact on any heavily glaciated hard rock surface, absolutely requires the presence of joints. The process operates by the freezing of water on to the rocks and the pulling away of sections of the rock as the ice moves away. The tensile strength of ice is so much less than that of any hard rock that such a process cannot be held to pull away lumps of rock, unless that rock is already jointed. In both the marine and the glacial erosion processes outlined above, the scale of the jointing is obviously a governing factor in the rapidity of the erosion, for the smaller the jointed block the more readily it will be moved by either of the processes cited.

Not only does the jointing govern the resistance to erosion directly through the way in which it helps or hinders the main erosive processes, but it also has a great effect on the permeability of rocks which in turn affects their resistance to erosion. It is true that in certain rocks the permeability is a function of the rate at which water can creep through the voids in the rock, but rocks of this sort are usually unconsolidated sands and gravels. It is the type of water movement which probably occurs on sand dunes, though even here there may be a concentration of water movement down such slightly easier paths as those provided by plant rootlets. A similar type of movement may also occur in unconsolidated and hence unjointed sands of earlier dates, for example the Eocene Bagshot Sands and the Cretaceous Folkestone Beds. Even in Chalk, which is uncemented and can readily absorb water in the mass of the rock, the main water movements are undoubtedly via the multitude of small joints which traverse the rock in all directions. This fact is familiar to those seeking a water supply in Chalk. If the rock were everywhere equally permeable as might be expected had water moved through the mass of the rock, all wells might be expected to produce comparable yields of water. But it is well known that certain wells are much more productive than others and that fissures require to be cut to ensure a large water supply. In some cases underground passages are cut in order to cut into a number of water-bearing fissures and so guarantee a constant profitable supply of water. In more cemented and consolidated rocks the role of joints becomes even greater. A rock such as Carboniferous Limestone, which is strongly cemented, is not permeable

PLATE 11. Widely opened joints, Barmouth Grits, Foel Penolau, near Trawsfynydd. Compare with Plates 9 and 10

through the main mass of the rock but only along the joints and other planes of weakness such as faults. Thus all erosive activity is concentrated along the joint planes and the more these are opened up the more permeable the rock becomes. The same is true of other rocks provided that they are sufficiently susceptible to chemical attack to have their joints opened by percolating underground water. The more the rock becomes permeable the smaller the amount of surface run-off and hence the smaller the susceptibility of the rock to surface erosion.

When one turns to the detailed relief forms of rocks, it is found that this depends just as much on jointing as does the general resistance of the rock to denudation. Where the whole rock is covered with a waste mantle and vegetation, the effects of jointing may be neutralised or obscured, but wherever bare rock surfaces are exposed, for example on sea cliffs, in all presently and recently glaciated hard rock districts and also in many deserts, joints mould the detail of the relief. This is more obvious in some rocks than in others. The large scale cuboidal jointing of the granite near Land's End or the polygonal columnar jointing of the basalts in the Giant's Causeway in Northern Ireland or in some of the islands off the west coast of Scotland are excellent examples. In sea cliffs the jointing plus the dip of the beds is important in governing the form of the profile. Other features in which the fundamental control is jointing come readily to mind, for example tors and inselbergs and sea stacks.

Thus, in this discussion of rock characteristics and denudation, it may be seen that, whatever processes of weathering and erosion are being considered, the influence of jointing can hardly be overstressed. If one thinks of the landscape in terms of a cycle of erosion the effects of jointing will be more pronounced in the early stages of the cycle, when lithological control as a whole is greatest, and will in some cycles decline with increasing age, when lithology tends to be obscured beneath the waste mantle. It should be noted, however, that some cycles, for example that which produces pediplains and inselbergs in semi-arid regions, involve a strong jointing influence right to the very end.

3
Igneous rocks and relief

Igneous rocks are those formed from the crystallisation of magma (a silicate melt) into rock-forming minerals. Although the origin of magma is a question far removed from the effects of igneous rocks on relief, it must be realised that molten material seems to occur at different levels in the earth's mantle. A variety of reasons have been suggested to explain the presence of molten material, but none of them stands up well to close examination. It has been suggested, for example, that very deep burial might result in fusion. Alternatively, local lowering of pressure within the earth's crust might so lower the melting point that fusion takes place. Again, excess heating, due possibly to greater concentrations of radioactive minerals, might produce similar effects. The incidence of these factors would be closely coincident with earth movements, and it is certainly true that many of the world's igneous rock masses seem closely related to such movements both in time and place.

The feasibility of the explanations of the formation of liquid magma has been discussed by Turner and Verhoogen (1960). Present ideas of the structure of the earth reject any suggestion that there exists a liquid core to the earth or a liquid layer at any depth. Although temperature increases with depth at a rate of approximately 30 degrees Centigrade per 1000 m (16 degrees F per 1000 ft) i.e. there is a temperature gradient which results in the outward flow of heat from the earth, the pressure due to the overlying rocks so increases that the melting points of all rocks at all levels are raised above the prevailing temperature. Surface flows of basalt usually appear at temperatures of about 1200 degrees C, and calculations suggest that lavas appearing at the surface at such temperatures are likely to have formed at depths of 50–100 km. At greater depths the melting points would be so high that, even allowing for heat lost during flow to the surface, the temperatures observed at the earth's surface would have to be much higher than they in fact are: for example, at a depth of 500 km the pressure is 150,000 bars (one bar is approximately the pressure of the earth's atmosphere) and the melting point of basalt would be in excess of 2000 degrees C.

Of the various possibilities the idea that magma is a residuum from a primeval state of the earth may be dismissed because it is uncertain that the earth ever was fluid and even had it been so it is incredible that liquid pockets could have survived. In connection with the relief of pressure hypothesis Turner and Verhoogen point out that, at the depths at which magma is generated, hydrostatic pressure conditions prevail so that local unloading does not lead to local relief of pressure, but to a slight general reduction. The hypothesis of deep burial seems to be vitiated by the occurrence of basalt flows in areas where there is no great load of rocks, e.g. the Hawaiian islands, and the absence of lavas from areas where the roots of mountains might be expected to be deeply buried, e.g. the Alps and the Himalayas. In the case of radioactive heat one would have thought that any initial irregularity in the distribution of radioactivity would have resulted in fusion of the rocks early in the earth's history and in the equalisation of the intensity of radioactivity. Thus some mechanism for ensuring a reconcentration would be required.

A possibility lies in the hypothesis of convection within the earth's mantle, an idea which has been used to explain a number of the earth's structural features, including both rifting and the development of fold mountains. If a mass of material could be transferred from depth without much loss of heat it might be brought into a part of the earth's mantle where its temperature remained above the pressure melting point at that level. It would, therefore, melt. But this is unproven: it is basing a hypothesis on a hypothesis. Turner and Verhoogen conclude that it may be that the temperature within the mantle is near to but usually slightly below the melting point and that slight variations in temperature, possibly caused by convection, may raise the temperature locally above the melting point, thus leading to the formation of magma.

Magma is not necessarily to be visualised as a mobile fluid readily penetrating the earth's crust. It is true that it can occur in this state, as is known from the outpouring of basaltic lavas which have formed such giant volcanoes as those of Hawaii and the great lava plateaus of the past. Further, certain residual fractions of the magma of some areas of igneous activity contain high concentrations of volatile constituents which act as fluxes and enable such material to penetrate widely into the country rocks (the rocks already present in any area where igneous activity is taking place), because the volatiles effectively reduce the freezing point of the magma. On the other hand it is fairly certain that much magma is a very pasty, viscous material which flows only slowly and which freezes or solidifies long before it has spread out into flat sheets. Indications of this are given by the form of the geologically young puys of the Central Plateau of France, which are often cited as their humpy form is due to the very viscous state of the acidic magma from which they formed. The same may well be true of many of the large masses of magma which crystallise deep within the crust and are only revealed as igneous intrusions by subsequent erosion. Under other conditions the magma may consist of a fluid portion containing and transporting a solid portion as a mass of mineral crystals, rather like a river, or perhaps better a glacier transporting its load. An example of this is the lava of Mount Vesuvius, which sometimes contains crystals of the felspathoid, leucite, and gives rise upon consolidation to a distinctive rock, formerly called leucitite but now termed leumafite by certain authors (Hatch, Wells and Wells 1961). This is not an exceptional occurrence because similar and related leucite-rich rocks occur in other parts of the world, e.g. among the igneous rocks related to the western branch of the East African rift valley. Not only may magma vary widely in its degree of mobility, its content of solids, but it may also have great variations in its bulk chemical composition and so give rise to a wide variety of rock types.

Magma becomes of geographical importance when its end product, consolidated igneous rock, is exposed at the surface of the earth. Some large part of this exposure may have been effected by denudation after the magma had consolidated, but it is mainly due to migration of the material towards the earth's surface. This may happen in two main ways.

The most obvious suggestion is that it migrates along fractures in the earth's crust, ultimately to consolidate as a mass of rock emplaced in those fractures or to appear at the earth's surface and spread out to a degree depending on the viscosity of the magma. The way in which many masses of igneous rock lie in relation to the country rocks is a clear indication that movement along fractures does take place, but the exact physical nature of the beginning of the process is not easy to imagine. The picture often invoked is one of rocks fracturing cleanly down to what may appropriately be called a magma chamber, where the change from liquid to solid is abrupt and complete. One would expect some transition from fracturable solids through semi-fluid to fluid materials, and clean injections into fissures are less easily understandable in such a context.

Again, the relation between earth movement causing fractures and igneous activity is not simple. Earth movements and igneous activity are not necessarily simultaneous, either wholly or in part. The Central Plateau of France provides an excellent example of this. This planed-off Hercynian massif was extensively fractured during the time of the Alpine folding. Movement along many of the faults had been taking place for a considerable time before this: for example, the marginal faults of the Limagne (the extensive basin in the middle Allier valley) were active during the first half of the Tertiary when the basin was filled with Oligocene sediments as fast as it subsided between the faults. At the height of the mid-Tertiary Alpine folding considerable movement took place along these and many other faults resulting in the shattering of the Central Plateau into a mass of uplifted blocks and rift valleys. If any time seemed propitious for igneous activity it was at this stage, but volcanic activity was much later, late Miocene, Pliocene and Pleistocene. It may be that the faulting is largely related to a period when compression prevails so that the fissures are not open, and that volcanic activity cannot occur until the compression is relaxed and the fissures opened.

The non-coincidence of major phases of volcanic activity and earth movements is quite common, however, and it is not always true even that earth movements are succeeded by vulcanicity. The volcanic areas of western Scotland provide an example of this. This area which formed part of a great Atlantic volcanic province is represented today by lava plateaus in Skye, Mull and Antrim, dyke swarms in Mull and Arran, and intrusive activity in Skye, Ardnamurchan and Arran. The lava flows are attributed to the Eocene from intercalated beds containing leaf impressions, though the dating of many other features is by analogy. Yet the main Alpine movements, which were presumably responsible for the uplift of the block mountains of Palaeozoic Britain, are later than this. In fact, the relations between vulcanicity and block uplift seem to be the reverse of those prevailing in central France.

Ancient examples of such non-coincidence are also known. The Lower Palaeozoic (i.e. the Cambrian, Ordovician and Silurian periods) was an era of geosynclinal sedimentation in Wales and ended in the Caledonian earth movements, which culminated between the Silurian and Devonian periods. There were earth movements of varying magnitude all through the Lower Palaeozoic just as preliminary movements to the Alpine folding occurred for millions of years before that orogeny. Yet the main folding period is not marked by the greatest outburst of volcanic activity, nor is the period immediately preceding it, the Silurian, which has few igneous rocks, whereas the period before that, the Ordovician, is one of great subaerial and submarine volcanic activity.

These examples stress the lack of coincidence in time between the climaxes of earth movements and outbreaks of igneous activity. The lack of coincidence is also evident in the spatial distribution of the two. The Atlantic igneous province of early Tertiary times, which included western Scotland, Antrim, the Faroes, Iceland, Jan Mayen island and parts of eastern Greenland, was far removed from the centre of tectonic activity in the Alpine geosyncline. In the Alps volcanic rocks are rare, as they are also in the Himalayas. But other fold mountain chains are characterised by intensive volcanic activity, for example the Andean Cordilleras, the East Indies arcs, the mountains of Japan and North Island, New Zealand.

Many of the areas of intense volcanic activity seem to have been in block-faulted regions away from the main areas of folding. The Atlantic area and the Central Plateau of France have been quoted above and to these may be added the Deccan of India, and many areas adjacent to the East African rift valley systems, for example in Kenya. But it must not be forgotten that there are many areas of block-faulted mountains not associated with volcanic activity, for example, Scandinavia, many of the Hercynian blocks of central Europe and large parts of Africa.

The connections between earth movements and vulcanicity do not seem to be simple, although most major intrusions of igneous rock appear in regions of intense folding where they may only be revealed after prolonged and severe denudation, e.g. in the Highlands of Scotland or in the Pre-Cambrian Laurentian shield of Canada.

Forms of igneous rock masses

The variety of forms which may be assumed by masses of igneous rock is very great. However, certain of these tend to recur, so that some sort of classification as an aid to memory may be attempted. One must enquire, therefore, what properties of igneous rock masses, other than the detailed properties of the rocks themselves, are relevant to a discussion of the relief formed on them.

Very easily forgotten is the mere question of size. Large intrusions are known to range up to sizes of approximately 1600 by 160 km (1000 by 100 miles), and the scale can decrease down to a volcanic plug eroded out from its surrounding ash beds and just large enough to have a chapel perched upon it with little space for further development, or to a minor dyke a few inches wide. From a geomorphological point of view the massive intrusion is large enough to dominate the pattern of denudation. Its permeability, allied of course with the local precipitation, will help to determine the density of the drainage pattern; its original surface relief coupled with its jointing pattern will govern the form of the drainage pattern. Compared with this, the volcanic plug is only an incident, however spectacular, in a landscape dominated by the effects of other rocks. Again, if one thinks in terms of lava flows, one tongue from a single volcano forms a little embroidery in another landscape, but plateau basalt sheets the size of those in the Deccan or in the Snake river basin of the north-western United States form the landscape as a whole. The question of extent of outcrop, so easily forgotten or ignored, is of fundamental importance in relief development not only in areas of igneous rocks but in all types of rocks. This may be appreciated by reference to limestone outcrops, where on large outcrops the special conditions required for the development of karst are more readily realised than on small outcrops.

A second main feature often used in classifying igneous rock forms is whether they are intrusive or extrusive. Extrusive rocks are those which are poured out on to the earth's surface; intrusive rocks are those which cool and solidify within the crust. The former may affect relief immediately, the latter only secondarily after they have been exposed by denudation. For this reason the division into intrusives and extrusives is very useful to the geomorphologist. The intrusive rocks themselves are often divided into major and minor intrusions, sometimes called respectively plutonic and hypabyssal rocks.

Another advantage of the division is that it reflects fairly accurately rates of cooling and hence crystal size, which in turn has a bearing on weathering.

The extrusive rocks generally cool the fastest and hence give rise to the finest-grained rocks, the extreme case being where cooling is so rapid that crystallisation is inhibited and glassy rocks result. Minor intrusions give rise to medium grain sizes, while the coarsely-crystalline rocks are found in the major intrusions or plutons.

Unfortunately this general rule about crystal grain size does not hold sufficiently rigidly for rocks to be differentiated easily on this basis. The size of the igneous mass largely controls its rate of cooling and complicates the effects of its being an extrusive or a minor intrusive, so affecting the crystal grain size. A lava sheet 12 m (40 ft) thick might cool at a rate much nearer to that of a 12 m thick sill than to that of a lava flow a mere couple of feet in thickness. Hence, the simple differentiation of extrusives from minor intrusives on grain size tends to break down. Both the lava sheet and the sill may have glassy margins where cooling has been most rapid.

Again, many minor intrusions and some lavas have a porphyritic texture in which large well-formed crystals (phenocrysts) of one mineral are set in a more-finely crystalline ground mass of other minerals. These phenocrysts, which tend in lavas to be of minerals crystallising at high temperatures and hence earliest in the cooling history, may represent slow cooling at depth followed by more rapid cooling leading to the formation of the ground mass when the magma is injected near to or on to the earth's surface. But this is not a complete explanation, for phenocrysts are absent from the glassy margins of many sills and they should be present on the above hypothesis. Such porphyries may be explained by the early crystallisation of one mineral followed by the more rapid later crystallisation of the rock as a whole when it reached a greater degree of supercooling (Turner and Verhoogen 1960). In these conditions the margins might be cooled so fast that there was no opportunity for one mineral to crystallise first. Whatever the explanation, it means that there is a class of generally medium-grained rocks, the porphyries, which also contain large crystals.

Within large intrusions there may be considerable variations between coarsely-crystalline fractions, perhaps due to the early crystallisation of one mineral which concentrated itself by settling under the effect of gravity, normal grain sized rock in the general mass of the intrusion, and finer-grained rocks where the margins of the intrusion have been cooled more rapidly. The last may well overlap the coarser parts of minor intrusions in grain size.

Finally, pegmatites, which are minor intrusions formed by the concentration of volatiles often occurring in the late stages of cooling, especially of granitic magma, and injected into the country rocks, are the most coarsely-crystalline rocks known. Huge crystals, often of rare minerals, e.g. muscovite, tourmaline, beryl, topaz, fluorspar, may be found. A beryl crystal, 18 ft long by 4 ft in diameter (5·5 by 1·2 m) and weighing 18 tons was found in Maine and muscovite (white mica) crystals up to 100 pounds

(45 kg) in weight are known from Pennsylvania (Dana and Ford 1932). Needless to say, these are exceptions and most pegmatites look like excessively coarse varieties of related plutonic rocks. Such rocks are often the source of gem stones either directly or indirectly through their erosion and the natural concentration of gem stones, many of which have high specific gravities, as placer deposits in river valleys. Fortunately for the welfare of the gem trade the largest crystals rarely have that degree of perfection required for the ornamentation of the human figure and are used as abrasive materials if they are used at all.

Although it is significant when discussing landforms developed, on modern igneous rocks to differentiate between lava flows and sills, simply because the one is completely exposed at the surface and the other is not, this does not hold for ancient igneous rocks where the lava flows may have been buried beneath later sediments and act like sills in the formation of relief. The differences are of interest to the geologist engaged in palaeogeographic reconstructions and may be inferred from the nature of the contact between the igneous rocks and the sediments containing them, but this has little relevance to relief development.

A third feature of igneous intrusion significant to the geomorphologist is whether the intrusions are concordant or discordant with the bedding of the intruded sediments or with the general grain of areas of more complex structure. The importance of this lies in the fact that concordant intrusions, provided that they are not very large, behave as harder or softer members of the existing rock successions and so accentuate but do not fundamentally change relief patterns. Discordant intrusions, on the other hand, produce an element out of harmony with the trend of the relief, for example a swarm of dykes or a rash of volcanic necks. In the case of intrusions so large that they dominate the relief the division into concordant and discordant is somewhat academic.

We may, therefore, produce a classification of igneous rock masses on the following lines:

I. Intrusions
   1 Major
     a Concordant: lopoliths.
     b Discordant: batholiths with associated bosses and stocks, diapirs.
   2 Minor
     a Concordant: sills, laccoliths, bysmaliths, phacoliths.
     b Discordant: dykes, ring complexes to include cone sheets, ring dykes and cauldron subsidence.
II. Extrusions
   1 Large: lava, ignimbrite and ash plateaus.
   2 Small: individual volcanoes, to include lava flows, cones, ash beds, volcanic necks or plugs.

It should be noted that the classification of extrusions is, bluntly, messy.

67

The lava flow is an extrusion: the feed, which later forms a volcanic plug, is strictly an intrusion: ash beds and plateaus may technically be regarded as aeolian sediments. But they are all so obviously related that they are best grouped as extrusions.

These various categories are described more fully below, but it must always be borne in mind that the categories are artificial and that natural forms of intermediate type occur.

Intrusive forms

LOPOLITHS

Some of the world's largest basic intrusive complexes affect a saucer-like form (Fig. 3.1), which is generally concordant in its relations with the country rocks. The generally concave-upwards surfaces may perhaps be due to sagging beneath the weight of the accumulating intrusive rocks. These intrusive complexes range in size up to features the size of the great batholiths.

The largest intrusion attributed to the lopolith class is the Bushveld Complex, which occupies a surface area of some 55,000 sq. km (20,000 sq. miles) in the Transvaal. It is essentially a layered intrusion of gabbros, norites and allied basic rocks, the central parts being concealed beneath a

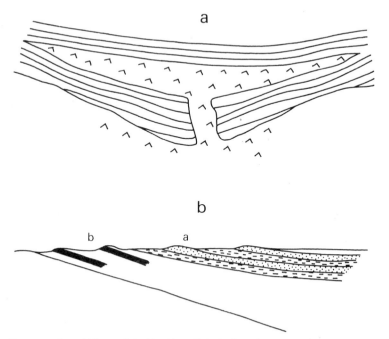

FIG. 3.1. Lopoliths: a. Idealised lopolith. b. Development of outward-facing scarps in the overlying sediment (a) and the layered igneous rocks of the intrusion (b)

roof of granitic rocks. Its thickness is estimated to range up to 8 km (5 miles). This sounds colossal until one remembers that the extent of the intrusion is perhaps 480 km (300 miles) by nearly 160 km (100 miles). Effectively it is a vast sheet. The same is true of the type lopolith, that of Duluth (Minnesota), which covers an area of 40,000 sq. km (15,000 sq. miles) and is estimated at nearly 16 km (10 miles) in thickness. Many lopoliths and related forms, including the Bushveld Complex, are layered and there is a tendency for the development of outward-facing scarps with variations in the resistance of the various layers. As the intrusions are essentially con-cordant this scarp development may be continued in the rocks lying above and below the main basic intrusions, for example in the quartzites which dip beneath the Bushveld Complex (Fig. 3.1).

Perhaps related to lopoliths but very much smaller in size are funnel intrusions. These taper downwards far more steeply than lopoliths and some might be conceived as emplaced in volcanic craters or calderas. Often they are layered, for example the Skaergaard intrusion of eastern Green-land. They differ from lopoliths not only in size but in the way in which the inward dip of the layers is far less than that of the sides of the funnel. Their layering is of course potentially likely to lead to a circular relief pattern of outward-facing scarps though this may be on a very small scale compared with that produced by a layered intrusion the size of a lopolith.

BATHOLITHS AND ASSOCIATED FEATURES

Batholiths or bathyliths together with lopoliths are the largest known masses of igneous rock. They are nearly always granitic, i.e. they cover the range granite–adamellite–granodiorite, in composition and are in direct contrast with the great lava flows which are predominantly basaltic. Batholiths are found mostly in areas which geologically are fold mountains whether they have since been planed down or not. But not all fold mountains contain batholiths. In Britain they are very frequent in the Highlands of Scotland and very rare in Wales. They probably attain their highest frequency in the world's Pre-Cambrian shield areas, possibly be-cause there was a greater frequency of granitic magma then or alternatively because these areas have been so denuded that the very roots of the mountains have been exposed.

An idealised batholith and its associated features are shown in Fig. 3.2. Batholiths themselves are often extremely large. For example, the Mesozoic batholith forming much of the Coast Range of Alaska and British Columbia is 1650 km (1100 miles) long and averages 160 km (100 miles) in width; the Sierra Nevada, California, batholith is 640 km (400 miles) long by 88 km (55 miles) average width. To smaller forms of the same general type a variety of names have been given. Turner and Verhoogen would use the term batholith for any intrusion with a surface outcrop exceeding 105 sq. km (40 sq. miles). This is really quite small and would mean that many of the British intrusions could be termed batholiths, e.g. Dartmoor, the north

Arran granite, the Mourne Mountains, the Wicklow Mountains, the Lands End granite and many Scottish examples including the Criffel, Cairnsmore of Fleet and Aberdeen granites. Just not large enough to qualify would be the St Austell and Shap granites.

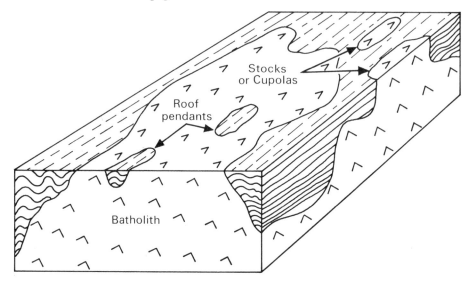

FIG. 3.2. Batholith and associated features

For smaller masses there are a number of available terms, stocks, bosses and cupolas. Stocks are defined by Hatch, Wells and Wells as features having surface areas less than 40 square miles. They may be small individual masses or merely unroofed portions of batholiths (Fig. 3.2). On the other hand, Hunt in his account of the Henry Mountains seems to use the term stock for discordant central intrusions formerly thought to be laccoliths. According to Hatch, Wells and Wells, bosses are small circular stocks. Of course, with some uparching of the strata above, these intrusions would not be radically different from the stocks of the Henry Mountains and, with fracture of the roof and forcing up of the magma in a fashion analogous to that of a salt dome, they become diapirs, which are scarcely recognisable except as a concept from bysmaliths (see p. 92). Cupolas are small protrusions of subjacent batholiths exposed at the surface. It was said above that the Devon and Cornwall intrusions of Dartmoor, Bodmin moor, Carn Menellis and Lands End are batholiths (because of size), and that of St Austell Moor is a stock. But all these are thought to be exposures of one large underlying mass—hence, the batholiths, on this hypothesis, are cupolas, too!

Although the student must be aware that the same words are used in slightly different senses by different authors, it is far better, because of the variety of natural forms, to keep a few simple terms and add qualifying descriptions; e.g. a small round stock possibly the top of a large subjacent batholith. What happens when every form has its own special term has been described by Hunt in his reference to the feeder to the Trachyte Mesa laccolith: to quote 'Because the form has certain resemblances to the woody structure of the cane cactus the name cactolith might be used and defined as a quasi-horizontal chonolith composed of anastomosing ductoliths whose distal ends curl like a harpolith, thin like a sphenolith, or bulge discordantly like an akmolith or ethmolith' (p. 151). It should be added that Hunt is not advocating but merely illustrating the possible terminology!

The origin of granitic batholiths presents a major geological problem. They are the commonest of the large intrusions and hence one would suppose that the spilling out of granitic magma at the surface would ensure that a comparable volcanic rock, i.e. a rhyolite, would be the most common extrusive rock, whereas rhyolites are in fact rather rare and local in occurrence. On the other hand, basalts, the most common volcanic rocks, might be thought to have some connection with granites, the most common intrusives, merely on the grounds that the most frequent of both types must surely be related. That the relation, if it exists, cannot be a very simple one is suggested by the fact that granites are confined to continental areas, where alone there is any great thickness of sial crust, while basic rocks with no associated granitic rocks are common in oceanic areas where little or no sial exists.

These difficulties and certain other observed features have been responsible for one view, that granites and similar rocks are not primarily magmatic but metamorphic in origin. They have been formed by the chemical and mineralogical changes associated with metasomatism, a form of metamorphism caused primarily by the replacement of original rock constituents by others usually in the presence of magmatic fluids. In an extreme form of this hypothesis it has been suggested that the changes take place by ionic diffusion through crystal lattices and along intergranular boundaries in a practically dry state, and are brought about by the upward migration of active ions (sodium, potassium, aluminium, silicon). In fact little is known about the feasibility of such processes and some authorities regard them as far too slow and limited to account for even small igneous bodies let alone batholiths. Again the presence of zoned crystals, ones in which bands of different composition alternate, and xenoliths, included masses of other rocks, is difficult to explain by a uniform ionic diffusion.

Yet certain features of the granite masses of ancient worn-down shields suggest no great evidence of intense heat associated with a liquid magma. These masses are often concordant with the enveloping metamorphic rocks and, if traced in the direction of increasing metamorphism, rocks which contain metamorphic minerals such as garnet may be seen to change into something much more granitic in composition with increased felspar content, as though a progressive metamorphic transformation of the rock had resulted in a granite. Furthermore, such granitic masses may not be surrounded, as are others at a higher level in the crust, by zones of progressive thermal metamorphism.

Other evidence, however, points more clearly to a liquid magmatic origin. The presence of rhyolites and acid dyke rocks shows that an acid magma must exist, though it does not prove a necessary connection with major acid intrusions. The general homogeneity of batholiths is not really consistent with a metasomatic origin, which should produce a much more general gradation of rock type. Further any local 'granitisation' observable at contacts may equally well have arisen from local changes due to the magma as from widespread metasomatic change. The general sharpness of batholith contacts and the finer-grained margins often observed are consistent with the freezing of a liquid magma. Finally, the progressive zones of thermal metamorphism found around some high level granites suggest that they were originally hot fluid material.

It may be that different granites have not all been formed by the same process, just as similar landforms, e.g. corries and spring heads, can be produced by totally different processes. There may be real differences in origin between deep-seated metamorphic granites and high-level magmatic granites.

Although granite magma has been regarded by some as a result of differentiation of a primary basaltic magma, there is no great evidence of this. In fact the lack of differentiation in oceanic areas would tend to be an argument against it. Some evidence consistent with differentiation is provided by tholeiites, basaltic rocks with interstitial acidic glass. If one could imagine interstitial material being squeezed out by some process an acidic magma might be produced, but only on a very small scale and, although perhaps sufficient to explain the local occurrence of rhyolites with basic rocks it would seem totally inadequate to explain batholiths.

It seems most likely that granitic magma may be produced by the fusion of the sial roots of continents and forced up through the crust in a fluid or semi-fluid form. The latter state is suggested by the preservation of flow structures in some of these rocks. A lack of mobility may help to explain one other problem, the relative scarcity of acid extrusive rocks similar in composition to granitic rocks. The injection of granitic magma into existing rocks may cause wholesale alteration by permeation of the injected rocks with the formation of mixed rocks, or migmatites, which are conceived as mixed igneous and metamorphic rocks of viscous, pasty character, capable of slow flowage.

Finally, there is the question of producing enough space for batholiths. If one reverts to the solid ionic diffusion idea or to the wholesale assimilation mentioned in the previous paragraph, much of what is now granite must have been something else and the space problem exists in a far less acute form. On the other hand, if batholiths are bottomless masses of granitic rock of magmatic origin, especially if they taper upwards as usually drawn, a tremendous disruption would inevitably seem to be involved and some signs of this should be visible. Some granites, e.g. that of Arran, appear to have been pushed up through the crust in a diapiric fashion and this can be seen in the outward dips of the adjacent rocks, but the degree of dislocation does not always seem to be sufficient. Several ways out of this difficulty have been suggested.

It may be that batholiths are primarily concordant, consisting of sheets, possibly several miles thick in places, below which country rocks reappear. This was essentially the idea of Cloos and Iddings. Geophysical work may finally resolve this question and it has already been shown that the Devon and Cornwall granites join underground and extend downwards for at least 16 km (10 miles). Alternatively the intrusion of the magma may be achieved by stoping, whereby magma is forced into the country rocks, lumps of which are broken off and appear in the magma as xenoliths and are finally assimilated in it.

The relief effects of batholiths may vary enormously both in type and degree. We might imagine an idealised batholith of large size and roughly circular or oval in form generating a radial drainage pattern either directly from its own slopes or by superimposition from the heaved-up strata above it. Development of drainage along joint directions might later control the pattern of many of the smaller tributaries while the exploitation of the joints by weathering would give rise to 'typical' granitic landscapes and in ideal examples to 'tors' (see below). More or less close approximations to these conditions may be provided by Dartmoor, Bodmin Moor and Arran, though the last is glaciated and has little obvious joint control on minor drainage pattern and a very imperfect radial general pattern.

But all sorts of departures can be expected so that the ideal or model batholith may be more or less a fiction. Many batholiths are found in worn down shields of crystalline rocks, e.g. the Laurentian Shield, the Fennoscandian Shield and the plateaus of Africa and peninsular India. All these areas have had long and complex erosional histories involving degradation probably over and over again by different agents at different times. Further, in these areas the granites are set in masses of crystalline metamorphic rocks, often essentially granite-gneisses similar in chemical and mineral composition to granites, perhaps also similar in origin if a metasomatic hypothesis is accepted for the formation of the granites, and very similar also in their joint patterns. Under these conditions the batholiths may have little overall relief effects. Such conditions may also be seen in the Highlands of Scotland where there are numerous granite intrusions, but where their effect on relief is very much less than that of the granites mentioned in the last paragraph.

On the other hand, where the granites are intruded into very different types of rocks the effects may be clearer, for example in the various intrusions of the Southern Uplands of Scotland, where the Criffel granite forms a dominant upland in the landscape, or in Eire, where the Wicklow Mountains are a fine example of a batholith forming a highland. However,

73

even the clear examples become less clear when looked at in more detail.

A comparison of the Criffel-Dalbeattie and the Loch Dee intrusions is instructive. Admittedly there are differences in the actual rocks of the intrusions, although both are broadly-speaking granitic; also, the former is intruded into Silurian sediments and the latter into Ordovician, but both of these are of roughly similar lithology. Both intrusions have a general overall effect on relief, but the effects are different and not entirely consistent in themselves.

Although Criffel stands out as a marked upland of comparatively simple form (Fig. 3.3), its extent is not coincident with the Criffel-Dalbeattie intrusion but only with the eastern end of it. The intrusion is granodiorite and composite but the relief does not seem to follow the different phases of intrusive rock either. The western end of the intrusion is much lower and more irregular in relief pattern than the Criffel highland. It is fairly obvious that there have been complications in erosion history, possibly both super-imposition of drainage and glaciation, which have blurred any simple relationships between intrusion and relief, so that, although the Criffel highland is granite, the granite is not coincident with the highland. Although there is an aureole from one half to one mile wide around the intrusion, this seems hardly to have affected the resistance of the Silurian sediments to denudation, so that there is nowhere any development of highland on the aureole.

The Loch Dee intrusion also shows a comparatively simple relation to relief (Fig. 3.4), but the relation is clear and completely opposite to that of the Criffel-Dalbeattie granodiorite. In the Loch Dee intrusion the igneous rocks are essentially eroded and their margin is generally coincident with a marked rise of relief. Further, differences of relief within the intrusion seem to reflect differences in the composition of the intrusive rocks. The central north–south ridge through Mullwharchar, reaching to above 600 m (2000 ft) and culminating at 680 m (2270 ft), is on granite, but the marginal parts of the intrusion are different, namely tonalite, a more basic rock with plagioclase felspar (Fig. 3.27). The latter has in general been eroded more than the granitic central ridge and forms a relative lowland about 300 m (1000 ft) lower in general than the central ridge. Surrounding the whole intrusion is a ring of highlands varying from place to place in elevation. They are at a maximum in the Merrick and the Rhinns of Kells respectively to the west and to the east of the intrusion, where long ridges rise above 600 m (2000 ft) and culminate at 830 m (2764 ft) in the Merrick and 800 m (2668 ft) in the Rhinns of Kells. To the north of the intrusion a lower ridge, culminating at a little above 510 m (1700 ft), is clearly in the same relative position as the eastern and western ridges. All three sides are on the aureole of contact metamorphism, which surrounds the intrusion in a similar way to that surrounding the Criffel-Dalbeattie complex. So far the general relations of the Loch Dee intrusion and the relief appear to be simple and systematic though opposite from those of the Criffel area.

But on the southern margin of the Loch Dee intrusion is yet another ridge of high ground reaching to 705 m (2350 ft). Unfortunately, although part of this is on the contact metamorphic aureole, it extends well south of the aureole as mapped and so spoils the simplicity of the general picture. Thus, these two intrusions show different relationships between zones of rocks and relief and in neither are the relationships consistently maintained.

Turning to the control of the detail of the landscape by granitic jointing, one finds that this, too, may vary considerably. The origin of joints and the general effects of joints on weathering and erosion have been considered in the last chapter. The effects of jointing on granitic landforms can be most pronounced. On the coast one may cite the Lands End cliffs for the marked control exerted by the cuboidal joint pattern and the same holds for some of the inland granites, for example in Arran or the Mourne Mountains, where fresh faces show this very well. The fresh faces have often been provided by geologically recent glaciation, especially by corrie formation, and probably the effects of joints are at a maximum in such places. Where, however, erosion has been less severe, the development of peat bogs on granites can effectively mask rock outcrops and any obvious joint control, as on the Shap intrusion in the Lake District or in parts of Bodmin Moor. Again the nature of the joint pattern may have a telling effect. In Arran there is a pronounced difference in detail between the coarse- and fine-grained granites in the main intrusion. This seems to be a reflection of differing jointing characteristics at least to a considerable degree. On the coarse-grained granite the bold joint pattern is responsible for the blocky east-facing cliffs of Beinn Tarsuinn and Beinn Nuis or the form of the ridge between Cir Mhòr and Goat Fell. Below the joint pattern level the granite seems to disintegrate into fragments the size of its individual mineral particles, i.e. to a fine gravel. The fine-grained granite seen in Beinn Bhreac and Beinn Bharrain on the west of the island breaks along a finer and more irregular joint pattern into pieces of intermediate size. The hills are more rounded, screes are conspicuous whereas they are absent from the areas of coarse-grained granite, and there is far less geometrical control of the detail of the landscape by joint patterns.

In different climates it is predominantly the effects of vegetation which control the extent to which joint influence is visible in the landscape. In glacial and arid climates the bare rock surfaces show the influences of joints most clearly especially in the inselbergs of tropical semi-arid countries. Where soil development is swift, as in the rainy tropics, joints may facilitate the penetration of deep weathering but are then masked by its effects. Even here, however, joint effects stand out very clearly where vegetation is minimal, for example in the sugarloaf mountains of eastern Brazil.

A special feature of many granitic landscapes is the presence of tors. On Dartmoor and Bodmin Moor, which may be taken as the type areas for these features, tors consist of small rocky hills usually varying in size up to

75

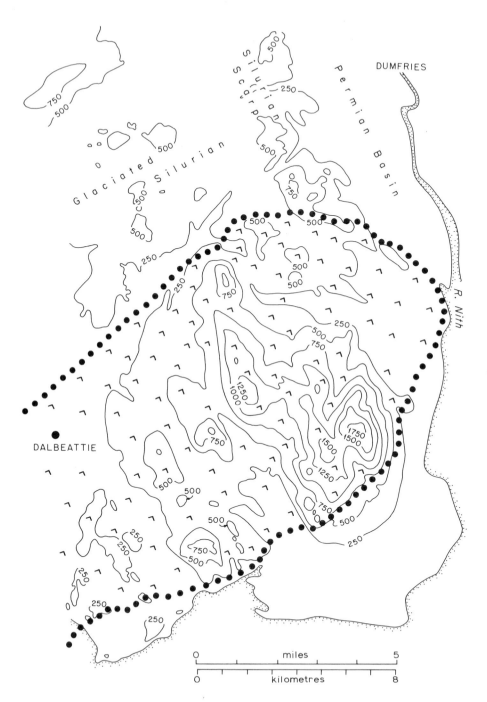

Fɪɢ. 3.3. The Criffel intrusion. Margin of intrusion shown by large dots

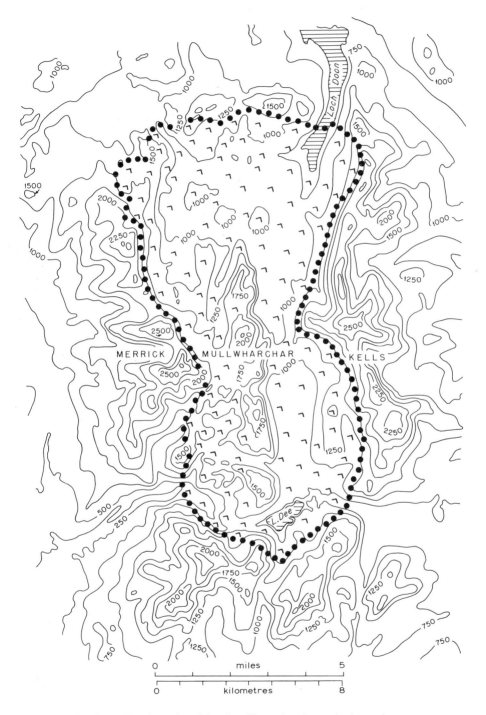

Fɪɢ. 3.4. The Loch Dee intrusion. Margin of intrusion shown by large dots

something of the magnitude of a small house. The detail of their form is obviously largely controlled by jointing. The term itself has been applied descriptively to comparable rock stacks on other outcrops, e.g. schists in New Zealand (Ward 1951), the features on the Stiperstones Quartzite in Shropshire (Plate 40), the dolerites of the Prescelly Hills and many others. In different areas tor has been used as a synonym for castle koppie (King 1948), which is a small variant on the inselbergs of tropical Africa. Although the term has mainly been used descriptively, it has been given a genetic connotation by Linton (1955), whose hypothesis of their origin is discussed below. A descriptive term is certainly needed. Otherwise the name of any given feature, if defined genetically, will have to differ when different hypotheses about its origin are applied by different geomorphologists. The term, rock-stack, which was mentioned above, was suggested by Linton, and, if it is possible to displace tor from its general usage, it seems a highly suitable one.

A hypothesis which has met with considerable approval is that which involves deep chemical weathering penetrating irregularly but generally guided by joint systems, so that masses of sound or nearly sound rock project upwards and are later stripped out mechanically from the waste mantle to form tors. This is the hypothesis used by Linton to explain the Dartmoor tors and the tor is defined by him as a feature produced in this way. The irregular penetration of the weathering, which is thought to have occurred under more tropical climatic conditions in the Pliocene, is explained by variations in the joint network (Fig. 3.5). Where the joints are close (Fig. 3.5a) weathering is more complete down to the water table and in these areas only small core-stones are left (Fig. 3.5b). Where the joints are wide apart the weathering is incomplete and the future tors are generated below the surface. Later mechanical removal of the weathered debris, perhaps by meltwater or by solifluxion in the Last Glaciation, revealed the tors (Fig. 3.5c). Some of the core-stones of the weathered closely-jointed granite may be left on the surface, i.e. the level of the former water table, or they may have been removed in the general mass movement of waste.

A similar idea had been used by Ward to explain the schist tors of central Otago in South Island, New Zealand. Tors here are so numerous as to form the main element of the landscape and to have suggested local geographical names, e.g. Raggedy Range and Knobby Range. These tors seem to be related to the exhumation of a Cretaceous peneplain developed on schist and greywacke. The peneplain was weathered to varying depths depending on the massiveness of the underlying schist. Where the schist is massive it rises from the general basement level towards the peneplain surface: where it is more jointed and hence less resistant it was weathered more deeply. Although the peneplain was later buried beneath 300 m (1000 ft) or more of non-marine sediments, exhumation has now removed the sediments and the weathered waste mantle of the schists, thus exposing the tors to view. Although the phase of formation and the phase of exhumation are farther apart than in Linton's hypothesis about the Dartmoor tors, the mechanism is virtually identical.

a

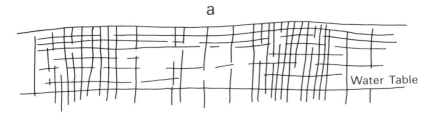

Water Table

b

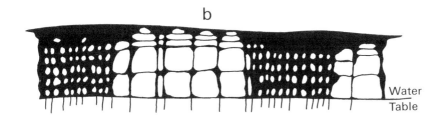

Water
Table

Original surface   c

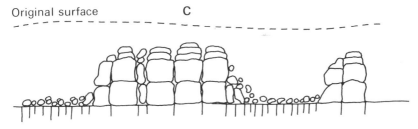

Fig. 3.5. Origin of tors *(after Linton)*. For explanation see text

A different interpretation of the Dartmoor tors has been proposed by Palmer and Neilson (1962). These authors find little sign of deep chemical weathering on the ridges with which the tors are associated. The maximum depth of rock alteration is found in the valleys and not on the ridges, where it should be if its depth is controlled by the level of the water table. Further, the alteration of the granite does not appear to have been controlled by the penetration of weathering agents from the surface, as in some places layers of altered and unaltered granite alternate. Such an arrangement is more consistent with pneumatolysis (see Chapter 4) of the granite, presumably along sheet joints, than with weathering. They quote modern geological opinion in favour of the hypothesis that the alteration of the

79

granites is mainly pneumatolytic, suggest that the valley directions are often controlled by zones of altered granite and emphasise the coincidence of tor formation with areas of sound granite. Where tors are found the granite debris is thin and has the appearance of being mechanically weathered for it contains few clay minerals. Excavations have failed to reveal any concentration of chemically weathered material between the joint boulders of the tors. Further, these authors do not consider that tors have the appearance of piles of sub-surface weathered core-stones, because, except at the edges of the tors, the joints are relatively unweathered. Where weathering has taken place at the margins it is said to be more readily explicable in terms of surface weathering. The upper surfaces of joint blocks tend to be more weathered than the lower surfaces as though due to rain. In addition, there tends to be a greater degree of weathering on the side facing the prevailing winds. Thus, the rounding of the joint masses of tors is regarded as a modification imposed subaerially on an original pattern of unrounded blocks.

Although there is much evidence inconsistent with the interpretation of tors as features of subsurface weathering, a positive hypothesis is still difficult. Palmer and Neilson suggest weathering under periglacial conditions. This would involve shattering along joints by freeze–thaw and the downhill sliding under periglacial conditions of boulders along joint planes. On steep slopes the amount of sliding would be great and the destruction of tors rapid: on gentle slopes sliding would be inhibited and the formation of tors retarded: on intermediate slopes one should expect the greatest development of tors. One finds it. Their precise location is thought to be controlled by variations in joint frequency, the more jointed areas being weathered most rapidly. In this last respect Palmer and Neilson are in agreement with Linton.

Whether the various types of tropical residuals, which are collectively called inselbergs, are related to tors is debatable. Their origin raises some of the major problems of geomorphology. It is true that they appear to be best developed on 'granitic' rocks, though not confined to them. Some authorities have believed in an identity of process regardless of climatic zone; others have not. The problems of scarp retreat and the formation of pediments are closely tied with the inselberg question. The majority view is that pediments have been formed not by lateral stream erosion but by sheetfloods. Or could it be that the sheetfloods are caused by the presence of pediments? The reader is referred to any good standard textbook on geomorphology for a discussion of the problems. Certain features, however, are to be much more directly attributed to the nature of the rock.

The first of these is the maintenance of the steep angle between the foot of the inselberg and the surrounding pediment. On King's hypothesis

PLATE 12. Glaciated granite, Pelvoux massif, French Alps. Dominance of process over rock type. Compare with Plate 13

inselbergs are the last remnants left by the parallel retreat of slopes away from rejuvenated river systems entrenched along major joint patterns. The maintenance of such a discontinuity must reflect either a change in process or a very marked change in intensity of process at that point. Basal attack, either by the sea or by lateral stream planation or by spring sapping, is often used as an explanation of such breaks in temperate climates but can hardly apply here. Instead a lithological explanation is often invoked. It is said that the granites break down into two calibres of debris, jointed boulders and mineral debris with material of intermediate size hardly existing or having a very short life. The change from the steep slopes of the inselberg to the gentle slopes of the pediment takes place where this calibre change occurs. It would be very interesting to have the results of detailed studies of the correlations between rocks, the sizes of their debris caused by weathering breakdown, and the angles of inselberg sides and pediments especially the sharpness of the transition between the two.

The detailed form of inselbergs is to a considerable extent a reflection of the nature of the jointing. The more massively jointed the rock the more massive the inselbergs, whereas castle koppies, which King has equated with tors, are on more closely jointed rock. Large massive inselbergs may develop their own joint systems parallel to their surfaces and so perpetuate their form due to progressive exfoliation. In this form they come very close to the sugarloaf mountains of Brazil.

Probably the most complete discussion of the possible relief effects of 'granitic' rocks is that by Wilhelmy (1958). Granitic relief is seen as an equilibrium between zonal and azonal effects. The zonal effects are provided by the latitudinal climatic belts and by altitudinal effects within a given climatic belt (compare Plates 12 and 13). The azonal effects are the landforms left from previously experienced climates and those caused by special microclimates within the main climatic zones. The interplay of these variable effects on rocks which may themselves vary subtly produces a great number of landform possibilities.

The general climatic variations show clearly. In the humid tropics deep weathering produces a reddish or kaolinised mantle in which core-stones are free in a mass of detritus. These may form block-fields if the fine material is removed. In tropical areas with a dominant wet season variously named domed hills and mountains may be formed by a massive shelling off of material. This may be facilitated by the action of salts crystallising below a zone weakened by expansion and contraction and it may in part be due to pressure release. Closely-spaced runnels may be formed on granites as they are also in the humid tropics. The tropical areas with dominant dry seasons are the type areas for inselbergs, although true

PLATE 13. Hercynian granite-gneiss block, Black Forest. Corrie developed at foot of Feldberg; otherwise generally smooth summit relief. Compare with Plate 12, which is some 2500 m higher and hence subject to different processes

inselbergs with an abrupt break of slope seem to require at least three months with over four inches (100 mm) of precipitation and four to seven months with less than two inches (50 mm). When the precipitation is less the wet season is insufficient for the removal of the debris mechanically weathered during the dry season: where the climate is wetter chemical weathering in the wet season produces more debris than can be removed. Both conditions lead to a concave profile instead of an abrupt break of slope. The wet–dry alternations lead to the breakdown of the granite into fine gravel. On the margins of the regions of seasonal rainfall and the arid regions night dews and diurnal heating lead to the capillary rise of dissolved salts, to a weathering of the interior of boulders and to the formation of surface crusts. Should weathering in shady situations breach the hard crust the whole boulder may become hollowed out to form taffoni. In truly arid areas taffoni do not develop, but hydration and the mechanical forces of crystallising salts lead to the formation of fine gravels and core-stones. Some of the features of the Mediterranean climatic areas resemble those of monsoon areas, e.g. the development of certain types of domed mountains. Deep weathering produces here masses of fine gravel and sound core-stones coated in limonite. The humid temperate regions show a great mixture of past and present forms. Periglacial processes and also normal erosion accelerated by forest clearance may reveal the core-stones, which were originally associated with deep tropical weathering in the Tertiary era, as block-fields or as tors. In the Arctic freeze–thaw produces masses of angular debris: this is the only area where the production of such debris is prevalent.

Altitudinal variations of the same type are also pronounced. In the Arctic these are obviously at a minimum, but in areas such as the Iberian peninsula and Corsica a zone of reduction to fine gravel occurs below about 1200 m (4000 ft). It is succeeded by a zone of formation of domed hills and block-fields, which in turn is replaced above 2400–2700 m (8000–9000 ft) by freeze–thaw action. Nearer the Equator, e.g. in Korea and Costa Rica, there may even be a complete altitudinal range from lateritic weathering to frost action.

Of the azonal effects those due to past climatic changes are most marked in the humid temperate regions and least marked in the Arctic and Equatorial regions. It seems that the combination of relics of past climates and altitudinal variations reaches its maximum in Corsica, which Wilhelmy regards as a laboratory for the study of granite landforms.

Azonal effects caused by local and microclimates may also be important. For example taffoni occur, outside the semi-arid zone, in maritime areas where strong sea winds cause arid microclimates due to high evaporation. Further away from the coast the necessary wetting and drying alternations do not obtain. Insolation effects and frost shattering are much more marked in continental than in maritime climates. Chemical weathering tends to be more marked near coasts than farther inland except in a few circumstances, such as in those areas where arid coasts are backed by high mountains giving increased precipitation and hence accelerated chemical weathering.

The possible combinations of all these effects make it very difficult to define granite relief in terms of the earth's main climatic zones. Micro-climates may overcome macroclimates, for example in the weathering of the bases of granitic monuments in Egypt where chemical weathering is more rapid on the shady sides. Local heating of bare rocks in the Arctic for short periods of the year may provoke the spheroidal weathering more typical of hot deserts, for example in western Greenland. The reduction to fine gravel of coarse granites and the isolation in this mantle of core-stones may be caused by a variety of chemical and mechanical weathering processes and so lead to a superficial similarity between one climatic zone and another, while the same products, relics from periods of Tertiary weathering, may appear in the humid temperate areas of Europe. As a final complication, it must be remembered that certain coarse massive gneisses, conglomerates and quartzites may produce landforms generally similar to those usually experienced on granite.

These generalisations given by Wilhelmy might not all be accepted by every geomorphologist. Some of them may be open to change in the light of either more detailed or more extensive work, but they do show the bewildering variety of forms liable to occur on rocks of similar type. Simple universal generalisations about the effects of the coarse granitic rocks of batholiths on relief are hardly possible, for the generally similar jointing patterns, mineral composition and degree of crystallinity react in very different ways to present and past weathering processes.

SILLS

Sills are minor intrusions approximately concordant with the bedding planes of the rocks. Their attitude should be defined in relation to rock stratification rather than in relation to the horizontal. The most marked relief effects occur when the sills and the rocks they are contained in are almost horizontal, but they can be at any angle to the horizontal either as the result of being injected into tilted strata or as the result of earth move-ments after emplacement. Vertical sills might thus be considered as identical in their effects on relief to normal dykes, but it will be realised that the entirely different attitude of the enclosing rocks will have also a marked effect on the relief.

Sills merge imperceptibly towards laccolith-like forms depending on the fluidity of the magma, the rate of cooling of the magma at the extremities of the sill, the resistance of the rock to the injection of the magma and the pressure of the overlying mass of rocks. Resistance to flow may result either in a dome-like form leading to a laccolith or to the bursting out of the magma as a fissure opens up and the movement later of the magma along an easier stratum. The latter form is called a transgressive sill, a self-defining term. A certain amount of load is probably necessary for the formation of sills, for, if magma came too near to the surface, it would certainly be easier for it to burst through to the surface than to force its way along bedding planes. The tendency for sills to be of basic rocks, very often dolerites, illustrates the importance of fluidity, whereas laccoliths are often of much more viscous acid magma.

Sills may vary considerably in thickness from a metre or so to a hundred metres or so (a few feet to a few hundred feet). Their relief effects are not greatly different from those of old lava flows which have been buried in the general later accumulation of rocks. They may be distinguished by a number of criteria. Size of crystals is not a foolproof method because this reflects rate of cooling. In this the size of the igneous body is probably of more importance than its exact origin. The rocks adjacent to a sill should be baked on both sides, whereas only below a lava flow will the rocks normally be baked, for the rock above would normally have been deposited after the lava had cooled and consolidated. The structure of the upper side of a lava flow will reflect its cooling and consolidation in the open, whereas a sill will be different. The upper side of a sill may include dislodged fragments of the rock above, but a lava flow cannot. However, these differences are all of more interest to geologist than to geomorphologist: the relief effects are not all that different.

Generally sills will act as harder members of the sedimentary or metamorphic series into which they have been intruded and often become pronounced scarp-formers, or, if more nearly horizontal, ledge-formers in plateau regions. They provide steeper sections if not actual waterfalls in the courses of the rivers they cross. At the same time they are not quite identical in effect to harder sediments because they usually have pronounced joint patterns of their own. The isotherms in a cooling sill are parallel to the marginal surfaces so that the jointing tends to be at right angles to the surfaces of sills. This tends to cause a pattern of columnar jointing similar to that observed in lava flows and accentuating the similarities between the two. Even where the jointing is not perfectly columnar the superior hardness of the rock and its well-developed jointing give in detail a distinctive mark to the relief.

Examples of the effects of sills on relief can be found in a variety of places in the British Isles, for this type of intrusion is found in a number of rocks of very different ages.

The youngest sills are of Tertiary age and occur in the igneous areas of western Scotland and Northern Ireland. Typical of this series are the sills of Arran, which are mostly intruded into Permo-Triassic sediments in the southern half of the island and usually lie almost horizontally. They range in composition from rocks of doleritic type to much more acid quartz-porphyry and pitchstone. The first two types of rock give rise to thick sills but the pitchstone sills are very thin. Many of the sills form scarps, though these vary somewhat in form. For example, the doleritic Clauchlands sill forms a north-facing escarpment south of Brodick. Quartz porphyry sills are important in the formation of Bennan Head overlooking the dyke swarm so well displayed on the southern shore of Arran and at Drumadoon where there is a splendid columnar-jointed sill in the south-west of the island. In Glen Ashdale in the south-east of the island sills form marked waterfalls in the stream course: a very good example is the Baoileig Sill which shows columnar jointed igneous rock lying between horizontal sediments. Holy Island near Lamlash rises abruptly from the sea to a height of 310 m (1030 ft) and is a triple sill formed from riebeckite trachyte. In fact, sills are generally the most spectacular relief-formers in southern Arran, much more so than the dykes with which they are associated.

Considerably older is the Whin Sill of north-eastern England, a quartz-dolerite sill of Carboniferous age. Intruded into the local Carboniferous sediments and reaching a maximum thickness of 60 m (200 ft), though 30 m (100 ft) is about its average, the Whin Sill shows many features typical of its class. It usually follows the bedding and so faithfully reflects the gentle folds and faults that have affected the Carboniferous rocks, but it is transgressive here and there and splits in places into two or more separate sheets. In many ways it behaves like the sills of south Arran. Locally, especially near Housesteads and Crag Lough (Plate 7), it forms a very bold, steep escarpment on which the Roman Wall is built. Where it cuts across stream courses it may give rise to waterfalls—the best is High Force in Upper Teesdale where vertical jointing effects are well displayed (Plate 14). Near the coast in north Northumberland it forms small scarps and crags often used for the sites of castles, for example Lindisfarne Castle on Holy Island, Bamburgh Castle and Dunstanburgh Castle. The Farne Islands are tilted slabs of Whin Sill dolerite, a fact which can be clearly seen even from the mainland coast. Finally, in the detail of relief the very rough polygonal jointing pattern is a feature of the dolerite and serves to differentiate it from the massive sandstones in which it is often exposed along the coast.

Quartz dolerite sills of Permo-Carboniferous age also occur among the igneous rocks of the Midland Valley of Scotland especially in Fife. In places they form marked hills and scarps. The crag on which Stirling Castle stands is part of one such sheet. It exemplifies the detailed control of relief by its well-developed joint pattern. These Scottish sills are of the same order of thickness as the Whin Sill.

LACCOLITHS

Laccoliths may well be illustrated from the area in which they were first described, namely the Henry Mountains of southern Utah (Gilbert 1880). In this remote semi-arid region dominantly diorite-porphyry has been injected into a series of mainly Mesozoic shales and sandstones of far less resistance to erosion. The igneous rocks in the form of laccoliths with associated dykes and sills have been exhumed to a considerable extent.

The complexity of the forms varies considerably. So does their interpre-

tation for the area has been surveyed in much more recent times by Hunt (1953), whose ideas and nomenclature differ from those of Gilbert.

Mount Ellsworth, a domed area associated with a swarm of dykes was thought by Gilbert to represent the typical simple large laccolith (Fig. 3.6a). The displacement here is of an area of about 6 by 5 km (4 by 3 miles) which has been updomed about 1500 m (5000 ft), but the exact area involved is difficult to measure because the dip flattens gradually. This probably accounts for the fact that Hunt later gave the diameter of the structural domes as 10 to 13 km (6 to 8 miles).

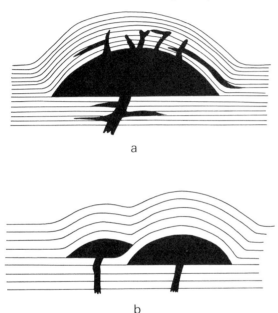

a

b

FIG. 3.6. Laccoliths: a. Simple. b. Compound
(after Gilbert)

Mount Holmes (Fig. 3.6b) represented to Gilbert the next stage of complexity, for two laccoliths were thought to produce this double, overlapping dome, each half of which is about the same area as Mount Ellsworth. The uplifts involved are smaller, about 90 m (300 ft) and 450 m (1500 ft) for the two domes of Mount Holmes.

Mount Hillers is the central element of a complex of laccoliths. Mount Hillers itself, in which the igneous mass is partly exposed, is a very steep form, 6·5 by 6 km (4 by 3¾ miles) and 2100 m (7000 ft) in vertical amplitude. It is asymmetric. Included in the same group, in which all the laccoliths are not at the same level, is the Steward laccolith (the Black Mesa

PLATE 14. Whin Sill dolerite, High Force, Teesdale. Dolerites, with rough vertical jointing, overlying Carboniferous Limestone and responsible for the fall. *(Photo by R. J. Sparks)*

bysmalith of Hunt). This is about 4 by 2·5 km (2½ by 1½ miles) in extent and almost perfectly exhumed from the soft sandstone of the Flaming Gorge group (the Morrison formation of Hunt). Within this laccolith complex Mount Hillers itself may be very thick but the Howell laccolith (the Trachyte Mesa laccolith of Hunt) provides an opposite example of a shallow form 60 m (200 ft) broad and a mere 15 to 30 m (50 to 100 ft) thick. Its affinity to a sill is thus obvious.

Within this laccolith region there is an estimated range of 1000 to 1 in the volume of the laccoliths and a considerable variation in their effects on relief. Where the laccoliths have not been reached by denudation they have little effect on relief, because the updomed sedimentary rocks have not been planed off. Where they are exposed their effects depend on their relative degree of complexity and the pattern of associated dykes and sills. The Steward laccolith is the most perfectly exhumed of the mushroom-like forms.

Some differences in the nomenclature of the features is obvious from the above, where both Gilbert's terms and Hunt's terms have been given. These differences go beyond names into interpretation. Gilbert's idea seems to have been that a laccolith complex consisted of one very large major laccolith (which could occur alone as in Mount Ellsworth) with associated minor laccoliths, sills and dykes, as shown in Fig. 3.6a. Hunt's interpretation is that the central major elements of each domed up area are in fact stocks, with which are associated dykes, sills, laccoliths and bysmaliths. Such an interpretation is shown diagrammatically in Fig. 3.7. The form

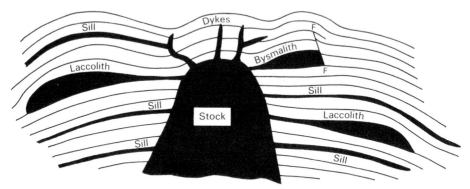

Fig. 3.7. Laccoliths and associated features

of the central intrusion is, of course, largely inferred, but all bodies known positively to be laccoliths are very much smaller and there is a difference in form between the conjectured main laccolith and the associated minor laccoliths surrounding it.

Although it is very difficult to get structural information in the vertical plane, Hunt points out that the central intrusions are rarely as concordant as they should be were they true laccoliths. Further, the small marginal bodies, some of which can be shown to be true laccoliths, are different in form from the main central intrusions. The smaller laccoliths are generally tongue-shaped bodies, while the conjectured central laccoliths would be more nearly mushroom in form.

Hunt's interpretation revives an earlier hypothesis of a plutonic plug put forward by Russell at the end of the last century (Harker 1909) to explain similar forms in the Black Hills of South Dakota. From the plug horizontal intrusions of viscous magma gave rise to sills and laccoliths, the whole very much resembling the so-called cedar tree laccolith (Fig. 3.8) except with regard to the source of the magma.

FIG. 3.8. Cedar-tree laccolith

Whatever the main magmatic source, a pipe or a stock, laccoliths have certain distinctive features. The magma causes a physical doming and breaching of the rocks. The updoming clearly shows that there is no question of the intrusion merely melting the country rocks and replacing them. In this respect the central intrusive member of a laccolith complex differs from the offshoots of batholiths. In fact, as Gilbert and Hunt both stress, the intrusion is the cause of the folding and not its effect. In this respect laccoliths are clearly distinct from phacoliths (see below). Laccoliths can only really exist in areas of simple structure with horizontal or near horizontal beds. The viscous magma—laccoliths seem to have been formed only from this type of material—effects a bodily uplift of the roof of sediments, whereas more fluid magma would have insinuated itself along the bedding.

It would seem that the lower parts of the intrusion of the central stock must be discordant, but as soon as there is a tendency for the intrusion to lift the overlying rocks lateral intrusion in the form of sills and laccoliths must take place and in turn aid the lifting. The lifting causes tensional cracks in the sediments above the laccolith and these are injected to form the commonly associated dykes.

In areas of complex rock structure lines of weakness, such as faults, shatter belts and the like, would tend to be followed by the magma and the ideal, comparatively simple laccolithic doming would not occur. Different igneous forms would be produced.

One question remains: as the stock intrudes and domes up the sediments one would have expected the spaces to be occupied by sills thinning away from the stock. Yet laccoliths seem to thicken away from the stock. This may be related to cooling effects which are further considered under bysmaliths below.

BYSMALITHS

Bysmaliths are only a faulted variant of laccoliths. In the intrusion of a normal laccolith one may imagine the magma, which is presumed to have been at a relatively low temperature, cooling and solidifying at its thin outer edge. Newly injected magma would increase the convexity of the laccolith until the rocks above fractured along faults (Fig. 3.9a).

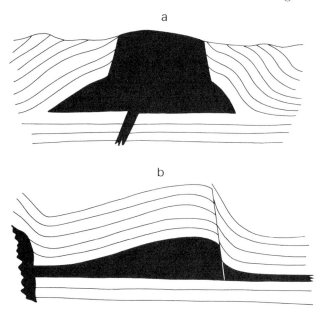

FIG. 3.9. Bysmaliths: a. Ideal bysmalith after erosion.
b. Asymmetric bysmalith with trap-door faulting

In Hunt's interpretation of the Henry Mountains the bysmaliths are not symmetrically faulted as shown in Fig. 3.9a but in a trap-door fashion as suggested in Fig. 3.9b. In this case one might assume that heat was greatest near the stock to the left, hence fluidity was at its highest there so that the magma penetrated the sediments easily. At the end of the intrusive tongue cooling would have been greatest, thus leading to greater viscosity and the piling up of magma, which eventually resulted in the trap-door faulting.

The intrusion on which the bysmalith concept was originally based by Iddings is Mount Holmes in the Yellowstone National Park.

Both laccoliths and bysmaliths may have considerable relief effects. It has been stated above that these forms are only possible in areas of simple near-horizontal sedimentary structures. Horizontal sediments of varying resistance, especially in the semi-arid American West, give rise to plateaus and escarpments ribbed by the outcrops of more resistant beds. In these conditions the domal uplift and exposure of laccolith complexes gives rise to higher mountains with radial dissection patterns, jointed cliffs where the sills, laccoliths and bysmaliths are exposed, and occasional local ridges where the dykes are more resistant to erosion than the sediments. In the particular example of the Henry Mountains the difference in resistance to denudation between the diorite porphyry and the Mesozoic sediments is so great that almost perfect exhumation has occurred in places. The effect of the intrusions extends beyond the igneous masses themselves. The sediments are uplifted in varying degree up to verticality. This introduces a departure from the plateau form in that the intrusions are fringed by inward facing scarps and hogback ridges depending on the angle of dip and the variation in resistance to erosion of the beds involved.

In Scotland the phonolite laccolith of Traprain Law 32 km (20 miles) east of Edinburgh is a pronounced hill (Plate 8b), but the adjacent trachyte laccolith at Pencraig Wood on the other side of the Tyne (not the Northumberland Tyne) has little effect.

PHACOLITHS

Phacoliths are lens-shaped masses occupying the crests of anticlines and the troughs of synclines (Fig. 3.10). They occur in such positions because of the tendency of any magma to migrate to positions where tension and weakness are present in the crust. At first sight they resemble laccoliths but they are the results of the folding and not the cause of the updoming of rocks as are laccoliths. Furthermore, laccoliths are approximately circular, but phacoliths are not, except in the rare case of their being formed in a dome. Usually they are elongated in the direction of the axes of the folds, so that, although in cross-section they resemble lenses, in plan they do not.

FIG. 3.10. Phacoliths

93

In their effects on relief they are not greatly different from laccoliths and sills. If almost circular the effect will be almost identical with that of a laccolith, e.g. Corndon Hill (Plate 15) in the Welsh Borderland (Fig. 3.11), but if elongated and breached by erosion they would tend to form scarps running along folds in roughly the same way as sills.

N.W.   S.E.

FIG. 3.11. Corndon Hill phacolith. Dolerite, black; ashes and andesites, dotted; sediments, pecked lines

DYKES

Dykes are probably the most common minor intrusions for dyke swarms are often associated with igneous centres. Essentially dykes are sheet-like intrusions at a high angle to the bedding or other stratification of the rocks into which they are intruded. They may range enormously in thickness from a few centimetres or inches to 60 m (200 ft) or more, but the majority are probably under 3 m (10 ft) (Tyrrell 1926). They are, therefore, on the average thinner than sills.

The facts that dykes are usually thin and that they often extend for considerable distances suggest that a fluid magma very rapidly invaded fractures in the earth's crust. The fractures may not have been in existence before the intrusion of the magma, for it is more likely that the pressure exerted by the magma rapidly opened fissures in an area already under tension. Had considerable pressure been required to open the fissures one might have expected much blunter wedge-shaped intrusions—rather like vertical laccoliths—whereas dykes on the whole are parallel sided. Once injected such fractures are obvious planes of weakness, so that it is not surprising that multiple intrusions of the same type of magma sometimes took place along the same plane, or that composite dykes have been formed of successive intrusions of different types of magma along the same plane.

The presence of dyke swarms clearly suggests regional tension in the crust. The curious thing is that dyke swarms are sometimes radial to a centre of igneous activity, as one would expect if the two were closely related, but in other dyke swarms the dykes are roughly parallel and apparently unaffected by the igneous centres. On the whole the Tertiary dyke swarm of western Scotland and Northern Ireland trends in a north-west to south-east direction and appears largely independent of the centres of intrusive

PLATE 15. Ordovician dolerite phacolith, Corndon Hill, east of Montgomery

activity associated with the dykes (Fig. 3.12). This is certainly true of Mull and Arran. Yet a radial dyke pattern occurs around the igneous complex

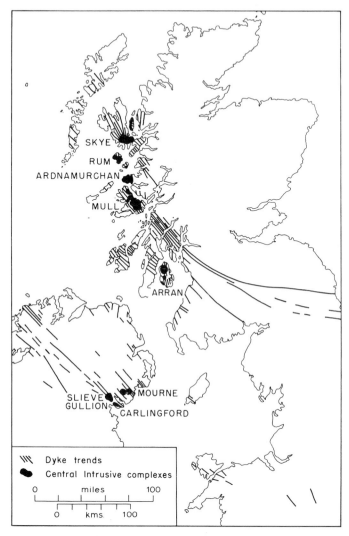

FIG. 3.12. Tertiary dykes and intrusions in northern Britain
(*after H.M. Geol. Survey*)

of the island of Rum (Fig. 3.13) which is a part of the same general igneous province. The number of dykes involved in this Tertiary dyke swarm is very large indeed: in south-east Mull 375 dykes occur in a belt 20 km (12·5 miles) wide, providing a total thickness of nearly 0·75 km (nearly half a mile) of igneous rock, each dyke averaging 1·75 m (5·8 ft) in thickness; in Arran 525 dykes occur in just under 24 km (15 miles), the average thickness being 3·5 m (11·5 ft) and the total thickness well over 1·5 km (1 mile). These figures represent considerable crustal stretching.

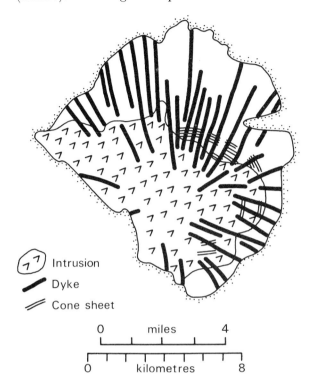

Intrusion

Dyke

Cone sheet

| 0 | miles | 4 |

| 0 | kilometres | 8 |

FIG. 3.13. Dykes, cone sheets and intrusion in Rum
(*after H.M. Geol. Survey*)

Although these Tertiary dyke swarms are probably the best known in the British Isles, they can be matched in Scotland by similar swarms of Old Red Sandstone age. In western Argyllshire and associated with the major intrusion of the Etive granite complex is a swarm of dykes, mostly formed of porphyritic microdiorites, which has a south-west to north-east Caledonian trend and, like much of the Tertiary swarm, is not radial about the intrusive centre. Indeed, its occurrence in a comparatively narrow belt between Loch Linnhe and Loch Awe suggests weakness in the crust above a subjacent intrusion aligned with the general Caledonian trend of the country. On the other hand the Old Red Sandstone volcanic rocks surrounding the granite intrusion of the Cheviot are intruded with an approximately radial pattern of dykes, probably of two generations and formed of microdiorite or andesite.

Obviously dykes must be expected to affect relief in a different way from sills, simply because of their different attitude in relation to the rocks they intrude. Being harder than the sediments, one might expect them often

97

to form wall-like features traceable for some distance across country. In fact they seem to form wall-like features far less frequently than sills form ledge- or scarp-like features. Several reasons probably contribute to this. First, dykes are often very thin compared with sills and so cannot be expected to have such an effect on relief. Second, it is usually true that the steeper the dip of a bed of given thickness the less effect it has on relief. This is merely a reflection of the fact that it represents a smaller area to be eroded away. Such a relation is clearly exemplified by certain sedimentary beds, notably the Chalk of the North Downs, where the continuously increasing dip westwards from the Mole Gap to the Hog's Back is accompanied by a decreasing area of outcrop and by a progressive lowering of the elevation of the crest of the North Downs (Sparks 1960). The reduced

FIG. 3.14. Tertiary dyke cutting Permian sediments on the shore in north Arran. Margins of dyke channelled

thickness and greater dip of dykes compared with sills probably both contribute to their decreased effect on relief. Thirdly, the isothermal surfaces in cooling dykes are nearly vertical and any resulting columnar jointing is approximately horizontal. In this position erosion is probably easier than where the jointing is vertical. Fourth, there sometimes seems to be a weakness where the contact between the dyke and the country rock occurs. One would suspect this from the occurrence of composite and multiple dykes. Erosion of the marginal contact seems to be taking place, for example, in some of the dykes on the northern shore of Arran (Fig. 3.14).

Arran shows well the differing effects of sills and dykes on relief. The south of the island has already been instanced as an area where the sills affect the coastal and the inland relief as well as the long profiles of streams.

FIG. 3.15. Ceum na Caillich, an eroded dyke, Glen Sannox, Arran

In the same area the dyke swarm shows up clearly on the foreshore, although this is as much due to differences in colour and rock composition as to differences in relief between the dykes and the Triassic sediments. On the '25-foot' beach lying behind the present foreshore the dykes have little effect—at most they are irregular masses of boulders—while inland they have usually very little effect indeed. They tend to occupy the cols between granite mountains and locally they form gullies rather than walls, the most famous example being Ceum na Caillich on the northern slopes of South Glen Sannox (Fig. 3.15).

It is often said that dykes may have three principal effects on relief: where they are harder than the surrounding rocks they project (Fig. 3.16a); where they are weaker than the surrounding rocks they form depressions (Fig. 3.16b); where they are harder they may stand higher than the

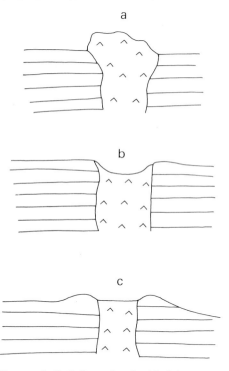

FIG. 3.16. Relief associated with dykes.
For explanation see text

surrounding rocks but be surpassed in resistance and elevation by the baked margins of the country rocks (Fig. 3.16c). In practice the relations are often more complex or more obscure. For example, the rocks on one side of a dyke may stand up and on the other side be planed off level with the dyke for reasons which remain obscure. Again, as mentioned above, the margins may be channelled out.

CONE SHEETS, RING DYKES, CAULDRON SUBSIDENCE

Features included in this class probably have more significance to the student of the mechanics of igneous intrusion than to the geomorphologist. Cone sheets are virtually circular dykes with an inward dip; ring dykes are circular dykes with a characteristic outward dip; cauldron subsidence, or, as it has been more graphically called, piston faulting, results in a ring of igneous material being injected round the margins of a subsiding cylinder of country rocks.

Cone sheets (Fig. 3.17) may be explained as intrusions along conical

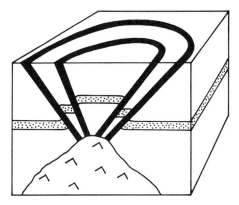

FIG. 3.17. Cone sheets, showing possible relation to intrusion

fractures consequent upon pressure from an underlying intrusion. They do not necessarily occur singly but in complexes of cone sheets. It will be apparent that the area within the conical fracture must be upfaulted to allow the intrusion of magma. Calculations based upon extrapolations of surface dips of cone sheets in the Scottish volcanic areas, where they are most common, suggest magma reservoirs at depths of 5–8 km (3–5 miles).

Ring dykes, on the other hand, involve the collapse of upward-tapering 'cylinders'. If the cylindrical collapse extends right to the surface (Fig. 3.18a) a caldera may be formed there: if it does not extend right to the

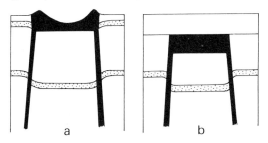

FIG. 3.18. Ring dykes: a. With caldera. b. With permitted intrusion

surface but it is terminated above by some plane of weakness an intrusion may be permitted (Fig. 3.18b). With very deep erosion it would be very difficult to distinguish between these two cases, for each would give rise to an intrusion of annular form. Intrusions such as this are usually thicker and coarser-grained than cone sheets. As with cone sheets, ring dykes occur in complexes. For example, there may be several successive ring dykes around a common centre or even overlapping series of ring dykes round several central igneous complexes. In the Scottish volcanic areas ring intrusions with thicknesses up to 1·5 km (1 mile) are known and it may well be that their origin is more complicated than described above. For example, if cylindrical faulting were associated with a zone of shattering magma might be injected into such a zone and assimilate the brecciated rock.

Annular intrusions in general are well exemplified in the igneous districts of western Scotland, notably in Skye, Mull and the Ardnamurchan peninsula. The reader is recommended to refer to the 1 inch Geological Survey (Scotland) sheet no. 51, Coll, to see the concentric arrangement of igneous rocks in Ardnamurchan around three separate centres (Fig. 3.19).

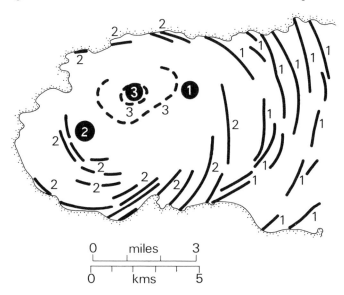

Fig. 3.19. Igneous centres, cone sheets and ring dykes of Ardnamurchan. Cone sheets 1 and 2 are associated with igneous centres 1 and 2 respectively; ring dykes 3 are associated with centre 3 *(after H.M. Geol. Survey)*

This shows up very clearly in the quartz dolerite cone sheets as well as in the thicker outcrops of the gabbros and eucrites (the latter are basic rocks with two pyroxenes, augite and hypersthene, and thus intermediate between gabbro and norite). A much more complicated example can be seen on sheet no. 44, Mull. Such natural complex examples must be studied as an antidote to the necessarily vastly oversimplified explanatory textbook diagram.

At their best ring complexes can give rise to circular patterns in the landscape and are comparable with the pattern ensuing from the dissection of a dome. In Ardnamurchan there is a central knoll of quartz monzonite and two rings of higher ground, the outer ring on the Great Eucrite being by far the more impressive. In Mull, too, there is a certain degree of parallelism between landforms and structural elements. For example, the general trough containing Loch Uisg and Loch Spelve follows the contact between ring complex and basalts in the south-east of the island, while the inner trough containing Glen Forsa and Glen More is roughly aligned along the division between Late Basic and Early Basic cone sheet complexes.

Extrusive forms

MAJOR FORMS

Just as the sheer size of intrusions is of great geomorphological significance so also is the extent of extrusions. Among the large extrusions are the great lava plateaus of the world, similar large expanses formed of ashes and ignimbrite, and the great shield volcanoes of Hawaiian type. The Hawaii volcanoes reach some 9000 m (30,000 ft) above the bed of the ocean in a series of great overlapping, interlocking domes, the largest of which, Mauna Loa, is on a base of 255 by 210 km (160 by 130 miles). In general terms the origins of these are readily comprehensible: the lava plateaus and shield volcanoes are formed by the consolidation of fluid basic lavas; the ash beds are literally the accumulation of vast amounts of volcanic ash; the ignimbrite plateaus were formed from nuée ardente eruptions of incandescent froth-like mixtures of gas and glassy fragments which moved with explosive velocities and on settling became in large part self-welded into flinty volcanic rocks.

Lava plateaus are found in various parts of the world and usually cover large areas. In the north-west of the United States the Columbia River basalt area covers well over 265,000 sq. km (100,000 sq. miles), while the adjacent Snake River basalts of more recent origin cover some 55,000 sq. km (20,000 sq. miles). The great outpourings of the Deccan 'traps' around Bombay are of about the same extent as the Columbia River basalts and attain a maximum thickness of 3 km (two miles) or so. A much older lava plateau area is the Atlantic Thulean province of early Tertiary age, which has already been mentioned. The lava plateaus still existing in western Scotland and Northern Ireland cover only about 5500 sq. km (2000 sq. miles), but other remnants are found in the Faroes, Iceland, Jan Mayen Island and Greenland. Vulcanicity lasted much later in some of these areas than it did in Britain and is, in fact, far from extinct in Iceland. Erosion and also possibly continental drift have combined to wreck the continuity of the lavas. The ring complexes of western Scotland described earlier may well have been the feeders of lava sheets long since eaten away by erosion.

There has always been considerable controversy about the origin of lava plateaus as extensive as these. This has revolved around the question whether the lava has been poured out of fissures or out of central complexes and coalesced to form a sheet. Whatever the hypothesis, the lavas comprising the plateaus are all basic, usually basaltic, in type and hence possessed of great fluidity. The old idea of a gigantic gaping chasm emitting a great upwelling of lava to form the plateaus has been abandoned, but fissure eruptions need not have been as simple in form as this. There seems little doubt that the emission of lava must have been rapid, but there is no reason why it should not have been effected from a whole series of fissures opening at different times in adjacent parts of the crust. It may even have been effected by localised centres of emission along such fissures and there may have been local centres of more pronounced flow where fissures intersected. Whatever the exact details, the theory of emission from large central volcanoes seems to be ruled out by the fact that individual lava flows rarely seem to have the degree of dip which such an origin would imply and they are too rarely seen to die out, as they must do if derived from central volcanoes. There seems little doubt that the flows of places like the Snake River plateaus spread out into still lakes of lava before they consolidated, because their surfaces are free of structures associated with flowage during cooling and because of the way in which they have filled canyon-like valleys and spread out to form thin but extensive sheets over the surrounding country. These facts argue large volumes of lava and a very high degree of fluidity.

This having been said, there remains the problem of why virtually indistinguishable lava types in other areas form vast low-angle cones of Hawaiian type. It is obvious from certain structural details of the surfaces of the lava flows, to be described below, that these lava flows did not spread out until they were horizontal, but cooled as they flowed. Perhaps in these great shield volcanoes the quantity of lava emitted in any one eruption is more limited: hence the distance flowed before cooling might be considerably less. It has been suggested that variations in iron content may have contributed, the greater the amount of iron the greater the fluidity. That there were considerable gaps between individual lava flows is shown, for example, in western Scotland by the presence of impure lignites, leaf beds and deep weathered horizons in places such as Mull between successive lava flows. Perhaps the answer is to be found in the exact place of eruption: for true plateaus the emission may have to spread over considerable areas by the successive use and abandonment of fissures. Where the place of emission remains more or less constant large low angle volcanoes are likely to form.

Ignimbrites and volcanic ash may both lead to the formation of widespread resistant expanses of volcanic rocks. The ignimbrites (or welded tuffs) are resistant at a much earlier age than ash beds, which may ultimately be compacted by lithogenic processes to form tuffs. Both rocks, usually formed of acid material for explosive volcanic activity is much more characteristic of acid than basic volcanoes, ultimately form resistant compact stony rocks resembling rhyolites in composition. Indeed, it may confidently be expected that re-examination of some of the older rhyolitic lavas and tuffs will lead to the suggestion that ignimbrites may be much more fully represented among them than had previously been expected. Alternate lava flows and minor periods of more explosive activity can produce interstratified lavas and ashes, which are prone to give rise to plateaus and scarps as the ash beds are generally more erodible.

The effects of these volcanic forms on relief may continue for a long time in the landscape. In the beginning it is the constructional form which dominates. Much later on when the rocks are exhumed as ancient lavas from their enclosing sediments they may again dominate the relief, as do the old Lower Palaeozoic volcanic rocks in parts of Snowdonia, for example, Tryfan and the Glyders.

Small scale initial features are provided by the nature of the surface of the cooling lava sheets. Two general types are recognised, both designated by Hawaiian words with more vowels than consonants, namely *pahoehoe* and *aa*. Both are characteristic of flows that cool during movement and are absent from those which become stationary before solidification.

Pahoehoe surface consists essentially of smooth forms, the bulbous ropy surface that one associates with very thick cooling toffee. It seems that the lava continues in motion beneath a glassy skin which, in the early stages at least, is elastic. Occasional ruptures when the surface is breached result in the welling out of further globules of lava, which themselves develop an elastic skin as they cool. In submarine eruptions the lava surface becomes chilled thoroughly, cracks, and the ensuing globules in turn become quickly chilled. This produces a pile of lava 'sacks' known as pillow lava and may be considered an extreme form of *pahoehoe* type (for a different explanation, see Chapter 7). Broad swellings in a *pahoehoe* surface may be due to the lava below bulging up, while tunnels within the lava mark the location of the feeding channels of the lava flow.

Aa surface is much rougher and more clinkery in texture. It is probably not caused by consolidated blocks of lava being rafted along in a lava stream, for most lava streams have such a high gas content that their density is less than that of solidified lava, which would therefore sink. It seems more likely that ejection of gases gives rise to rapid solidification of the surface and to the formation of the incredibly rough, scoriaceous *aa* surface. Such an explanation demands acceptance of the very great importance of gases at high temperatures in maintaining the fluidity of lava flows in spite of their having such a large cooling surface in contact with the atmosphere.

The other important initial feature of lava flows is the polygonal joint pattern, which has its greatest effects on relief at the edges of lava flows and is very similar to that produced in sills.

In a lava plateau the presence of discrete lava sheets will cause the tendency for stepped effects in the landscape, especially where the flows are separated by weaker ash beds or where the surface of one flow has become weathered before it was covered by the next flow. Stepped landforms of this sort have been found in the plateau basalts of Eigg and Skye, where the slaggy tops of the lava flows are more erodible than the basalt sheets proper. There seems to be a general tendency for the jointed cliffs at the edge of basalt sheets to retreat parallel to themselves and in the final stages remnants of lava flows form isolated, flat-topped, steep-sided mesas and buttes. These forms are well known from the western United States, the mesa being a larger remnant than the butte.

Somewhat comparable effects may be produced, with an associated inversion of relief, in the dissection of lava flows occupying former valleys, especially when the lavas invade areas of not very resistant rocks. Lavas originally flow into valleys (Fig. 3.20a), but, because of the resistance of the lava, the intervening ridges may be denuded leading to inversion of relief (Fig. 3.20b) and finally to the formation of mesas and buttes (Fig. 3.20c). One of the most famous cases of this type is the Montagne de la Serre south of Clermont-Ferrand. Basalt flows from the volcanic areas overlying the crystalline complex to the west flowed across the faulted margin of the crystalline block into valleys in the soft Lower Tertiary sediments occupying the Limagne. Later the relief was inverted to leave the Montagne de la Serre as a scarped, flat-topped interfluve.

The tendency for the edges and not the surface of a basalt sheet to be eroded may be accentuated by the permeability of the rock. The predominantly vertical joints of basalt must swallow a large part of the precipitation, while any slaggy, cavernous or ashy layers will further increase the general permeability.

The permeability of lava flows may be part of the reason why valleys in basalts tend to be broad, flat-floored and steep-sided. When the cross profile is smoothed it tends to be very like the U-shaped form said to be characteristic and indeed diagnostic of glacial troughs. This tendency will be accentuated when the beds underlying the lavas are less resistant ashes and breccias as in the Cantal in the Central Plateau of France. Here glaciation may have helped in the production of the U-shape, as it is one of the few glaciated areas of central France. In the island of Oahu in the Hawaiian Islands peculiar, deep U-shaped valleys with amphitheatric heads have long attracted attention. They resemble glaciated valleys in form but must obviously have another origin. Their mode of origin remains obscure but may be connected with the very wet, tropical climate, which causes intensive chemical rotting with mass wasting and the production of long concave valley-side profiles.

The tendencies for preservation of the plateau surfaces, inward retreat of escarpments and development of U-shaped valleys suggests some affinity between basalt plateaus and limestone plateaus. But the analogy must not be pushed too far. It probably arises from comparable joint patterns and permeability, and the generally horizontal alternations of

a

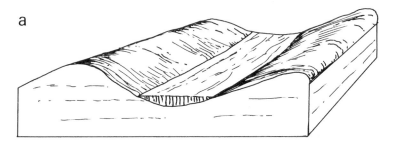

b

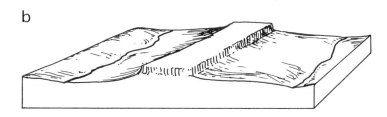

c

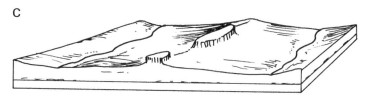

FIG. 3.20. Inversion of relief on lava flow

harder and softer beds. Beyond that the whole susceptibility of limestones to chemical weathering and the ensuing landforms make them greatly different from basalt plateaus.

Any drainage pattern initiated on basalt plateaus will develop under the influence of the initial slopes, varying resistance and jointing of the beds, so that its form may be very irregular and complex. On large central volcanoes, however, a radial pattern may well develop and lead to the dissection of the volcano ultimately into a series of triangular interfluvial areas (Fig. 3.21). These are usually referred to as planèzes from the name of such a triangular facet on the Pliocene volcano of Cantal, which must be regarded as the type of such dissection and which is in approximately the

state shown diagrammatically in Fig. 3.21 with the centre reduced to a series of narrow radiating ridges (Plate 16). The reduction of the centre will be aided wherever weaker beds lie under a hard capping of lava and where the surface gradient of the former volcano is greater than that of the

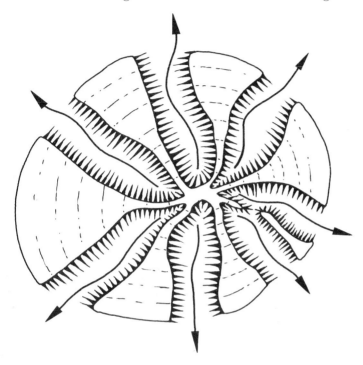

FIG. 3.21. Dissection of a volcano into planèzes

streams in equilibrium. In these conditions the stream will be entrenched more deeply in the central parts of the volcano than in the outer parts, the valleys near the centre will tend to be wider and the centre heavily denuded producing the planèze form.

Individual lava flows may locally modify drainage. If the flows are thin, temporary lakes, later to be filled in with sediment, may be produced in valleys. If the flows are much thicker diversions of drainage akin to glacial diversions may occur. Such is thought to have happened to the Truyère (Fig. 3.22). This stream is believed to have been a tributary of the Allier via the Alagnon. Evidence of this is provided by the open form of the upper Truyère valley where it flows across the foot of the Margeride and the presence of alluvium of the former Truyère, containing Jurassic flint nodules derived from the Causses region well to the south, almost as far as

PLATE 16. Dissected Pliocene volcano, Cantal, Central Plateau, France. View west from Puy Mary near the centre, where the interfluves have been reduced to narrow crests

St Flour. This course is thought to have been blocked by lava flows from the Cantal so that the Truyère was diverted via its present gorge-like steeply-graded valley westwards between Cantal and Aubrac into the Lot and ultimately into the Gironde estuary.

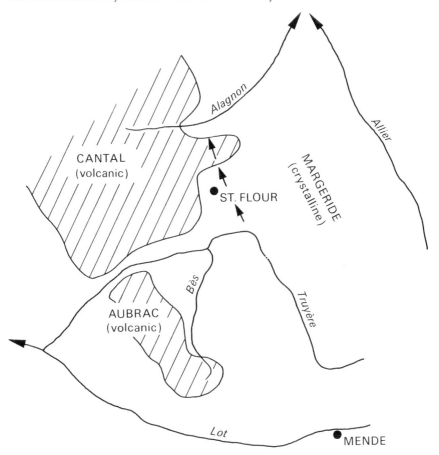

FIG. 3.22. The diversion of the river Truyère

On a much larger scale it is possible for regional superimposition to occur from lava plateaus. An interesting example of this has been described by West and Choubey (1964) from the area round Sagar in the Vindhya Hills, which form the watershed between the westward-flowing Narmada (Narbada) and the drainage flowing north into the Jamuna (Jumna) and Ganga (Ganges). The area consists basically of the Vindhya Series, a set of unfossiliferous sandstones, limestones and shales believed to be most likely late Pre-Cambrian in age. These rocks are not greatly disturbed: indeed in large part they are almost horizontal but are locally affected by sharp folds. A maturely-dissected surface developed on the Vindhya rocks was in the late Cretaceous invaded by the Deccan Trap lava flows which buried the pre-existing relief and are thought to have once covered the area with great thicknesses of basalt, for the highest outcrops of the basalts are at higher levels than the Vindhyan summits. The Traps here are made up of nine lava flows and the upper surface of each is either vesicular or weathered. The result is that these lava flows are stripped back from each other along these planes of weakness giving rise to typical terraced land-forms. The authors believe that virtually no erosion takes place on the near horizontal surface of the flows which are reduced entirely by the back-wasting of the 20 degree escarpments separating them, although in places the chemical weathering of the basalts is so pronounced as to have com-pletely laterised the top two flows, a thickness of 55 m (180 ft). The final stage before a flow is completely consumed is a small inselberg-like hill bounded by 20 degree slopes.

On the regional scale a tendency for the basalts to have a northward dip may be responsible for the general northward drainage into the Jamuna. Locally the drainage is being superimposed on to the Vindhyan rocks, for example near Rahatgarh (Fig. 3.23), where the Bina flows in a gorge 30 m (100 ft) deep through a Vindhyan inlier, and again near Bilaspur, where the river neatly divides a low Vindhyan hill in spite of

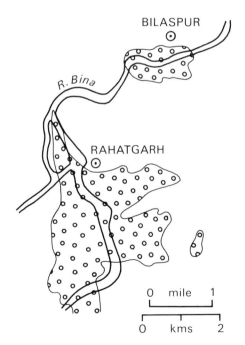

FIG. 3.23. Superimposition of river Bina from Deccan lavas (unshaded) on to inliers of Vindhyan Series (circle shading) *(after West and Choubey)*

the lavas on either side being lower than the hill. The quartzitic sandstones of the Vindhya series are much more resistant to chemical denudation than the lavas, so that ultimately the Vindhyan relief is exhumed as the lavas are removed sheet by sheet and so comes to dominate the basalt plateaus. Some adaptation to the Vindhyan structures is already taking place, partly because there is a relief developed on the Vindhyan surface beneath the lavas and partly because the Vindhyan series has considerable lithological diversity.

Most of the preceding discussion has been concerned with lava plateaus and volcanoes still occurring at the surface. But many series of volcanic rocks have been buried and later exhumed by denudation. Some of these may assume an entirely spurious volcano-like conical form, for example the Pre-Cambrian volcanic rocks of the Wrekin and the eastern side of the Church Stretton fault, where the Lawley and Caer Caradoc have splendid conical forms. As an aside it might be mentioned that the genesis of conical forms is an interesting problem of denudation: they occur on some quartz-ites in the metamorphic rocks of the highlands of Scotland, on volcanic rocks as stated above in the Welsh Borderland, on granites and other intrusives in the Lleyn Peninsula, and on Old Red Sandstone in Sugar Loaf Mountain near Abergavenny. Not all of these could possibly be extinct volcanoes.

The influence of older complexes of volcanic rocks is much more indirect than this. Snowdonia is an area where rhyolitic rocks of Ordovician age are well developed and exposed in such mountains as Snowdon itself, Tryfan and the Glyders. The area is often cited as one in which highland glacial forms are supremely developed and, indeed, the corries, arêtes, shattered rock faces etc. are well developed, but how much of this is due directly to glaciation and how much to the preservation of these forms on the well-jointed, splintery rhyolitic rocks? Farther to the south, Cader Idris, with its scalloped north-facing escarpment and the superbly-developed Llyn Cau (Plate 17), one of the best corries in Britain, is at the same time an area of intensive glaciation and one of a great development of Ordovician igneous rocks, both intrusive and extrusive (Fig. 3.24). Areas of comparable height and similar in position relative to former ice centres, e.g. the Berwyns some 40–50 km (25–30 miles) east-north-east of Cader Idris, have apparently far less pronounced glacial features, but they also have far less well-developed suites of igneous rocks. One begins to suspect that even in glaciated country rock type is very important indeed in its controlling effect upon relief. Very similar conditions obtain in the Lake District where most of the best glaciated relief is developed on the Ordovician Borrowdale Volcanic Series, a mass of andesitic lavas and tuffs, some 3500 m (12,000 ft) in total thickness.

PLATE 17. Llyn Cau and Cader Idris. The corrie lake is on mudstones and the southern ridge of Cader Idris to the left on acid volcanics

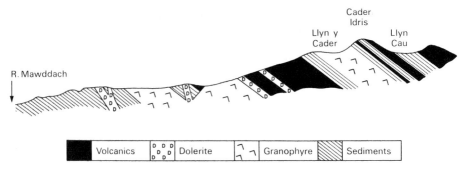

Fig. 3.24. Sketch section of Cader Idris *(after H.M. Geol. Survey)*

Minor forms

Minor may seem to be an inappropriate term to apply to such volcanoes as Vesuvius, Stromboli, Mt Pelée and Etna, but, however great their individual interest, they do not exert a regional influence on landscape comparable to that produced by lava plateaus and the giant basic central volcanoes. Complicated classifications of such volcanoes may be made when one's attention is focused on details of the volcanic activity, a famous volcano being taken usually as the type of a certain type of activity. The nature and relative proportions of the volcanic products will considerably affect the form of the volcano emitting them. Broadly speaking, apart from gases, there are lavas and fragmentary or clastic products. The latter may be fine (ashes), coarse (lapilli) or moulded as they cool in passing through the atmosphere (bombs). The lavas may range from acid to basic and hence vary greatly in viscosity. Different combinations of products may be produced by different volcanoes, while these combinations may change during the history of activity of any one volcano. Accordingly, the tendency towards certain types of forms with predominance of certain types of products, but not all the possible combinations, will be discussed.

Basic lava, basalt primarily, is very fluid and leads to the development of forms broad in proportion to their height; in the case of widespread extrusion to the lava plateaus and shield volcanoes already discussed.

Acid lavas, usually much more viscous, flow with much greater difficulty. In intrusive forms they tend to produce laccoliths rather than sills: in extrusive forms they give rise to steep, humped, convex hills of which the Auvergne puys (Plate 18) are the type examples, e.g. the Puy de Dome and the Grand Sarcoui. These are mainly trachytic in composition.

Ashes, cinders and clastic debris in general produce cones, which are often regarded as true volcano shapes. The degree of symmetry of such cones depends mainly on the prevailing winds: where there is an approxi-

PLATE 18. Chaine des Puys, Central Plateau, France. A series of ash and cinder cones with some acid lava volcanoes of Pleistocene age

mately even distribution of wind direction around the compass the symmetry will be most perfect; where there is one dominant wind, as in the
trade wind belt, the chances of pronounced asymmetry will be greatest.

Alternations of lavas and ashes may result in a type of armour-plating of
the ash form and its greater resistance to erosion. Ashes lend themselves to
erosion when unconsolidated, especially in the heavy convectional showers
which may be associated with volcanic eruptions. On the other hand they
are very permeable and may be characterised by dry valleys, a lack of
surface run-off and hence a relatively low degree of erosion in areas where
heavy intense rainfall is absent.

Much larger craters, explosion craters or calderas, may be formed by
explosion or collapse. If the various lava vents and feeders in a volcano
become solidified, sufficient pressures may be built up to cause a paroxysmal
eruption or explosion. Such pressures are likely to be built up if either sea
water or ground water gets into contact with the molten magma at depth.
This may have happened in the case of the great 1883 explosion of
Krakatoa in the Sunda Strait between Java and Sumatra. Alternatively
great craters may be built by collapse (Fig. 3.25). A lowering of the level

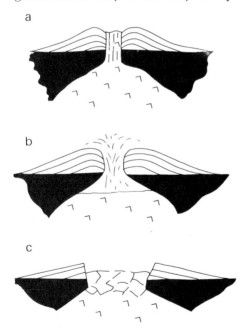

FIG. 3.25. Formation of caldera by collapse

of the magma below a volcano may cause both the overlying cone and
parts of the country rock to collapse. This may later be blown out by
explosive activity as clastic debris and pumice. It is probable that collapse
rather than explosive ejection is responsible for most calderas, because

there seems to be too little clastic volcanic debris in the surrounding land-scape to support the explosion theory.

With ancient volcanoes it is the skeleton which is of the greatest import-ance in forming the relief. Within any cone the consolidated plug of lava in the crater and any associated dykes and lava flows will form the resistant elements in the relief when the whole is exhumed by erosion.

Volcanic plugs are very important in the Midland Valley of Scotland (MacGregor and MacGregor 1948), where they are mostly of Carbon-iferous or Permian age. Although some may be definitely dated, there is considerable doubt about the precise age of others, because the affinity of rock type within an igneous province is no guarantee of identity of age. The rocks filling the vents vary. They include basaltic agglomerate, for example in Arthur's Seat at Edinburgh, basalt in Castle Rock at Edinburgh, and in Dumbarton Rock on the river Clyde, phonolitic trachytes in North Berwick Law (Plate 19a) and Bass Rock, and varied volcanic agglomerate (an aggregate of lumps of volcanic and country rocks) in two great groups of necks. The first of these, which includes about sixty examples, lies in the Central Ayrshire coalfield between Irvine and Galston to the north and Dalmellington and New Cumnock to the south. The other group, a hundred or more strong, lies in east Fife. Inland many of the necks form prominent hills, while many of the eroded necks on the coast expose excellent plans of the beds, for example on the south side of the Firth of Forth near North Berwick and Dunbar and again in eastern Fife.

Volcanic necks are also superbly developed in the Central Plateau in France where they are composed mostly of trachytes or phonolites. The examples of the ideal, and hence probably exaggerated effect on relief, are provided by the necks of le Puy. Crowned with chapels and exhumed from non-resistant Tertiary sediments, the ancient plugs are the dominant elements in the landscape. But there are many more examples of the same effect in the landscape of the surrounding region. On an altogether larger scale is the dissected volcano of Mont Dore, where trachytes are the dominant rocks (Fig. 3.26) (in this respect it may be contrasted with Cantal). It forms the highest point in Hercynian Europe, 1886 m or a little

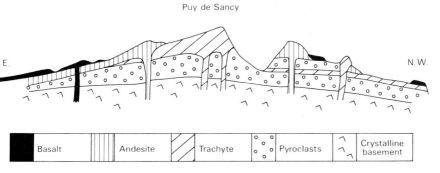

F IG. 3.26. Sketch section of Mont Dore *(after de Martonne)*

PLATE 19a. Carboniferous volcanic neck, North Berwick Law, East Lothian. The rock is phonolitic trachyte

PLATE 19b Fault-guided valley, Carboniferous Limestone, Gower Peninsula, South Wales

above 6000 ft, although this elevation is only reached by much younger volcanic rocks piled on the Hercynian massif proper. It is now considerably dissected, but its general form and detail of its various dyke sheets are still both prominent in the landscape (Plate 20).

A classification of igneous rocks

Classifications of natural objects are usually arbitrary systems designed to facilitate handling a range of phenomena at the risk of some distortion of the truth. There is no one correct classification; some classifications are better than others for certain purposes; some may be better than others because, being more elaborate, they inevitably approach the continuous gradation which is probably the natural truth.

If we were fundamentally interested in igneous rocks from the chemical point of view, then a chemical classification would be best; if mainly concerned with the association of rocks in different suites then a genetic classification of some sort would be preferred. But with our fundamental interest in rocks as relief formers, the classification must be moulded around those properties which most closely affect landforms. They are:

(*a*) The nature of the igneous mass. The importance of this has been fully discussed and no rock classification can completely cover it. But grain size, which is one criterion of the classification proposed below, does reflect to a large degree whether the rock was formed as part of a major intrusion, a minor intrusion or an extrusion, although the classes of coarse, medium and fine grained rocks do not, as was stressed earlier, correspond exactly with the form of intrusion or extrusion.

(*b*) The jointing, which cannot easily be brought into a rock classification.

(*c*) The mineral composition. In the last chapter, when considering the effects of minerals on weathering, it was shown that the nature of the felspars, the nature of the ferromagnesian minerals and the abundance of quartz were all important factors.

Accordingly we have four factors, grain size, felspar type, ferromagnesian type and abundance of quartz to try to fit in as bases for classification. Fortunately, the second and third of these usually vary sympathetically, so that the classification is reduced to a problem in terms of three variables.

The common rock-forming minerals are fortunately few in number in spite of the large number of minerals known. They can be arranged into two main groups:

(*a*) Light-coloured minerals. These include quartz, felspars, felspathoids and white mica.

1. Quartz is simply crystalline silica and is one of the minerals least affected by weathering. It forms hexagonal crystals though in igneous rocks this form is not apparent as it usually crystallises late and fills in the

gaps between other crystals. It appears as a vitreous substance with an irregular fracture for there are no cleavage planes along which the mineral splits.

2. White mica, or muscovite, has a perfect cleavage—hence its property of flaking into sheets parallel to one of its crystal faces. Cleavage is basically related to planes within a mineral in which the atomic bonding is weakest. Thus, mica is a great contrast to quartz in which the bonding is equally strong in all directions.

3. The felspars include two principal types; potash felspar or orthoclase and the soda–lime felspars or plagioclase. Orthoclase is a mineral of fixed composition but the plagioclase felspars form a continuous series between the extremes of soda felspar (albite) and lime felspar (anorthite) these two being miscible in any proportions. It is customary to recognise half a dozen plagioclase felspars, albite, oligoclase, andesine, labradorite, bytownite and anorthite with increasing lime : soda proportions. They are usually greyish or whitish porcellanous minerals, with a good cleavage but not nearly as perfect a one as in micas. They may, however, vary considerably in colour, pinks and light brick colours being often produced by small amounts of impurities. Mixed crystals of plagioclase and orthoclase can occur but they are rarer than plagioclase mixtures. Reference to the table in the last chapter (p. 12) will show that there is a definite order of susceptibility to chemical weathering ranging from orthoclase through soda plagioclase to lime plagioclase.

4. The felspathoids are in many ways very much like felspars in their structure, but have a lower silicon : aluminium ratio, that is they have been formed from magma deficient in silica. Thus they cannot normally coexist with quartz for the latter implies an excess of silica in the magma whereas felspathoids imply a deficiency. They are much rarer minerals than felspars but give rise to a great variety of specialised rock types.

(*b*) Dark coloured minerals. These include the amphiboles, the pyroxenes, the olivines and the dark mica, biotite. They are often referred to as ferro-magnesian or mafic minerals and are primarily silicates of iron, magnesium and calcium.

1. Dark mica or biotite possesses similar cleavage properties to muscovite, but is more susceptible to chemical weathering.

2. The main differences between amphiboles and pyroxenes are complex structural ones reflected in the systems in which they crystallise. The principal amphibole found in rocks is hornblende, a blackish-brown mineral with a fairly good cleavage and more resistant to chemical weathering than the pyroxenes.

3. The two main pyroxenes are augite and hypersthene in that order.

PLATE 20. Puy de Sancy, Mont Dore, Central Plateau, France. Dissected Pliocene trachytes. Note lava sheets to left and rough columnar jointing to right

Their cleavage is about as well developed as in the amphiboles but they are less resistant to chemical attack.

4. The olivines are a series of iron and magnesium silicates with no cleavage but usually with a well-marked fracture pattern, along which the mineral, which is suceptible to chemical attack, is readily altered (Fig. 2.1). In some rocks where it is fresh, for example in some young olivine basalts, the olivine shows as clear light green crystals. Like the felspathoids, it is not found associated with quartz in rocks under normal conditions.

Turning now to the classification set out in Fig. 3.27, it can be seen that the primary classification in one direction is based upon grain size, and hence to a great extent on mode of formation, and that within each grain size three divisions are recognised depending on the presence or absence of quartz and felspathoids. These three states are referred to as oversaturated (with free quartz), saturated (with little or no free quartz and felspathoids) and undersaturated (with felspathoids). Thus grain size and one aspect of mineral composition are taken into account vertically on the diagram.

In the other direction the main division is based on the nature of the felspars. It should be understood that the vague term soda felspar used must not be taken to mean pure albite, but refers to the sodic end of the plagioclase series: similarly with the term lime felspar. From the table of relative susceptibility to chemical weathering (p. 12) it will be noted that there is a progressive range in this attribute from one side of the classification to the other. The second aspect of mineral composition used in the horizontal direction, because it is largely sympathetic with the felspar change, is the content of ferromagnesian minerals and, perhaps more important, their nature. In rocks to the left of the table the ferromagnesian mineral is likely to be biotite or more rarely hornblende or an associated amphibole, while to the right of the table pyroxenes become more important and, well to the right, olivines. There are exceptions to this general trend: for example, one of the ferromagnesian rocks in the last column, hornblendite, does not conform with this general range. Nevertheless, the shift does hold in general and involves increasing susceptibility of the ferromagnesians to chemical weathering from left to right just as it does with the felspars. Thus chemical weathering susceptibility as a whole increases from left to right and also from top to bottom within each of the grain size divisions.

In the classification the rocks to the left are those which are often called acidic while those to the right are called basic. Between are the intermediate rocks, which are sometimes divided into sub-acid and sub-basic. These divisions are really based on silica percentages within the rocks and often form the bases of elementary classification. The classification used here, perhaps a little more complicated, is more closely related to mineral content and is thus more directly related to weathering potential.

The table does not pretend to include all igneous rocks. It contains the names a geomorphologist is likely to meet and attempts to show how they differ from one another. Thus, if we take the three rocks alkali-

		Mainly potash felspar	Potash and soda felspars	Soda-lime felspars	Mainly lime felspars	Predominantly ferromagnesian minerals
COARSE GRAIN	With quartz	ALKALI-GRANITE, PEGMATITE	ADAMELLITE	GRANODIORITE, TONALITE	QUARTZ-GABBRO	SERPENTINE, PERIDOTITE, PICRITE, PYROXENITE, HORNBLENDITE
	With little quartz or felspathoids	SYENITE	MONZONITE	DIORITE	GABBRO, NORITE, ANORTHOSITE	
	With felspathoids	NEPHELINE-SYENITE			THERALITE, TESCHENITE	
MEDIUM GRAIN	With quartz	MICRO-GRANITE	MICRO-ADAMELLITE, QUARTZ-PORPHYRY	MICRO-GRANODIORITE, MARKFIELDITE	QUARTZ-DOLERITE	
	With little quartz or felspathoids	MICRO-SYENITE e.g. RHOMB-PORPHYRY (APLITE)		MICRO-DIORITE	DOLERITE (DIABASE)	
	With felspathoids	NEPHELINE-MICRO-SYENITE				
FINE GRAIN	With quartz	RHYOLITE, FELSITE, *OBSIDIAN*, *PITCHSTONE*	TOSCANITE	DACITE	THOLEIITE	
	With little quartz or felspathoids	TRACHYTE, KERATOPHYRE	TRACHY-ANDESITE	ANDESITE	BASALT, SPILITE, *TACHYLYTE*	
	With felspathoids	PHONOLITE			OLIVINE-BASALT, TEPHRITE, BASANITE	

Ferromagnesian minerals increasing →

FIG. 3.27. Classification of igneous rocks. For explanation see text

123

granite, granodiorite and diorite, the following information can be gleaned from the table without recourse to more detailed descriptions of the rocks. They are all coarse-grained rocks. Alkali-granite resembles granodiorite in containing essential quartz but differs in the nature of the felspars, orthoclase in the former and soda-lime plagioclase in the latter. At the same time more ferromagnesian minerals might be expected in a granodiorite. Diorite differs from granodiorite simply in the reduction of the quartz content. Thus a good deal of knowledge of significance to a geomorphologist is immediately to hand, but complete descriptions of the rocks cannot be inferred from the table. Many alkali-granites contain muscovite as one of their main constituents, but there is no way of telling this from the table; nor can anything be inferred about the nature of the accessory minerals; nor could it be guessed that alkali-granites are usually somewhat coarser than diorites. It is necessary to recognise both the usefulness and the limitations of the table. Again, if we had no idea what a basalt was, we could infer that it was a fine-grained rock without quartz and with lime plagioclase and a considerable amount of ferromagnesian mineral as its main constituents, which from comments made above might be deduced to be pyroxenes perhaps with some olivine. The inference would be reasonably correct, but if the same inference were made about spilite because of its position in the table, it would be far from correct in one important aspect, the nature of the felspar. Again, the table is only a useful key, not a complete replacement for detailed reading about the rocks. It will put into rough position unfamiliar rock names met with in reading.

On the table an underlining system has been used to give a coarse and largely subjective idea of the relative frequency of the different rocks at the earth's surface. The range from no underline through single and double underline to triple underline denotes increasing abundance. The frequency is assessed in terms of physical abundance and not in terms of the frequency with which the names are likely to be met, because some names (e.g. granite) are used with a rough meaning which extends far beyond a petrographic definition. Rock types inserted in italic capitals are those which are wholly or in large part glassy.

Certain features of the rocks require further elaboration. Considering first of all the coarse-grained rocks, it should be noted that the term 'granite' is often used very loosely. At times it has been applied to a whole range of rocks including true igneous rocks and coarse gneisses which go to make up many ancient crystalline massifs. The term granite-gneiss is suitable for such an assembly of rocks which geomorphologically are very comparable with each other. More frequently the range from alkali-granite through adamellite to granodiorite has been collectively treated as granite. Much depends on the origin and date of the description concerned.

Alkali-granites are usually characterised by a very low content of ferromagnesian minerals. Apart from orthoclase and quartz the two micas, muscovite and biotite, are the only other commonly occurring minerals.

In adamellite and granodiorite, apart from the different character of the felspar, dark minerals become more important, usually biotite but occasionally hornblende. This is in accord with what has been said above about the increase in frequency and change of character in the dark minerals from left to right across the table. Granodiorite is probably the most common 'granite' and prevails widely in the giant intrusions of the Western Cordilleras of North America. In the British Isles the Donegal and Mountsorrel, Leicestershire, intrusions are granodiorites, the Shap intrusion in the Lake District is an adamellite, while alkali-granite is well developed in the intrusions of Devon and Cornwall. An alternative name for adamellite which may be met is quartz-monzonite; the reason for this should be apparent from the table.

Tonalites, which are put in the same 'box' as granodiorites, differ from the latter in their quartz content. In granodiorites quartz may be regarded as essential: in tonalites, which have sometimes been called quartz diorites in the past, the quartz is subsidiary, but the two rocks grade into each other, nevertheless. The use of the term tonalite by some American authors is somewhat different as it refers to all rocks with quartz, plagioclase and ferromagnesians. It thus covers tonalite plus granodiorite—but not all American authors use it in this way (e.g. Turner and Verhoogen 1960).

Three other rocks in one box are gabbro, norite and anorthosite. Gabbros and norites have the same type of felspar, but differ in that the usual ferromagnesian mineral is augite in gabbro and hypersthene in norite. Olivine may also occur in substantial amounts in either. A rock formed of lime plagioclase and olivine with no pyroxene is a troctolite, a learned translation of its ancient name, troutstone, which referred to its speckled appearance. In Britain gabbro intrusions occur in the Cuillin Hills of Skye and in Carrock Fell, Cumberland. Norite and associated rocks make up the giant Bushveld complex of southern Africa. Anorthosite is an almost pure lime felspar rock with very little dark mineral content. It is found in great layered intrusions such as the Bushveld complex, where it is associated with norites and various purely ferromagnesian rocks. It is the most widespread of all the monomineralic rocks. In the present classification it is included in its present position solely because of the nature of its felspar content, which is usually labradorite or bytownite and not anorthite as its name might suggest.

Theralite and teschenite, both uncommon rocks, indicate the difficulty of arbitrary classification. They belong to a group of rocks which shares some of the characteristics of syenites and gabbros—the equivalent fine-grained rocks are called trachy-basalts. Their generally high content of dark minerals defines their general position in the table, but they contain some alkali felspar as well as the lime felspar one would expect. Both theralite and teschenite are undersaturated rocks, the difference lying in the nature of the felspathoid. They occur in the Midland Valley of Scotland, for example teschenite in the basic sill of Salisbury Crags, Edinburgh.

The group of coarse-grained rocks composed mainly of ferromagnesian minerals is a series of not very common rocks. Peridotite is olivine plus other dark minerals but no felspar; pyroxenite and hornblendite are self-explanatory terms; picrite is very similar to peridotite but contains some felspar (lime plagioclase). Serpentines are altered versions of these rocks, usually peridotite and picrite but also to some extent pyroxenite. In them the minerals have been hydrated. The occurrence of these rocks in layered complexes such as the Bushveld with norites has already been mentioned. In Scotland peridotites are associated with gabbros and norites in the Huntly complex near Banff. Serpentine is well exposed on the western side of the Lizard peninsula at such places as Mullion and Kynance coves.

Finally something must be said about pegmatites and the aplites which are closely associated with them. The table shows that these two terms cover more than one of the principal rock types. Pegmatites are formed from very volatile and mobile magma which penetrates beyond the margins of 'granite' masses to form irregular minor intrusions. The very low viscosity of the magma results in very well formed and often very large crystals, some of them being of quite rare minerals. Aplites which are usually fine- or medium-grained mixtures of quartz and felspar, may have been residual parts of granitic magma poor in fluxes and hence solidifying rapidly. They occur in very thin intrusions often associated with pegmatites, a combination difficult to explain. Both pegmatites and aplite vary in composition with the parent intrusion, hence the range shown on the table.

The nomenclature of medium-grained rocks is chaotic. The simplest terminology simply adds the prefix micro- to the name of their coarse-grain equivalents. Unfortunately many of these rocks are porphyritic and thus there has arisen a range of names of porphyries. Quartz-porphyry is a common term, but it covers a range as indicated in the table. Quartz may not be the only mineral forming phenocrysts in it, for felspar is common and ferromagnesian phenocrysts are also known in some micro-granodiorites. Non-porphyritic microgranites also occur; they are near to aplites but usually contain minerals other than quartz and felspar in greater quantities. Rhomb-porphyry is an old name for a range of rocks from the Oslo district of Norway. In them the felspar, which is of unusual type, occurs as large rhomb-shaped phenocrysts. Some of the rocks, which are distinctive and fairly common glacial erratics in eastern England, fall into the micro-syenite category. Markfieldite is only included as a Charnwood Forest example of a rock which is basically a porphyritic micro-diorite.

Dolerite in English usage is the same as diabase in American usage. Unfortunately diabase has been used in England as a synonym for old and altered dolerites. Hence, confusion over the meaning of the term could arise unless the nationality of the author is known. Dolerites are common sill rocks in north-east England and in central Scotland.

The largest number of entries in a single box in the fine-grained rocks is associated with rhyolite. Rhyolite, like granite, has been used to cover a range of rocks, in this case rhyolite, toscanite and dacite, and older descriptions of rhyolite areas may well involve such a range of rock types. Alkali-rhyolite would strictly appear to be the best name for the narrow range corresponding with alkali-granite. The word, rhyolite, refers to the common presence in these rocks of flow structures. Rhyolites are common in the Ordovician of the Snowdon area, where their hard flinty character has preserved model glacial erosion features.

Obsidian is the equivalent volcanic glass, black in colour as a rule and fracturing in the same conchoidal manner as flint. It is a rare rock. Pitchstone is also mainly glassy, but may contain phenocrysts and also incipient crystals known as microlites. These rocks are well developed in the Scottish volcanic districts, notably in Arran and in Eigg, where a mass of pitchstone forms the Sgùrr. In time glass devitrifies and both obsidian and pitchstone crystallise into a hard, compact, fine-grained stony rock, known as felsite. Obsidian and pitchstone are only found among young volcanic rocks of Tertiary or even younger age, while felsites are found among the older volcanic rocks, for example the Ordovician of Wales.

Trachytes are also likely to have suffered considerable alteration, especially where they are older—the Ordovician of Wales will again provide an example—and in this state the coloured minerals are liable to have been altered. Such rocks are known as keratophyres, which may also include altered trachy-andesites, the difference being in the nature of the felspar.

In the normal British usage andesite and basalt are clearly differentiated, but the Americans have grouped certain oceanic lavas, which are basaltic in all respects apart from their content of oligoclase felspar, as oligoclase-andesites.

Among the basic fine-grained rocks two points require amplification. Tachylyte is basaltic glass, the equivalent of but rarer than obsidian and usually found near the chilled margins of lava flows. Glass also occurs in the ground mass of many basalts. Spilite is an awkward rock to classify. Its silica percentage makes it the equivalent of basalt, but its felspar is the soda plagioclase, albite. In addition its dark minerals are usually somewhat different from those of basalts. Spilite usually occurs as pillow lava erupted under the sea in geosynclinal belts. Its origin is a matter of dispute, especially the question whether it has suffered a low grade form of metamorphism. Tephrite and basanite are again difficult to class. Mineralogically they belong to the group of trachy-basalts. Tephrite is the normal form and basanite is an undersaturated form containing olivine. They correspond with the coarse-grained theralite and teschenite mentioned above. Olivine-basalt, a self-explanatory term, is a normal undersaturated basalt.

In these notes, those rocks whose composition may be reasonably closely inferred from the table have not been commented upon, but only those with certain unusual features. For more detail of all types the reader is referred to some of the reference books listed.

4
Metamorphic rocks and relief

Metamorphic rocks are, as their name implies, changed rocks, but the changes must have taken place under a certain range of defined conditions. In a sense most rocks are changed rocks. Granites may be formed from the fusion and assimilation of pre-existing rocks in a molten magma. Most sedimentary rocks are formed by the weathering of other rocks, redeposition of the weathered products and the lithification of the unconsolidated sediments into rock. But neither of these two changes would normally be called metamorphism.

Metamorphism implies a series of mineral and texture changes which take place in an essentially solid medium. Thus new rocks formed by complete fusion and recrystallisation are ruled out. Of course the boundary is not completely sharp, for there must come a stage when traces of the pre-existing rocks remain although most of them have been melted in an advancing magma: such rocks are called migmatites.

The factors involved in metamorphism are essentially heat, pressure and chemical alterations, the last induced by the presence of chemically active volatile materials of which steam is a common example. Some surface processes might be included logically in these changes. Weathering is a form of chemical alteration. Compaction and consolidation of rocks under near-surface conditions leads to a degree of alteration due to pressure. But such processes are excluded from metamorphism by most geologists, although there have been exceptions. Nevertheless, it is possible to get closely comparable rocks, one by an ordinary sedimentary process and the other by a metamorphic process. A good example is provided by the series of rocks known as quartzites, which essentially consist of silica grains bonded together by interstitial silica. It is possible for the silica to have been deposited between the grains by an ordinary sedimentary process, thus giving rise to a sedimentary rock. It is also possible for the sediment to have been crushed so that heat is generated and fusion promoted at the contacts between the grains, so producing a metamorphic rock. But the two would have virtually the same effect on relief and soil development, so that from a geomorphologist's point of view their precise modes of origin would not have much relevance.

Although it is possible to speak of the effects of the different agents of metamorphism separately, these effects rarely occur isolated in nature. The reasons for this are easy to see, if one takes, for example, the effects of heat. Heat in varying degrees may be supplied by igneous intrusions, very locally at its margins in the case of a small dyke and widely through a mass of rock in the case of a slowly-cooling giant granite batholith. The first example probably illustrates the effects of heat alone, but in the second it is very unlikely that there will not be a general chemical soaking of the rocks in addition to the effects of heat. Thus the bulk composition of the rock will be altered either by the import of material by the volatile elements or by the export of material taken away in solution. Such a process, involving some alteration in the general composition of the rocks, is called metasomatism and is associated in varying degrees with almost every type of metamorphism. Thus the metamorphism will not be purely thermal.

Again heat increases with the depth of the burial of rocks beneath overlying material. But so also does the pressure: in fact at a depth of about 20 km (12 miles) the pressure is approximately 6000 atmospheres. Under these conditions rocks may start to flow, so that the textural and mineralogical changes will reflect both the very high temperatures and the very high pressures, the two being hardly separable.

Or, thirdly, the case of rocks subjected to intense shearing by directed pressure near the surface, for example in earth movements along major planes of dislocation, may be quoted. The primary effect here is that the rocks are shattered and milled out—the process produces a type of rock called mylonite—but the very milling process may produce so much heat that fusion takes place with the formation of glassy streaks within the rock.

Thus the factors of temperature, pressure and chemical change can rarely if ever be considered as acting in isolation.

Furthermore, one cannot leave pressure, unqualified, as an element on its own. There are two types of pressure which need to be separated in understanding the process of metamorphism. On one hand there is the condition when the pressure in all directions is equal: on the other there is the condition when pressure in some directions is greater than in others.

The former type of pressure is the only type that can occur in a liquid and so it is usually referred to as hydrostatic pressure. It occurs at considerable depths in the crust when rocks are deeply buried. Its main effect is towards a reduction in volume and if mineralogical changes occur within a rock they should be such that minerals of higher specific gravity are formed. Such minerals are sometimes called anti-stress minerals. In theory pressure can only be the same in all directions in a homogeneous (or isotropic) material, for example a liquid, a gas or a glass. A rock, such as granite, is not isotropic because it is composed of a mosaic of crystals of different minerals which yield differently in different directions because of directional properties such as cleavage. On a large scale the directional properties should cancel out, but on a small scale certain degrees of shearing within minerals would be visible.

Directed pressure or stress can only exist within solids and not in liquids and gases. Its result is not a reduction in volume but an alteration in shape. Hence, minerals capable of existing in elongated and sheet-like forms will be favoured by the presence of such shear stresses. These conditions obviously exist in the uppermost layers of the crust during orogenic processes. Very locally intense directed pressure may operate along faults especially in thrust zones, while on a large scale major mountain building movements will cause large regional areas of stress to be set up, resulting in a widespread metamorphism of a type involving primarily directed pressure.

But the temperature, pressure and chemical environments are only one set of major factors affecting the nature of the consequent metamorphic rocks. Equally important are two other factors. The first of these is the nature of the parent rocks from which the metamorphic rocks are being produced. The second is the duration and intensity of the metamorphic processes: combined these latter two result in what is termed the grade of metamorphism.

First, the nature of the parent material must be considered. The degree of the metamorphic change will depend to a considerable extent on the range of temperatures and pressures under which a given rock remains stable. Thus, rocks which have been formed under high temperatures and pressures will not change as readily as those formed in equilibrium with temperatures and pressures characteristic of surface conditions on the earth. This means that a rock such as granite, formed by cooling under considerable pressure well within the earth's crust, will not change nearly as readily as an argillaceous limestone formed as a result of surface weathering processes and biochemical deposition. In this connection it should be noted that, although metamorphism may progressively increase once a marginal or threshold set of temperature and pressure conditions have been exceeded, the reverse does not hold. As the temperatures and pressures reduce the rock does not return through a series of progressive changes to a form which is stable at the surface of the earth. A moment's thought will show that, if this were so, no metamorphic rocks could exist at the surface of the earth. It is true, however, that some changes towards equilibrium may take place, for example the pressure-release joints discussed in Chapter 2.

Different classes of rocks vary in their susceptibility to different types of metamorphism. The purely superficial mechanical forms of metamorphism —the strong shearing and milling known as cataclastic metamorphism— will be determined largely by the mechanical properties of the rocks involved, particularly their strength and brittleness. Rocks capable of flowing will change differently from rocks which shear and splinter, the one forming new flow structures, the other a ground-down mass of fractured particles in which the degree of recrystallisation reflects the degree of heat generated.

Metamorphism involving temperature and pressure increases with deep burial will be governed in its effects by the chemical stability of the original material under a range of pressures and temperatures. Several broad classes of rocks may be distinguished in this connection. Firstly, there is a group of quartzo-felspathic rocks, which includes a variety of apparently different types. Among igneous rocks there are the granites and their relations which are usually grouped as granitic rocks. In the metamorphic rocks gneisses and various granulitic rocks, including such things as leptites, all of which are approximately equigranular quartzo-felspathic rocks (see below), should be included here. Among the sediments the whole range of sandstones, grits, conglomerates and arkoses (see Chapter 5) may be lumped together in this class. It might be thought that the sedimentary rocks would be stable only under surface temperatures and pressures and would alter with changes in these. But many of them have been derived from the unweathered residues of igneous and metamorphic rocks of the types mentioned earlier, which have remained unchanged because metamorphic rocks of a high grade do not suffer retrogressive metamorphism with falling temperatures and pressures. These quartzo-felspathic mixtures are stable under a wide range of conditions and will, on the whole, only show changes with an intensive (or high grade) metamorphism. This does not apply to cataclastic metamorphism for these brittle minerals suffer this type of change readily. Indeed, in a mass of varied rocks at a given degree of metamorphism the quartzo-felspathic rocks may show only this type of change while other types of rocks, e.g. argillaceous or calcareous sediments, may have undergone profound alteration. It must be borne in mind that the extent of the change will depend greatly on the purity of the original quartzo-felspathic mixture. The greater the purity the greater the intensity of the metamorphism required to exceed the crushed and milled stage: the greater the admixture of 'impurities' with the quartz and the felspar the more progressive will be the series of metamorphic changes observed. Where the metamorphism is mainly effected by heat the same will generally hold good. If the rock is pure quartz all that can be expected is a recrystallisation resulting in a quartzite: the greater the clay and calcareous impurities the more varied and progressive will be the metamorphic changes.

In contrast to the quartzo-felspathic rocks one may take the argillaceous rocks, the clays, shales, mudstones and the various hybrids between these and calcareous and arenaceous rocks among the sediments, together with the already slightly metamorphosed rocks such as slates. Whereas quartz and felspar exist at depth in the earth's crust as igneous and metamorphic rocks, the minerals collectively known as clay minerals are all usually formed by the action of chemical weathering on alumino-silicate minerals under surface conditions of temperature and pressure. Thus clay minerals are very closely linked with such conditions and are not stable under the range of conditions under which quartzo-felspathic rocks are stable. Under conditions of directed pressure and rise in temperature (dynamothermal metamorphism) argillaceous rocks exhibit a progressive series of changes, firstly into slates, then with increasing development of mica into phyllites and schists (see below). Again, in the metamorphic aureole of a cooling acid intrusion the progressive gradation of change consequent upon temperature and chemical gradients is best known in argillaceous or slaty rocks.

Calcareous rocks form a third, clearly-defined class of rocks. Most of these are formed at the surface of the earth by chemical weathering followed by deposition due to chemical action or the intervention of animals, which use the calcium carbonate in their skeletons and so allow it to accumulate when their dead remains are swept together. A relatively small class of primary calcareous rocks, known as carbonatites, are associated in some areas with alkaline igneous rocks, notably in the area of the East African rift valleys. But compared with the great mass of sedimentary limestones these are geomorphologically insignificant. Limestones are chemically very reactive rocks and very susceptible to metamorphism. If pure limestone is heated in the presence of air it forms quicklime as it dissociates—this is the normal process of lime manufacture. However, under pressure, dissociation is prevented and crystalline calcium carbonate or marble is produced. This reaction occurs readily but there is little more that can happen to the rock. There is, then, really only one change and not a progression of metamorphic changes. The result is the formation of pure saccharoidal (the word refers to the sugar-like consistency of the rock) marble of the type used in memorial tablets and statuary. Carrara in Italy is one of the most famous localities for the exploitation of marble, but there are many other sources in and near Greece, for example the island of Marmora. But limestones are not all pure, and the extent and nature of the impurities affect the metamorphic changes. Impure limestones suffer change with a relatively low intensity of metamorphism. A whole series of metamorphic changes occurs with the formation of various types of impure marble.

The final class of rocks includes the basic igneous rocks. One might be excused for assuming that the intrusive rocks of this class, e.g. gabbros and norites, which were formed under conditions of high temperature and pressure, would be stable, like the quartzo-felspathic rocks, under a wide range of conditions. This is not so. Admittedly, many of the basic igneous rocks are extrusive and therefore cooled under atmospheric temperature and pressure conditions: in fact there are probably more basic than acid igneous rocks which fall into this category. All these basic igneous rocks are much more susceptible than the acid igneous rocks and exhibit a whole range of changes as metamorphism proceeds. It is curious that the basic igneous rocks are not only much more susceptible to chemical weathering but also to metamorphism than the corresponding acid igneous rocks.

After the effects of the parent material on the nature of the ensuing metamorphic rock there remains for consideration the questions of the metamorphic grade and facies. The first is really a measure of the maximum intensity of the metamorphism, provided that the intensity was maintained long enough for the associated metamorphic changes to take place. Facies, in a metamorphic sense, refers to a particular group of rocks which have suffered a similar degree of metamorphism. The rocks will not have been originally identical but will develop, with a given grade of metamorphism, into a suite of rocks with their own characteristic metamorphic minerals. The minerals will reflect the prevailing metamorphic conditions, the bulk composition of the original rocks and any additions or subtractions made during metamorphism. As yet there is no complete subdivision of metamorphic rocks into a pigeon-hole classification of grades and facies but certain facies have been recognised.

For example, one well-known metamorphic facies is the greenschist facies. Typical greenschists are albite–chlorite–epidote schists with calcite or actinolite (the nature of these minerals may be obtained by reference to a textbook of mineralogy) formed from basic igneous rocks under relatively low temperatures and moderate directed pressure. But the greenschist facies contains very different rocks formed from different parent materials under metamorphism of the same general type and grade.

By contrast an eclogite facies refers to a very dense type of rock composed principally of a pale pink garnet (almandine-pyrope) and a pyroxene (omphacite). In general bulk composition it is very similar to a basic igneous rock such as gabbro and is usually thought to have been formed at a very high temperature and pressure, partly because of the very high density of the rock. This represents an extremely high grade of metamorphism, if the rock can be considered as metamorphic at all. It has a very uniform composition and is unstable in the presence of water, which is usually present in most metamorphic environments, in the range of temperature and pressure covered by metamorphism. Its high density, uniformity and proneness to retrogressive metamorphism suggest that it may have been formed by crystallisation from a basic igneous magma at great depth.

It will be noted that the eclogite facies represents, if it is indeed metamorphic, a much higher grade of metamorphism than the greenschist facies—but it is a high grade of a different type of metamorphism. It is also possible to imagine an extreme grade of shearing of a basic igneous rock in a thrust zone, so that the rock was milled out and recrystallised. This would again be metamorphism of a high grade, but of yet another type.

Thus, we can start with the same rock and assume it to be involved in different types of metamorphism, varying themselves in intensity. The result will be a whole range of derived metamorphic rocks. It is also possible to start with different parent materials and by metamorphic processes produce closely similar metamorphic rocks. Further, just as in meteorology abstract discussions involve adiabatic temperature changes, which are entirely due to pressure variations and changes of state, and eliminate heat exchange with the surroundings, so in metamorphic geology one talks about changes without the additions and subtractions characteristic of metasomatism. But in practice any parcel of air rising over mountains probably exchanges heat with the land and few rocks are metamorphosed without some alteration, however slight, in their overall chemical composition.

Hence metamorphic rocks will be treated generally under several main classes, the main types and textures will be defined and some attempt made to relate these characteristics to weathering and erosive effects. Two of the main classes of metamorphism discussed below, namely contact metamorphism and dislocation metamorphism, are primarily local. The third obviously is not from its very name, regional metamorphism, but it includes a great variety of effects produced by different causes and ideally should be subdivided. Finally, metasomatism, which some geologists would separate from metamorphism, is separated somewhat artificially though in practice it is bound up with the other types of metamorphism.

Contact metamorphism

Contact metamorphism may be defined as the alteration of rocks at the margins of intrusion or beneath extrusions. It is fundamentally a form of thermal metamorphism, but is not synonymous with it as heat enters largely into deep-seated regional metamorphism as well as playing a role in dislocation metamorphism when the shearing is intense. Obviously neither of these can be classed as contact metamorphism. Contact metamorphism differs from other forms of metamorphism not only in being due primarily to heat but also in owing very little to pressure effects. A certain degree of shear may be exerted at the margins of a large intrusion as it becomes emplaced in the country rocks and so may give rise locally to schistose rocks, but these play a very minor part in the products of contact metamorphism.

It is apparent that the degree and duration of the heating will vary greatly with the nature of the intrusion. A thin dyke or sill will cool quickly —indeed, the fine-grained and even glassy textures shown at the margins of some such features are evidence of this—and, although the initial temperature may be high, alteration is confined to a relatively narrow zone, a metre or two or even only a few centimetres, adjacent to the intrusion. Where heating is continuous, or often repeated, as in a volcanic pipe, the alteration may be greater. But for alterations on a broad scale, with the formation of contact aureoles a mile or two in width, one has to turn to granitic batholiths and bosses. Here the heating is prolonged and the rocks are soaked, to a degree depending on their permeability, by the emanation of volatile reactive materials from the granites. In this case contact metamorphism involves a degree of metasomatism. Examples of contact aureoles are illustrated in Chapter 3 (pp. 73–5) in connection with the granitic intrusions of the Southern Uplands of Scotland.

Some authors use special terms to describe the different types of metamorphism which may be conceived as included in the range mentioned above. Tyrrell, for example, largely confines contact metamorphism to the graded changes seen in the contact aureole of an intrusion of some size and describes as pyrometamorphism that in which a very high degree of dry heat—but short of fusion—is involved, and as optalic metamorphism the burning and fritting which occur locally at the margins of dykes and lava flows.

Whatever the parent material the rocks produced by contact metamorphism differ from most other metamorphic rocks in being equigranular and having the grains unorientated, except where contact schists have been formed by the emplacement of an intrusion. Such equigranular textures in metamorphic rocks are called granoblastic or, more simply, hornfelsic from the main type of rock formed by these processes. Because of the lack of deformation rocks produced by contact metamorphism retain features of the original rocks better than other metamorphic rocks. If the rocks affected are slates their slaty cleavage often remains visible; if clearly bedded sediments traces of the bedding may remain; if very fossiliferous then the fossils may still be recognisable. To the geologist these marks from a 'previous existence' are vital in reconstructing the history of rocks. Such a superimposition of the textures and/or the minerals from two or more phases is known as palimpsest structure, a term originally used to describe a manuscript partly erased and later written over.

The effects of contact metamorphism are probably best graded when the intrusion affects pelitic rocks, i.e. those deriving from argillaceous sediments. For example, in contact aureoles in slates a not uncommon rock succession starts at the outer edge of metamorphism with a zone of spotted slates. In these the major part of the slate is unchanged but spots occur due to the segregation of carbonaceous matter. Towards the intrusion there is a change to a zone of spotted hornfels, in which the rock has recrystallised and the slaty cleavage has been eliminated, while the spots are recognisable as crystals of andalusite or cordierite. Both of these metamorphic minerals are aluminium silicate minerals, the latter including iron and magnesium as well. Andalusite is the form of aluminium silicate produced under fairly high temperatures and low stress; hence it is the form characteristic of contact aureoles. As a hornfels the rock is a lot harder than in the form of a spotted slate. Still nearer to the intrusion the rocks have usually been recrystallised to a pure, granoblastic andalusite hornfels. At the contact itself there may be locally even coarser rocks or foliated rocks if stress has been important.

A very high degree of baking, analogous with brick and tile manufacture, may result in the formation of a very hard material called hornstone or porcellanite.

Different effects will be caused by the contact metamorphism of calcareous rocks. If the original limestone was pure, then a white grano-blastic marble is the usual result. Small amounts of impurities result in colour streaking of the marble and add to its commercial value. With metamorphism dolomites suffer dedolomitisation as the double carbonate is broken down. Argillaceous limestones react to form a series of lime-bearing silicate minerals, and the resultant rocks are known as calc-silicate hornfels, if they are derived direct from a parent impure limestone, and as skarn, when they are formed from the addition to a pure limestone during metamorphism of large amounts of silica, aluminium, iron and magnesium.

The results on arenaceous or psammitic rocks are different again. The intense baking of a pure sandstone will result merely in a granoblastic quartzite, but impure sandstones may lead to the formation of more varied rocks. The Silurian rocks forming the metamorphic aureoles of the Southern Uplands granites (Chapter 3) were flagstones and greywackes and the result has been the formation of a series of biotite-hornfels, some of them bearing cordierite and others sillimanite. Sillimanite is a form of aluminium silicate stable at higher temperatures than andalusite and also in the presence of a certain amount of stress. It is thus formed in the very inner-most zones of metamorphic aureoles in some localities.

Basic igneous rocks also undergo alteration in contact metamorphism. At some distance from the intrusion hornblende-hornfels may be pro-duced, the pyroxenes of the original rocks being changed to amphiboles. Nearer the intrusion, where the temperature approaches that of the original basic magma, these hornblende-hornfels become altered to pyroxene-hornfels, i.e. they revert to a composition nearer to that of the original rock, which is only to be expected because of the close similarity of temperature conditions.

Dislocation metamorphism

Just as contact metamorphism is a simple form involving only one main agent, temperature, so dislocation metamorphism also involves only one main agent, shear or directed pressure. It is found predominantly along thrusts and shear-zones and is very local in occurrence. Its main mani-festation is mechanical rupture, hence its alternative name, cataclastic metamorphism.

Under low temperatures mechanical breakage at various stages may occur. Crystals of individual minerals may first of all become strained, an effect which may be recognised in thin sections under a microscope by peculiarities in their optical properties. These crystals may shear along planes of weakness in their lattices or along surfaces parallel to planes of high shear stress. At a higher level the rock as a whole may be ruptured and sheared.

There is also a chemical side to these processes of deformation. In the presence of stress there is some transfer of material from stable to unstable grains and also the transfer of matter from the less stable surfaces to the more stable surfaces of individual grains. This transfer of matter is brought about by diffusion and results in the formation of elongated or platy crystals, which are oriented at right angles to the direction of principal stress. This type of deformation is known as crystalloblastic and normally takes place when the temperatures are higher than those prevailing during cataclastic deformation. Not only is it characteristic of higher temperatures but also of the presence of active pore fluids.

It cannot be completely divorced from cataclastic deformation because high temperatures may be realised during this process and because different rocks react differently in the presence of stress, some undergoing crystalloblastic deformation at temperatures at which others are still being subjected to cataclastic deformation. Thus, in a series of mixed sediments it is possible for the metamorphic process to change from bed to bed. In the case of alternations of silica-cemented sandstone and shale, the sandstone may undergo primarily cataclastic deformation and the shale a crystalloblastic deformation resulting in a schistose rock.

Perhaps the simplest way to approach the changes due to cataclastic deformation is to think in terms of a possible succession. Early stages in a thrust zone may involve the slicing of the rock by a series of thrust planes into what is termed imbricate structure. Further shearing in the immediate vicinity of the thrust planes may ensure the slicing of the rock by close and repeated planes of movement into a breccia (see Chapter 5). Whether a breccia produced by mechanical stress is a metamorphic rock or not is a nice point. If movement ceases, the rock, composed of angular shattered material, may be cemented with quartz or calcite into a fault breccia. Such rocks are fairly common in the Silurian sediments inland from Aberystwyth, where they form coarse breccias of grey-green mudstone cemented with quartz.

If the deformation proceeds beyond this stage, the rock is going to be systematically sheared on a minute scale. Larger and smaller fragments may be formed, partly due to the shearing itself and partly due to the different susceptibilities of the different minerals to shearing. Felspars are generally resistant to shearing and it is possible for rocks to be formed in which lens- or eye-shaped pieces of felspar are enclosed in a fine-grained, foliated ground mass. Such rocks have been called augen gneiss or flaser rocks, though not all such rocks may be formed by the processes described. Flaser is an adjective derived from a German word used for lenticles and is used in such combinations as flaser-granite and flaser-gabbro. The large felspars, which might have been called phenocrysts had they occurred in an igneous rock, are termed porphyroclasts in metamorphic rock terminology. A third term for this group of metamorphic rocks, in which the nature of the parent material may still be ascertainable from the included porphyroclasts, is cataclasites.

With a more intense milling and grinding, or with a parent material more susceptible to these processes, the result will be a rock of tiny angular sheared particles with dust in the interstices and perhaps with occasional porphyroclasts. These rocks are mylonites. They are most commonly quartz-felspar rocks as these two minerals are stable under a wide range of temperature and pressure conditions, which would cause changes in minerals chemically less stable. If the temperatures are higher or there are more chemically active pore fluids a certain amount of recrystallisation may take place in the mylonites. Such rocks, which are known as ultra-mylonites or pseudotachylite (a reference to its resemblance to basic igneous glass—see Chapter 3) show streaks of dull flinty or glassy material some of which, but apparently not all, may have been produced by fusion.

A final type of rock to be developed by dislocation metamorphism is phyllonite, a product of a high degree of dislocation acting on originally coarse-grained rocks. One aspect of dislocation may be the formation of a series of very closely spaced slip surfaces. Indeed, slates may be formed in this way and owe their fissility to what some authorities term strain-slip cleavage, whereas normal slaty cleavage, which is described below, may be termed flow cleavage. In phyllonites considerable movement along the slip surfaces and the development of some mica and chlorite give the rocks a close resemblance to phyllites (see below).

Although rocks caused by dislocation metamorphism are very significant geologically, for their study may help to reveal the degree and intensity of movement in thrust zones, they are local rocks of no great regional significance and hence less important to the geomorphologist than the other classes of metamorphic rocks.

Regional metamorphism

Regional metamorphism covers a wide range of processes because the areas involved are so large. It is essentially linked with great intensity of mountain building processes, as may be inferred from the present distribution of regions of metamorphic rocks, e.g. the highlands of Scotland, Norway, the Central Plateau of France and the Himalayas. In such areas directed pressure will reach very high but variable values and will be accompanied by high temperatures, due partly to shear, partly to the deep burial of the rocks and partly to the heat supplied by intrusions which commonly but not invariably accompany mountain building. Locally heavy loads of overlying rocks may give rise to high levels of hydrostatic pressure.

It will be realised that the ensuing distribution of metamorphic agents may be very complex in detail but one might expect to see certain major trends, viz.

(*a*) A distribution of stress decreasing from the centre of orogenesis.
(*b*) A decrease in the relative importance of stress downwards as high temperatures and hydrostatic pressure become increasingly dominant.

Thus, theoretically, it would seem that envelopes of equal metamorphic intensity might be fitted over an orogenic belt. Such an idea lies behind certain divisions made early this century, which may be illustrated by that of Grubenmann who distinguished three zones:

(*a*) An uppermost or epizone in which stress was very important, temperatures no more than moderate and the rocks produced were slates, phyllites and low-grade schists.

(*b*) An intermediate or mesozone with stress still high but with higher temperatures and uniform pressures, the rocks being predominantly high-grade schists.

(*c*) A deep or katazone with low values of stress but very high temperatures and pressures, the rocks being essentially gneisses and granulites.

Unfortunately, although there may be a tendency for metamorphic zones to be distributed in this way, there are so many local variations in the intensity of the agents of metamorphism that the model is of no great use. High-grade metamorphism, for example, is known to have occurred in many areas at quite shallow depths.

Another form of zonation was proposed by Barrow for the Scottish Highlands in 1912 and somewhat modified by later workers. This was based not on the general facies of the metamorphic rocks but on the appearance of certain index minerals as the grade of metamorphism increased towards a central area of sillimanite-gneiss, a zone in which the rocks are really migmatites as they contain numerous local occurrences of granitic material. As mineralogical changes reflect the nature of the parent material as well as the grade of metamorphism, the grading was based on changes in pelitic rocks only. Six index minerals were used to give zones of increasing metamorphic intensity. They were: chlorite, biotite, almandine (the common iron-magnesium garnet), staurolite (an iron-aluminium silicate characteristic of certain types of metamorphic rocks), kyanite (the form of aluminium silicate stable under moderate stress) and sillimanite (the form of aluminium silicate stable at high temperatures and under a certain amount of stress and hence characteristic of high grades of regional metamorphism). The associated rocks are slates in the chlorite zone, slates and phyllites in the biotite zone, schists of various grades in the other zones with gneisses in the innermost zone. In different materials the mineral changes may not always be the same, while even in the same parent material the order is not always normal. Further, in different areas, even though one takes the right type of parent material, the succession may prove to be different. This is due to the fact that in nature stress and temperature conditions, even apart from original rock composition, are so rarely identical that universal models rarely apply to individual examples. Even within the highlands of Scotland Read later showed that, a little farther to the north-east, Barrow's succession did not apply exactly.

Nevertheless, although the distributions may not be ideally symmetrical, it is possible to distinguish between low and high grades of metamorphism and between the predominance of shear and of heat plus uniform pressure. Thus three zones may be established: low-grade metamorphism due mainly to stress, medium-grade metamorphism still due mainly to stress, and high-grade metamorphism due to the heat and pressure of deep burial.

Low-grade regional metamorphism

The typical rocks of low-grade regional metamorphism, effected in the presence of stress and at temperatures from below 200 to about 400 degrees C, are slates, phyllites and certain schists. In terms of ideal metamorphic zones they should be found in the epizone.

In the transformation of mudstones and shales, which consist of very finely divided and therefore chemically reactive material, the first stage is the production of slate, a rock in which recrystallisation has only proceeded to a minor degree in that, although white mica and chlorite are formed, they occur in minute crystals with a preferred orientation which give to slate its often perfect fissility. The flakes are said to align themselves normal to the principal stress direction partly by differential growth in different directions of the individual minerals and partly by physical movements. Rocks of this type actually tend to flow under metamorphic conditions. The evidence used to support the idea of physical movement is provided by garnet crystals (often found in the schists which result from increasing metamorphism of the slates), which have spirally arranged inclusions so demonstrating that they were rolled as they grew (Fig. 4.1). As well as cleavage produced by reorientation of minerals, other cleavage may be

FIG. 4.1. Diagram of garnet crystal with spirally-arranged inclusions

produced parallel to slip and shear planes, as indicated in the section on dislocation metamorphism above. Some authors believe this to be very important in the production of most slates.

Although slates are very fine-grained, recrystallisation means that they are not as fine-grained as the material from which they were derived. The further stages of metamorphism result in increasing coarseness as the individual crystals, especially mica and chlorite, grow and become visible to the naked eye. The next recognised stage is phyllite in which the mica crystals give a lustre to the cleavage surfaces and in which the full perfection of slaty cleavage is lost. Slates may be split into roofing material of a degree of fineness not to be achieved with phyllites.

The last stage of low-grade regional metamorphism of pelitic rocks leads to the development of schists. Schists are essentially recrystallised rocks with parallel arrangements of tabular minerals, mainly mica and chlorite, but they are formed under a wide range of metamorphic conditions and high-grade schists present mineral compositions different from those of the low-grade schists included here. The main constituents of low-grade schists are muscovite mica, quartz and albite felspar. Any initial potash felspar tends to be transformed into mica plus quartz.

The series, slate–phyllite–low-grade schist, is characteristic of pelitic rocks, and other rocks under the same grade of metamorphism give rise to different metamorphic products. Thus quartzo-felspathic parent material, such as sandstone, may give rise to quartzo-felspathic schist in which the two named minerals are much more important than mica and chlorite. The ensuing psammitic schist will exhibit less perfect schistosity than the pelitic schist described above. Calcareous parent material may produce various types of marbles and calc-schists. Basic igneous rocks are transformed into greenschists in which the main minerals are chlorite, albite and epidote; the last is a metamorphic mineral, a complex basic silicate derived from the alteration of ferromagnesian minerals. A final type often produced in low-grade regional metamorphism is glaucophane-schist. The mineral is a bluish sodium-aluminium silicate and probably needs a degree of metasomatism with soda-rich solutions to be present for its formation.

Medium-grade regional metamorphism

This rather arbitrary class is designed to include the high-grade schists but not the products of deep-seated regional metamorphism, in which stress plays a subordinate part to uniform pressure and high temperature.

In the high grade schists included here are those of the almandine, staurolite, kyanite and sillimanite zones described above. The rocks are thoroughly recrystallised and any traces of original bedding, constituent pebbles and fossils have disappeared. Both the main micas, muscovite and biotite, are present, while the index minerals of the zones concerned tend to appear in large crystals: in igneous rocks they would be phenocrysts but in metamorphic rocks they are called porphyroblasts. A common rock is a garnetiferous mica schist.

All the rocks mentioned in the previous paragraph are pelitic schists for Barrow's mineralogical zones are applied to rocks of this type. They belong to what is called the amphibolite facies of metamorphism. Amphibolite itself is a rock derived not from pelitic sediments but from basic igneous rocks and impure limestones and dolomites. It is essentially a hornblende-plagioclase rock which has not been subjected to great deformation. Indeed, it tends to resist shearing and mixed rocks containing amphibolite often show broken lumps of amphibolite in a matrix of rock that has much more evidence of flowage. When amphibolites have been deformed and foliated they change into hornblende-schists.

High-grade regional metamorphism

At high pressures and temperatures rocks become plastic and, instead of showing the strongly schistose textures of the lower grades of regional metamorphism, they tend to recrystallise completely. The directional tendencies characteristic of the schists are reduced in importance so that the rocks are much more equidimensional in mineral development.

Various names are used for rocks of this class and these names are by no means used in the same sense by all authors. A common term is gneiss, a rock which is normally coarsely crystalline and only roughly banded. Gneisses may be derived from igneous rocks, when they are sometimes termed orthogneiss, or from sedimentary rocks, when they are called paragneiss. Most of them are essentially quartzo-felspathic rocks with a certain amount of mica, which may give them a rough schistosity. Such rocks, which some authors would call gneiss, are sometimes called quartzo-felspathic schist, e.g. by Turner in a modern American text, while the terms gneiss and schist are almost interchangeable in some of the older writing about the Highlands of Scotland. The reader needs then to consider carefully not only the name given to a rock but also its description. In gneisses associated with the amphibolite facies the usual dark mineral, if one is present, is hornblende, but at a higher level of metamorphism in the granulite facies it is usually the pyroxene, hypersthene. The higher grades of metamorphic rocks are formed in an environment closely approaching that in which igneous rocks are formed so that it is to be expected that the two would tend to converge in mineralogical content. Gneisses of the granulite facies are in fact very close in composition to norites among the igneous rocks (see Chapter 3).

Granulites themselves are rocks of a high grade of metamorphism in which, although the texture tends to be equigranular (or granulitic), there is usually a rough banding due to a certain amount of layering of different chemical composition and the elongation and flattening of coarse quartz. If a dark mineral is present it is usually hypersthene and with a high content of this mineral the rock passes into a pyroxene granulite. Fine-grained granulite is sometimes referred to as leptite.

A peculiar rock of this granulitic facies is charnockite, a mixture essentially of quartz, orthoclase and/or soda plagioclase, and hypersthene. These rocks are possibly to be regarded as metamorphic, possibly as igneous in origin.

Finally there is eclogite, a curious rock composed of the magnesium garnet, pyrope, and a soda pyroxene, omphacite, often with kyanite as the other main constituent. As mentioned above various features have led eclogite to be referred to both igneous and metamorphic classes of rocks.

Metasomatism

It has been stated earlier in this chapter that in practice most metamorphic changes involve the addition or subtraction of matter from the rocks being metamorphosed, so that metasomatism is inextricably linked with metamorphism. Ramberg (1952) has listed a number of lines of evidence which testify to the prevalence of metasomatism. Frequently, bodies of ore, such as sphalerite (ZnS) or galena (PbS) are found in metamorphosed limestones in such conditions that it is inconceivable that they have been injected as a fluid ore melt or that they were originally deposited with the limestone—they must represent a replacement of the limestone. Certain metamorphosed sediments change along the strike towards an acid intrusion without there being any evidence of an original lithological change in the sediment: again progressive metasomatism related to the intrusion seems to be the only answer. The bulk chemical composition of some bodies of metamorphic rock differs from those known to be caused by igneous activity and sediment deposition. There seems to be a relationship between the bulk chemical composition and the degree of metamorphism, thus suggesting a progressive addition of material as metamorphism proceeds. Where rocks with igneous appearance are emplaced without disruption and deformation of the adjacent rocks it seems likely that replacement by diffusion, i.e. metasomatism might well have been the effective cause.

Exactly how metasomatism takes place is difficult to understand. The different distances to which metasomatism may take place in different rocks indicates that permeability to the emanations varies from rock to rock, although it is not clear whether pore fluids pass physically through the rocks. Such movement is quite conceivable in poorly-cemented sediments and the concentration of tourmalinisation (see below) adjacent to major fissures is further evidence of the bodily transport of fluids or gases through open pathways in the rocks. On the other hand rocks may behave as semi-permeable membranes and allow single molecules or ions to pass through without their having any permeability to mass movements of liquid. One can see this easily in the case of a clay. It is well known that clay is impermeable to water but the fact that clay becomes saturated and plastic must mean that it absorbs water by some process of diffusion.

Both liquids and gases may permeate rocks and these volatile constituents are much more important associates of acid intrusion than of basic intrusions. Usually such action follows metamorphism. Very often the volatile constituents are in a gaseous state and then their action is known as pneumatolysis. Apart from water the chief agents of pneumatolysis are the borates, fluorides and chlorides. Emanations of borates lead to the process of tourmalinisation, so-called because the main mineral is tourmaline. All stages of alteration are shown in the regions adjacent to the granites of Cornwall. The tourmaline may be present as rosettes of minute radiating crystals as in the altered granite, known as luxullianite, or the whole rock may be altered to a quartz-tourmaline aggregate known as schorl, which is found, for example, in Roche Rock near St Austell. Fluorine and steam lead to the process of greisening, which results in the alteration of the felspar to white mica, while topaz is usually formed in addition. Kaolinisation, which causes the wholesale alteration of the felspars to kaolin and was responsible for the china clay deposits of St Austell Moor and Dartmoor, is a process which demands principally superheated steam plus some small quantities of the halogen elements.

Pneumatolysis and metasomatism, important as they are as geological processes altering rocks, are, like many other rock forming processes, of more interest to the geologist than to the geomorphologist.

The limits of metamorphism

It has been said that the higher the grade of metamorphism, e.g. in the case of pyroxene-granulite and eclogite, the closer the rock approaches an igneous rock in general composition simply because the two processes converge. Indeed, the product of an extremely high grade of metamorphism may well be an igneous rock. This is well illustrated by the example of granite. It has always been a curious fact that the most common extrusive rock is the basic basalt while the most common intrusives are acidic rocks of granitic type. Perhaps the latter are the results, at least in some localities, of a fusion of pre-existing rocks. Many examples are known of mixed rocks, which contain a host material, always metamorphosed to a considerable degree in view of the high temperatures to which it has been subjected, and an invading granitic material. Such mixed rocks are called migmatites. In this stage of ultrametamorphism the whole rock may become pasty and viscous and quartz veins injected into such material exhibit folding of the utmost sinuosity, termed ptygmatic folding. In some migmatites the granitic magma has been injected along various planes of weakness giving rise to lit-par-lit injection.

In other rocks alteration may proceed in a solid state by the metamorphism and metasomatism of the material, a process known as granitisation. Of course, at these high temperatures melting may take place and if this has occurred the process becomes palingenesis. The net results, the formation of granites, are virtually indistinguishable. But not all granites have been formed in this way: others are true intrusions consisting of granitic magma whatever its origin may have been.

Metamorphic rock types

The processes of metamorphism and metasomatism obviously lead to a great variety of rock types, converging in some ways and diverging in others. Any geomorphologist needs a basic appreciation of what is involved, but the practical question for him centres on the nature of the metamorphic rocks and the ways in which they will react to weathering and erosion. Hence out of this general discussion of metamorphic processes and intensities it is necessary to abstract a certain number of rock types.

HORNFELS These are essentially baked rocks, the products of thermal contact metamorphism. The degree of baking will obviously reflect the duration and intensity of the heat. Characteristically they show no preferred orientation of mineral grains, which are generally equidimensional. Their nature will also reflect the parent material, and this may be important in their denudation. For example, calc-silicate hornfels will be more susceptible to chemical action than ordinary hornfels.

CATACLASITE, MYLONITE AND PHYLLONITE These rocks, the products of dislocation metamorphism, are areally less important than many of the other classes. In cataclasites the alteration has been caused by shattering without chemical action. With greater milling these become mylonites, again without there being any great degree of chemical change. Finally, at a higher degree of dislocation the degree of chemical reconstitution may lead to phyllonites with films of mica smeared through the rock thus giving an appearance similar to a phyllite.

SLATE Slates are usually formed by low-grade regional metamorphism of fine-grained sediments. Their distinguishing feature is the possession of slaty cleavage. It is possible that there are different types of cleavage, some due to very closely spaced slip surfaces, some due to the reorientation or growth in a preferred direction of mineral grains. The perfection of cleavage also varies: Welsh slates, which at their best have a very fine cleavage enabling them to be split into very thin grey or purple sheets, were used for roofing much of Victorian England; less perfect cleavage either makes the slate commercially unusable except for the production of crushed aggregate (the waste tips at old slate quarries show how much imperfect material there is) or results in the production of coarse, but sometimes handsomely coloured slates euphemistically termed rustic. Unlike many other metamorphic rocks there is no segregation of minerals in slates and the individual minerals cannot be recognised in hand specimens.

PHYLLITE   This is really intermediate between slate and schist. Phyllites are coarser than slates, because recrystallisation has proceeded further as is evident from the small plates of mica and chlorite. The perfect slaty cleavage has gone and with it the commercially useful quality of slate.

SCHIST (Plate 21)   Like most of these metamorphic rock terms the word really refers to a texture rather than to the general composition of a rock. As we have seen different types of schist may be formed under different degrees of regional metamorphism from different parent materials. They are coarser than phyllites for they have undergone a higher grade of metamorphism. The perfect schist is the highly micaceous or highly chloritic one in which minerals of platy habit lie in parallel planes giving rise to the texture defined as schistose, while there are often developed segregated layers of different mineral composition. But not all schists are so perfect. Larger contents of quartz and felspar give rise to rougher schists, some of these quartzo-felspathic rocks being as coarse as the gneisses of other authorities' terminology. Apart from the variation of metamorphic minerals resulting from the degree of metamorphism, schists vary in composition with the rock from which they were derived. Highly-sheared amphibolite may result, for example, in hornblende-schists, a very different rock from the weathering point of view from a quartzo-felspathic schist.

AMPHIBOLITE   Regional metamorphism of much more basic rocks leads to the development of hornblende-plagioclase rocks. Although the hornblende prisms have a preferred orientation, this mineral is not nearly as micaceous in habit as the micas themselves so that the schistosity is far less marked. The rocks are also on the whole coarser than schists.

GNEISS   This is usually much coarser grained than schist and is produced by a high grade of regional metamorphism. Micaceous minerals are subordinate so that the rock has only a coarse and imperfect schistosity. The division between gneiss and schist is not absolutely clear and different authors use the terms in different ways: this applies to old descriptions as well as to modern writing so that care must be exercised in interpreting the term. Gneiss formed from igneous rock is sometimes distinguished as orthogneiss, that from sedimentary rock as paragneiss. Paragneiss, which is common in the Moine Series in the highlands of Scotland, is usually developed from coarse sandstone or arkose.

GRANULITE AND LEPTITE   These are rocks produced by the highest grade of regional metamorphism in which the texture is equigranular, not schistose, and any layering present is caused by the alignment of lenses of quartz and felspar. Leptites may be a little finer grained than granulites but the terms are virtually synonymous.

PLATE 21. Weathered schist, San Bernardino Pass, southern Switzerland

MARBLE   Metamorphosed calcareous rocks may have a great number of forms. The perfect marble, i.e. statuary marble, is produced by contact metamorphism of a pure limestone, but various sorts of marble may also be produced in regional metamorphism. Usually there is no great development of schistose structure because minerals with a platy habitat are in a minority. Varying types of marble will reflect the varying degrees of purity of the parent material.

QUARTZITE   This is essentially a recrystallised silica rock, which is extremely resistant to denudation. The use of the word for silica-cemented sandstone of sedimentary origin should also be noted.

Weathering and erosion of metamorphic rocks

From the general discussion of weathering in Chapter 2 it will have become apparent that it is a function of two sets of characteristics. In the first place there is the overall chemical composition of the rock which finds its reflection in the mineral content. In the second place are the physical characteristics, the degree of crystallinity, the openness of the texture (important in sedimentary rocks as will be shown in Chapter 5), the nature and frequency of the jointing and so on. The discussion of the weathering of igneous rocks was to some extent facilitated by the fact that the classification used had a chemical or mineralogical basis, so that the name of an igneous rock fairly closely defines its composition, and also incidentally its texture. It should be obvious from the discussion above that most, if not all, of the names of metamorphic rocks refer to the textures, with only a few, e.g. marble, referring to general composition. Therefore, the nature and rate of weathering will not necessarily follow the textural classification given.

   Where rocks of different classes, igneous, sedimentary and metamorphic, approach each other in mineral content their reactions to denudation probably also converge. One must not expect their precise form of origin greatly to affect the landforms produced on them. This can be most clearly illustrated from the coarse acid gneisses. These rocks are primarily quartz-felspar mixtures with the development of some micaceous layers. Thus their crude mineral composition is very similar to that of granite. In fact if they were called foliated granite it might lead to a quicker appreciation of their likely reactions to denudation. Their similarity to granite may be judged by the fact that the typical inselberg landscape is best formed on granitic rocks, a bulk term which with many writers includes both coarse acid igneous and metamorphic rocks. Furthermore, as gneiss is a product of a high grade of regional metamorphism and was thus formed at high temperatures and pressures, it is quite likely to develop similar types of jointing to granite when it is relieved of the pressure of the overlying rocks. As a result granite landscapes may be scarcely distinguishable from gneiss

148

landscapes. This does not mean that all gneiss landscapes are similar to granite landscapes. It is possible to have much more basic gneiss, e.g. pyroxene-gneiss. Such a rock will have a general mineral content similar to that of a basic igneous rock, such as a gabbro or a norite, although there may be differences among the accessory minerals. It will weather like such igneous rocks and not like the acid gneiss which resembles granite. In Ghana different forms of inselberg type seem to develop on granite-gneiss and on basic gneiss. Such different profiles reflect the nature of the jointing and probably the calibre and composition of the debris produced by the disintegration of the two forms of gneiss. The general convergence of forms on acid gneiss and granite reflects a degree of convergence in the composition of the rocks, which is itself probably a reflection of the fact that at least some granites may be little more than ultrametamorphic stages of gneiss.

Granulites and leptites should also be included here, for these also are equigranular quartzo-felspathic rocks. Their mineralogical composition and deep-seated origin ally them to gneisses and granites, so that they too can be included within the general term, granitic, in the sense in which the word is used when speaking of the crystalline rocks, partly igneous and partly metamorphic, of ancient shield areas, such as the Laurentian Shield, the Fennoscandian Shield, or the surviving parts of Gondwanaland. But just as with gneiss one must beware of including all granulites, because there are rocks of basic type, such as pyroxene-granulite, which, like pyroxene-gneiss, should be grouped with the basic igneous rocks. With these it would also be best to include eclogite with its high content of ferromagnesian minerals.

To compare the similarities, a well-cemented arkose from the sedimentary rocks might be considered together with the granite, the gneiss and the granulite. The chemical composition is very similar so that the chemical weathering might be expected to be similar. But differences might well develop because arkose, a rock formed at the earth's surface, would have a stratification and a type of jointing different from the other rocks of deep-seated origin. This can be seen in parts of the Torridonian Sandstone in north-western Scotland, on which the landforms developed are very different from those on the adjacent Lewisian Gneiss, which is predominantly an acid gneiss (Plate 1).

Although it represents a different facies and grade of metamorphism amphibolite could well be considered with the pyroxene-gneiss and pyroxene-granulite. They are all rocks composed of ferromagnesian minerals and plagioclase felspar and hence are fairly susceptible to chemical weathering. Amphibolite has hornblende and not pyroxene and is thus the equivalent of diorite among the igneous rocks: indeed, some are so difficult to distinguish from igneous diorites that the term epidiorite is used for them.

Schists, as is apparent from their large range of mineral content, must vary enormously in their susceptibility to weathering and erosion. The coarse quartzo-felspathic types, often classed as gneiss by some authors, will approximate to granitic rocks in their weathering. On the other hand, hornblende schists will be very similar to the amphibolites from which they have been derived. In tropical weathering conditions basic rocks of this type will weather to a mass of rusty clay and mica flakes, for the muscovite mica is very resistant to chemical weathering. Many of the characteristic metamorphic minerals are very resistant to weathering, as is evidenced by their presence as heavy residual minerals in the sedimentary rocks formed from the decomposition of metamorphic rocks, but as they form only a minor element in the metamorphic rock they cannot add greatly to its resistance to denudation. Fortunately the common geological practice of naming a schist in terms of its minerals facilitates an assessment of its likely weathering characteristic, e.g. an andalusite-muscovite-biotite-cordierite-quartz schist or an almandine-biotite-plagioclase schist. There are simpler varieties, of course! These considerations of the weathering of schists are obviously based solely on their mineral content: the effect of the rock texture will be considered below. In a discussion of denudation the phyllites can be generally grouped with the schists.

Slates are somewhat different in type. They cannot be approximated to any known type of igneous rock for the degree of metamorphism is insufficient. They may more nearly be compared with the pelitic or argillaceous sediments from which they have been derived. But this is not completely true. Through the metamorphic process they have become very much harder and more resistant and form uplands and plateaus rather than the lowlands characteristic of their parent shales, though it should be noted that well-indurated shales and mudstones, such as those of the Silurian and Ordovician of Central Wales, do not seem to be much inferior in resistance to the true slates of North Wales. Slates retain the characteristic impermeability of their parent rocks and are, therefore, areas of surface drainage. But they do not absorb water as do clays and are hence much more resistant to freeze–thaw which is the greater destroyer of clays and shales. This is obvious from the way in which slates, less than a quarter of an inch in thickness, stand for a century before they decay sufficiently to need replacing in the roofs of houses. Their impermeability was also used before the advent of bitumen felts and various plastics for providing the damp courses of the main walls of buildings and also between the sleeper walls and the floor joists resting on them. Ultimately slates weather and then freeze–thaw probably aids their final destruction.

The products of cataclastic or dislocation metamorphism, the cataclasites, mylonites and phyllonites, are after all only shattered and ground rocks and in the case of the first two should be compared with the parent material from which they have been derived. The phyllonites may be compared with the phyllites which they closely resemble in properties if not in origin.

Quartzites, whether cemented by secondary deposition and hence sedimentry in origin, or recrystallised by metamorphic processes, are probably the most resistant rocks known. Although silica is often contained in solution in groundwater it is probably formed by the breakdown of silicates and not from quartz which is virtually unassailed by chemical action. Quartzites, whether metamorphic or sedimentary, often form the residuals on erosion surfaces.

Marbles should be compared with the older and more crystalline lime-stones, such as the Carboniferous Limestone of Great Britain. Because they have been crystallised they have lost the ability to absorb water and transmit it through the mass of the rock, a property possessed by many younger limestones of Mesozoic and Tertiary age, such as the Chalk. The transmission of water through the rock will be confined to joints and along such features weathering will be concentrated as it is in the older harder sedimentary calcareous rocks. In detail the attack of weathering will be along the cleavage and twinning planes of the constituent calcite crystals.

Finally, there is the series of baked rocks of which hornfels is the main example. Their nearest equivalents are in neither igneous nor sedimentary rocks but in man-made rocks such as bricks and tiles. It is probably true $\longrightarrow$ $P70$ that most baked sediments are more impervious and more resistant than the materials from which they have been formed. The degree of fusion, itself a function of the temperature attained and the duration thereof, must affect their resistance to denudation. Whether they form prominent relief depends on their relations to the adjacent rocks: in Chapter 3 the example of the Rhinns of Kells and the Merrick were quoted as examples of a contact aureole forming the highest land within a region, but as was there pointed out not all contact aureoles have this relationship with the surrounding areas. Sometimes, too, the baked margins adjacent to dykes may be hardened to a degree that makes them both more resistant than the dyke itself and more resistant than the rocks into which the dyke was intruded. But it is impossible to generalise about this.

So far metamorphic rocks have been approximated to rocks in other classes with virtually the same mineral content, but such comparisons fail to take into account the characteristic metamorphic structures, the slaty cleavage, foliation and schistosity. It works well enough with the meta-morphic rocks with equigranular texture, the hornfels, the marbles, the quartzites and the granulites, and also with those such as amphibolite and gneiss in which the directional tendencies are less well marked.

This leaves the slates, phyllites and schists, in which the cleavage and schistosity planes are obvious weaknesses. The behaviour of these rocks to denudation will to a considerable extent depend on the angle at which the planes of weakness lie to the land surface.

Where the slaty cleavage or schistosity is vertical any agent grinding the surface with a traction load, especially an ice sheet, will tend to produce a differential striation of the surface particularly when the direction of movement is the same as the orientation of the planes of weakness. It might

also be thought that the vertical inclination would facilitate the ingress of water and hence of freeze–thaw shattering in slates and schists, and so accelerate denudation, but it is doubtful whether it does lead to faster denudation, although it may well add to the superficial fine-scale fretting of the surface.

Where the cleavage or schistosity is approximately parallel to the surface any joints are likely to be vertical and plucking due to hydraulic action by the sea, rivers or ice may well be easier than where the planes of weakness are vertical, although the joint pattern of individual examples would always have to be considered. In slates and schists water seepage will tend to be along planes of cleavage: some evidence of this is provided by the oxidation along cleavage planes which adds to the charm of the rustic slates described above. If water is present along horizontal surfaces and the rock is subject to freeze–thaw, the amount of destruction is likely to be much greater than where the surfaces lie vertically because whole sheets of rock are likely to be flaked off. The importance of the direction of the grain is well known in the case of sedimentary rocks used as building stones.

Whatever the truth in individual cases, the attitude of the cleavage and the schistosity will have a marked effect on the relief in all areas where such rocks are virtually bare, i.e. in coastal areas and in areas of heavy Pleistocene glaciation (Plate 21). The effects, however, will be confined to the detailed embroidery of the relief rather than reflected in the major lines of the landforms which probably reflect overall resistance and permeability (Plate 22).

PLATE 22. The upper Hérault valley, Cevennes, southern France. Close-textured dissection and straight slopes on crystalline schists of Hercynian age

5

Sedimentary rocks and relief

Sedimentary rocks are secondary rocks in that they are derived from other types of rocks, but it is not possible to treat the three main classes of rocks, igneous, metamorphic and sedimentary, in absolutely logical order because both of the latter two classes may be derived from members of their own class or of either of the other two classes. Unlike igneous and metamorphic rocks, both of which are formed at temperatures and pressures very different from those prevailing at the earth's surface, sedimentary rocks are formed in equilibrium with surface temperatures and pressures, though later deep burial may lead to a degree of lithification not consistent with surface conditions. Similarly, the minerals forming sedimentary rocks are usually stable under surface conditions.

Sedimentary rocks are formed by the deposition of the weathered residues of other rocks. Where deposition takes place depends on the size and shape of the fragments in relation to the forces moving them. Hence, sedimentary rocks are characteristically sorted. As conditions of deposition at a given place are hardly likely to stay uniform indefinitely but vary either rhythmically, for example seasonally, or in a more random fashion, the deposition at any one place usually shows regular or irregular changes. The rocks are thus characteristically bedded. The process may be illustrated by the type of deposition known as varves, which takes place in proglacial lakes. The amount of sediment fed into a proglacial lake varies directly with the amount of melting of the ice sheet, which will reach a maximum in spring and summer. Apart from the coarser material deposited in the bed of the influent stream or in the delta where it flows into the lake, there will be a range of material in suspension from silt to clay. The silt settles first forming a coarse lower layer, while the clay settles very slowly forming a finer upper layer. Such a pair of layers is a varve. Deposition as slow as this is only possible in freshwater because the electrolytes in sea water promote flocculation and the near simultaneous deposition of all particles. During the following year spring melting brings a further access of sediment with the consequent formation of another varve. This deposition produces bedding so fine that it is usually called lamination. Much coarser rhythmic succession of beds, the origins of which are often not known, are called cyclothems. Ordinary bedding will occur when there are pauses or changes in deposition. Each bed, however thick or thin it may be, represents one phase of virtually homogeneous deposition.

But not all sedimentary rocks are sorted and bedded. It is customary to include in sedimentary rocks those residual deposits which form as a result of weathering without being transported at all. The lack of transport virtually excludes the development of sorting, while the mode of origin excludes the formation of bedding. Even some transported sediments are not sorted, or are so poorly sorted as to justify the description unsorted. These include solifluxion deposits and the ground moraines of some glaciers and ice sheets. Not all transported sediments need be visibly bedded. If the transporting agent is highly selective with regard to the fraction transported and if it acts continuously over a long time, a homogeneous, apparently unbedded deposit of considerable thickness may develop. A good example is the aeolian sediment, loess.

Sediments also possess a number of other diagnostic features, which, however, do not have any great effect on landform development. These include such things as ripple marks, sun cracks and so-called rain pitting, which may be due to molluscs or to the escape of gas bubbles formed from decomposing organic matter in the sediment.

There is one type of rock, which is strictly sedimentary, but which is not normally included in the class. This includes the beds of ash and cinder deposits formed by volcanic activity. These rocks may become welded through being deposited in an incandescent state (see Chapter 3), but if deposited at lower temperatures they are strictly sediments.

The formation of variety in sedimentary rocks

The nature of a sedimentary rock will reflect primarily the original material, the way in which it has been broken down by weathering and the way in which the fragments have been distributed and sorted in transport. To this may be added the way in which the unconsolidated sediment has been converted into rock. Obviously, a great deal of variety is possible so that an exhaustive treatment of all the possibilities is out of the question, but the processes may be illustrated in the following way.

Let us take the breakdown of a muscovite-biotite granite by weathering in so-called normal, humid temperate climate. The main minerals of the rock are quartz, felspar (let us say orthoclase: should it be plagioclase it may affect the rate, but not the type of decomposition product), biotite and muscovite, together with such accessory minerals as apatite, magnetite and zircon. The weathering of these will produce three classes of products:

(a) Unweatherable materials produced by the decay of the bulk of the rock. The most important of these will be quartz particles and muscovite flakes, but any zircon present will fall into this class and some of the magnetite may also survive.

155

(*b*) Clay minerals produced from those minerals susceptible to chemical weathering. In the example under consideration these will be produced mainly from the weathering of the felspars and to a smaller degree from weathering of the biotite.

(*c*) Soluble products formed from the chemical weathering of the felspars. These will include a variety of salts and a certain amount of silica which will go into solution.

These products may be subjected to different modes of transport and deposition.

It is possible that there will be no transport. In this case the material will accumulate slowly as a residual deposit. The soluble constituents will be lost because the rain water will take them away in solution. Residual deposits on granite are not common in many humid temperate areas because the vicissitudes of the Quaternary Ice Age with the advent of glacial or periglacial conditions have removed them, but they are much more common in the Tropics, for example laterite. One does not need to go as far as the Tropics: many of the 'granitic' parts of Portugal, especially the north, have such extensive decomposition of the granites that the rock is quarried as a general purpose coarse sand. This weathering may be a relic of much more tropical conditions in the recent geological past.

On the other hand the weathered material may be subjected to the transporting, sorting and depositional activities of normal fluvial erosion. Any mechanically weathered fragments will be transported for a distance determined by their size, shape and density. Thus selective deposition could give rise to a series of conglomerates, grits and sandstones (see below for fuller descriptions of these rocks). Not only will the size of the fragments be reduced by attrition but any material easily breakable (for example, felspar along its cleavages) or weatherable (again felspar may be quoted) will be progressively eliminated as the residues are transported towards the sea. Concentrations determined largely by density give rise to natural placer deposits of metallic fragments, especially the valuable high-density ones such as gold and platinum, and to gem gravels: these two are not to be derived from the type of granite under consideration but they may be conveniently mentioned here. Alternations of transporting conditions could give rise to alternations of quartz grains and mica flakes, so producing a sediment which would ultimately form a flagstone. Should the broken fragments, often called clastic material, reach the sea, further attrition and sorting could be caused by marine action. The clay minerals will largely be removed as suspended material in the streams, and are likely to be transported to the sea, where they will be flocculated by the electrolytes in the sea water and deposited, ultimately giving rise to clays, shales and mudstones. The material in solution will also be transported to the sea and may be precipitated in a number of ways. It might be precipitated directly if the concentration of a particular salt becomes high enough. Precipitation of calcium carbonate (again, let it be noted, not to be derived from the acid igneous rock under consideration) may take place in very thin layers around nuclei so producing a concentration of ooliths, which will ultimately form oolitic limestone. It may also be abstracted to form the external skeletons of invertebrate animals and the accumulation of these after death will produce a variety of limestones, coral, shelly, crinoidal, foraminiferal and so on. Thus, even under fluvial conditions, the variety of rocks which may be produced by the denudation of a granite is very large. Some of the rocks described above, especially those with large clay or calcium carbonate contents, could not be produced as readily from the denudation of granites as from that of much more basic igneous rocks such as gabbro.

Let us now imagine a more violent form of transport tending to cease suddenly so that fragments of different sizes and shapes are transported and deposited almost simultaneously. This could be produced by solifluxion occurring in periglacial climates in spring. The weathered debris would then be moved en masse and redeposited to produce a very poorly sorted rock without bedding, a very different end product from the range of products described above.

If we imagine mechanical shattering to be much more important than chemical weathering in the decomposition of our granite, as might occur in Polar or desert conditions, there may well be a higher proportion of small rock fragments rather than individual mineral grains, and also of unweathered fragments of weatherable material such as felspar. If the degree of sorting during transport is fairly high, a felspathic sandstone, or arkose, will be formed. If the material remains largely unsorted, a type of sandstone known as a wacke (see below) will be formed. If the soluble constituents are leached out in deserts their final fate may be to be deposited by evaporation as salts in some interior drainage basin: such rocks are called evaporites.

The stress in these theoretical examples has been on the variety of lithological types which may be produced rather than on the many environments in which they may be deposited.

Environments of deposition

The main factor governing the reaction of sedimentary rocks to weathering and erosion will be lithology, but the environment in which deposition took place may play a minor role by affecting the detailed textural features of the rock.

Broadly speaking, there are two main environments of deposition, continental and marine, together with a smaller intermediate group which includes marginal types of deposition, principally estuarine and deltaic. If, as the whole of geomorphology suggests, the processes of denudation are fundamentally engaged in wearing the land down and transporting the waste to the sea, marine deposition is the final stable form, many continental deposits representing in the long view breaks in the overall seaward transport of material.

CONTINENTAL DEPOSITS

Continental deposits are mostly relatively thin and very variable in lithology from place to place. There are a few exceptions, which will be discussed below, to this generalisation.

In the first place there are fluvial deposits. Depending on the regime of the river, there will be a large variation in the nature of the deposition. A river with the smallest difference between maximum and minimum discharge will tend to have the most evenly graded deposits, while those with large seasonal or longer period variations will show much more variability in their deposits, for example streams fed by melting ice and snow from a mountain region may lay down very coarse material in early summer and very much finer material in winter. In the aggradation of a floodplain, when coarse material is deposited at the sides of the river while the finer material settles out from the flood water spread over the floodplain, the pattern of the deposits will be very complex in detail, especially as the meandering will not only give a shifting pattern of deposition but will also channel deposits already laid down. The thinness and the variability mean that such deposits are of no great significance in rock formation.

Lacustrine deposits, although again very variable, may be considerably thicker than river deposits. The final fate of a lake, assuming that it is not emptied by some renewed tectonic movement or drained by the downcutting of the stream issuing from it, is to be filled with sediment. The type of sediment may vary from coarse waste, such as would be hurled into a tectonic trough lake by vigorous streams draining adjacent fault scarps, to fine organic matter, known as detritus mud if it is transported into position, in a widespread shallow lake into which sluggish streams drain through fens and bogs. Where clastic sediments have accumulated in considerable thicknesses and been converted to rock, those lacustrine rocks will have the same general degree of resistance as marine rocks of similar types.

Glacial deposits, of which only boulder clay or till sheets are widespread, are again very thin when compared with the thickness attained by marine sediments. In East Anglia, where tills are well developed, thicknesses of over 60 m (200 ft) are known, but the average is very much less than this. Many of the other glacial deposits, such as eskers, kames and outwash trains, are really a special class of fluvial deposits showing great lateral variation and very unlikely to form thick homogeneous beds when lithified.

Aeolian deposits, principally accumulations of sand dunes and loess (see below) are again liable to great variability in thickness, though loess can form substantial widespread deposits when the conditions favouring its deposition remain in operation for long periods, for example where steady winds blow dust into an area of increasing precipitation, as seems to be the case in north China. Desert deposits, although varying considerably and

158

not all aeolian, may reach great thicknesses in the filling of subsiding tectonic basins. Ideally one would expect some such arrangement as that shown in Fig. 5.1 with conglomerates widespread at first as active denudation acts on high relief and with streamfloods the main transporting agents.

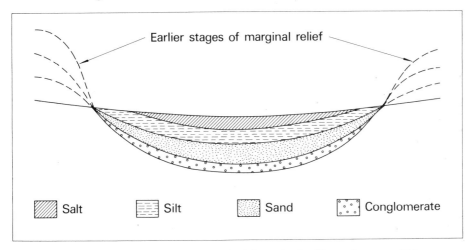

Fig. 5.1. Successive stages of fill of generalised desert basin. Time planes would be horizontal and lithological boundaries diachronous

The conglomerates would tend to become increasingly restricted to the sides as the basin filled up. In the middle parts of the basin belts of dunes would separate these deposits from the playa lake deposits of the very centre of the basin. The final stages in the extinction of the playa lake could well be the formation of an evaporite succession. Examples of thick infillings of former desert basins are known from rock systems such as the Old Red Sandstone (Devonian) while modern examples of the process may be found in the tectonic basins of Central Asia.

Chemical deposits could be taken out as a separate class largely because of their lithological similarity. They occur under a variety of conditions: the evaporation of playa lakes mentioned above, as indurated horizons in former soils, and as the filling of evaporated lagoons marginal to tropical seas.

MARINE DEPOSITS

Generally and theoretically the deposits on the sea floor grade from coarsest near the land to finest in the ocean depths, but as soon as one has said this exceptions come to mind. This is probably due partly to the fact that in shallow water reworking and sorting of deposits by wave and current action may lead to irregular concentrations of coarser and finer material, and partly due to the fact that many of the shallow areas were dry land in the Pleistocene, when they had glacial deposits spread upon them which have never yet been completely reworked and assimilated by marine action.

Other factors such as turbidity currents on continental slopes have also probably upset the regular pattern.

Beach and shallow water deposits can be a variety of things. They can be organic deposits such as coral reefs or they can be coarse pebbles such as are found in some of our coastal features, e.g. Dungeness and Chesil Beach. Rarely do they attain any great thickness because they are confined to a fairly restricted zone of shallow water and once that depth is exceeded in a transgression the character of the deposit changes. They are also liable to considerable lateral variation as observations on any modern beach will show. When cemented into resistant conglomerate they may become important in relief development even though of no great thickness.

Continental shelf (or neritic) deposits may be found down to depths of 200 m according to some authorities, to 600 m according to others. The depth is important because this is a zone of mixed, generally rather coarse, sedimentation liable to resorting by the action of waves, the active base of which is generally though not greatly to exceed 200 m. Current-bedded sands may be expected in this zone as sandbanks are shifted about by wave action; also bio-mechanical deposits, such as various types of shelly limestones, as broken shells are swept into concentrations by current action. Sediments on the whole are thin and variable. Consequently, the relief formed on areas of rock derived from continental shelf sedimentation is liable to rapid lateral changes. The Jurassic of Britain (Chapter 11) affords some fine examples.

Continental slope (or bathyal) deposits occur where material is brought to the edge of the continental slope and dumped over. The deposits on the whole are finer than shelf deposits but coarser than true deep sea deposits. They include mainly silts and muds. As on the edge of any tip or delta the equilibrium slope is likely to be exceeded, so leading to slumping and in favourable conditions turbidity currents. These are dense suspensions of mud and sand in water which may flow rapidly down the continental slope guided by its relief, reaching speeds of a mile a minute and, according to some authorities, capable of eroding submarine canyons and disrupting submarine cables. Such currents transport relatively coarse material out on to the ocean floor and deposit it far from the sites where it would normally be found. The slumping and turbidity currents are reflected in the bedding. This is often graded bedding (Fig. 5.2) in which the coarser material is at the bottom and the finer at the top, a reflection of the different settling velocities of the various grades of material. Rocks with structures of this type are known from the marginal areas of the old Palaeozoic geosyncline of Wales. In the past their sometimes heterogeneous composition has led to their being confused with fossil till deposits.

Deep sea (or abyssal) deposits are very thin fine oozes, mixed with volcanic dust, any siliceous or calcareous organisms which may reach the ocean floor without dissolving, and such resistant organic matter as sharks' teeth and the ear bones of whales. Their geomorphological significance is slight, as examples of rocks definitely formed from abyssal oozes are very rare.

Because the environments of deposition impress upon the beds interpretable lithological and fossil features, preoccupation with this aspect is to be

FIG. 5.2. Graded bedding

expected primarily from the palaeogeographer seeking to reconstruct the geography of past periods and from the palaeontologist in his attempts to interpret the ecological meaning of assemblages of fossils found in rocks. The geomorphologist is more interested in the lithological nature of sedimentary rocks and this may be virtually the same in different environments of deposition, but the geography of the past controls the pattern of deposition and this may affect the relief pattern developed millions of years later (Chapter 6). For geomorphological purposes a simple and fundamental division may be made between stable and unstable environments. Stable environments would include those in which conditions of deposition remained relatively unchanged over long periods, thus giving rise to great thicknesses of relatively homogeneous rock, e.g. geosynclinal and other deep water conditions. Unstable environments would cover continental and shallow water conditions of all types in which variable deposition conditions produce characteristically variable rocks.

Consolidation and cementation of deposits

Before proceeding to a classification and description of sedimentary rocks some attention must be paid to the way in which originally unconsolidated sediments become lithified, i.e. become rocks in the non-technical sense of the word. Provided that the pressures and temperatures involved do not rise to the levels at which metamorphism takes place, these processes are referred to as diagenesis.

The processes of diagenesis may begin almost as soon as sediment is deposited on the sea floor. The solid and liquid phases of the sediment may not be in equilibrium with each other or with conditions prevailing in the environment. Thus chemical changes leading to the formation of minerals stable under prevailing conditions may take place. When sediments are buried and subjected to pressures, even though they may still be on the ocean bed, initial or connate water may be forced to migrate through the sediment transporting and redepositing dissolved minerals, while compaction with the rearrangements of the grains may also occur. It is clear that these are processes which probably operate almost simultaneously. Later, when the rocks are uplifted, not only may further compaction take place as the removal of water is facilitated, but the zone of rocks above the water table, in which all water movement, apart from some capillary reversal near the surface, is downwards, becomes a zone of weathering. Below the water table there is theoretically a zone of deposition or secondary enrichment. This must not be visualised as a layer of increasingly concentrated solutions like an underground playa lake, because it is really a zone of water movement with seepage from above and outflow at spring lines.

Compaction is a process affecting mainly fine-grained sediments. In the case of rudaceous and arenaceous sediments some degree of compaction may be effected by the rearrangement of the particles, usually quartz, but the expulsion of water from clays results in a compaction of far greater degree. The degree of compaction depends on the depth of burial, and, as this will usually increase with the age of the rocks, it is generally true that the older the argillaceous rock the more compacted it will be. Some idea of the effects of compaction may be derived from the fact that soft surface mud may be over 90 per cent water, that surface clays may have a porosity and therefore a potential water carrying capacity of about 50 per cent by volume, while shale at a depth of 2000 m (6000 ft or so) may have its porosity reduced to about 5 per cent of its total volume. If no more than compaction has taken place then the rocks are breakable by long soaking in water, plus any measures needed to disperse the material if it is colloidal. For example, in breaking down Pleistocene sediments in the laboratory it is often necessary to soak them in solutions of sodium hydroxide or a sodium zeolite, such as Calgon, to produce the deflocculation characteristic of sodium clays. Shales cannot usually be broken in this way for a certain degree of mineral change has taken place as well as straightforward compaction.

If cementation takes place without the introduction of any outside material the process is known as authigenesis. It takes place as the result of the re-establishment of equilibrium between the environment and the solids and liquids present. Some of these changes may be effected by bacteria: for example sulphate-reducing bacteria may produce sulphides which in turn react with carbon dioxide to form carbonates, e.g. calcium carbonate, which may be precipitated as cement in the rocks. Many depositional environments are characterised by conditions which reduce any ferric compounds deposited there as a result of subaerial denudation. Clay minerals also react with iron and magnesium to form chlorite minerals which are common cementing material in many old impure arenaceous sediments. These are only examples of minerals which may be produced by chemical reactions in sediments: in addition silica, felspars, calcite, dolomite, clay minerals and chlorites are commonly produced while isolated examples of much rarer minerals have been reported.

In addition to the formation of new minerals existing minerals may be dissolved and redistributed. Many sediments contain soluble matter, for example organic phosphatic matter and calcium carbonate. Phosphatic material contained in calcareous sediment, for example teeth, bones and excrement, may be taken into solution and redeposited through the rock giving rise to phosphatised limestones. Solution of calcium carbonate and its redistribution may be aided by pressure, which results in solution where the particles are in contact with each other and redeposition in pore spaces. Simple solution and redeposition is somewhat different from authigenesis in that no new mineral formation is involved. Simple examples of redeposition of material are provided by Pleistocene gravels. Many of these show leached surface zones with redeposition of earthy calcium carbonate as a soft cement below. Also, accumulated iron and occasionally manganese layers occur in these sediments at or about water level.

The form of the redeposition and its strength as cement will be discussed more fully in connection with the description of rock types below. Meanwhile, it should be made clear that everything about the processes of rock cementation is by no means completely understood. The minerals which are either soluble, or in a colloidal state, or chemically reactive, are most liable to be dispersed through the rock as cement. It is not clear why minerals are taken into solution and then redeposited: perhaps the chemical reactions are much more complex than straightforward deposition. Further, there is the opposite tendency for segregation to produce concretionary bodies during the process of redistribution and crystallisation. Calcium carbonate, calcium sulphate (gypsum) and pyrite concretions are all common. Some of these concretionary sediments may be the result of a two-stage lithification: an early phase with insufficient matter for complete lithification producing an irregular concretionary cementation, while a later phase cemented the main mass of the rock either with the same or with a different cement. Yet some concretions are so regular and so localised, for example the spherical concretions of the Permian Magnesian Limestone of Durham, that something still needs to be discovered to provide a complete explanation.

Lithification may also be facilitated as a result of a sediment being

163

saturated by solutions derived from an external source. Like the metamorphic processes promoted by similar saturation, this is primarily a process of replacement and is known as diagenetic metasomatism. Theoretically it differs from metamorphic metasomatism as it takes place at much lower temperatures, but in practice it is very difficult to draw a sharp distinction between the two. The characteristic products of the process show their original structures, although the original materials have been completely replaced. Where the original material consisted of finely divided sediment and it has been completely replaced it is obviously very difficult if not impossible to demonstrate that replacement has occurred; for example, it is conceivable that certain quartzites might be silicified, chemically-precipitated limestones. Where, however, fossils are present, or where the rock had a characteristic original texture, such as an oolitic limestone, it should be possible to demonstrate the metasomatic origin of the rock. Thus among the metasomites, as such rocks have been called, silicified limestones, including silicified oolites, phosphatised limestones, dolomitised limestones, calcified siliceous conglomerates, as well as limestones converted to ironstones, have all been recognised.

It is often difficult to understand whence are derived all the chemical solutions for large-scale alterations. There is evidence for solution within beds in a number of forms. Generally speaking the composition of the suite of accessory heavy minerals in sediments changes with time, the most weatherable being much more common in younger sediments. In addition, it has been noted in certain concretionary limestones that the concretions contain higher percentages of the more weatherable minerals than the matrix; in an example cited by Pettijohn (1957) of certain sandstones with calcareous concretions in California, hornblende formed 40 per cent of the heavy mineral assemblage in the concretions but only 5 per cent in the matrix. Again, especially in pure quartzites and limestones, there are sometimes structures known as stylolites: these consist of irregular surfaces like barograph or anemometer traces along which the beds above and below are locked by teeth-like projections (Fig. 5.3a). They are usually parallel to, but occasionally normal to, the bedding and most likely represent surfaces along which solution has taken place by pressure, although there is no consensus of opinion regarding the detailed operation of the process. The amount of solution may be estimated by the degree of interpenetration of the teeth and sockets, or, when a vein cuts the surface diagonally, by a simple geometric reconstruction (Fig. 5.3b). Micro-stylolitic contacts between grains of sediments have also been cited as evidence of solution within a bed due to pressure. While such processes might suggest a means by which a fair amount of material may be taken into solution, certain metasomites involve such vast quantities of silica or dolomite in solution that they cannot be explained by derivation from anything other than some outside source.

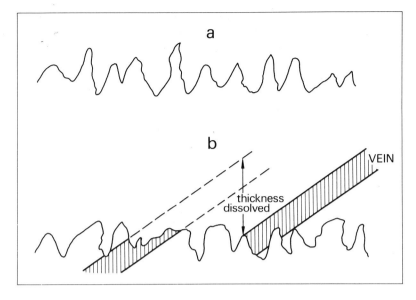

FIG. 5.3. a. Stylotites. b. Method of estimating thickness dissolved from displacement of vein

Classification of sedimentary rocks

Although the environment of deposition and the processes of diagenesis may make minor differences to the relief-forming properties of rocks, they are overwhelmed by the overall lithology of the rocks. Hence the classification used below (Fig. 5.4) is based primarily on lithology: it is essentially

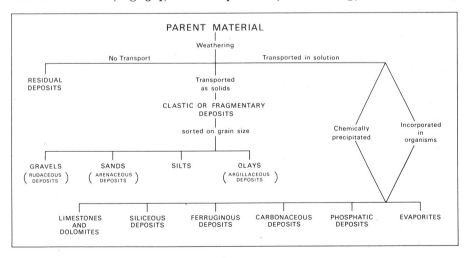

FIG. 5.4. Classification of sedimentary rocks

165

a modification of that used by Tyrrell (1926). Tyrrell described the chemical and organic deposits separately. Here, they are recombined under lithology simply because this dominates relief development.

Residual deposits

By definition residual deposits are those which result from weathering, but which undergo no transport and hence accumulate *in situ* on the parent rock from which they have been formed. It follows that residual deposits are not bedded and are characteristically related in composition to the parent rock from which they have been derived. This relationship holds in both chemical and physical aspects. The grain size of the unaltered fragments, which are characteristically angular, must be in keeping with that of the parent material: the products of chemical weathering must be insoluble products derived from the underlying rock by demonstrable chemical processes, which may, however, involve alteration in two stages. It does not follow, however, that all rocks which are unbedded and/or related to the rock on which they rest are residual deposits. Any extensive rock outcrops will provide ample scope for transport and redeposition without the material crossing to another outcrop, e.g. the Chalk of southern England or the Keuper Marl of the Midlands. Similarly, there are transported rocks which are unbedded or virtually unbedded. Many of these are glacial or periglacial deposits such as till or solifluxion gravels, e.g. the Coombe Rock of the Chalk areas of southern England, but they also include slumped sediments on geosynclinal floors and any material transported by an agent very selective as to grain size and deposited virtually without pauses, e.g. loess.

Soils, of course, are residual deposits and one of the factors required for the full development of climatic or zonal soils is lack of movement, but it is not proposed to go into the whole subject of soil development here. Fossil soils are known in the stratigraphical column and may be very important in interpreting palaeogeographical conditions. In Pleistocene studies fossil soil developments are used as indicators of warmer interglacial or interstadial conditions. Older soils are also known, for example the fire clay 'seat-earths' in which the fossil roots of the Coal Measures vegetation are found, and the dirt beds of the Purbeck (Upper Jurassic) Beds of the Dorset coast. None of these fossil soils has any real significance in landform development.

Yet when one thinks of the various types of soils it becomes apparent that there are certain conditions in which more indurated beds may be formed. The process of podsolisation leads to the accumulation of iron and more rarely humus pans in the B-horizons of such soils. Some authorities would regard the tropical process of laterisation as analogous. These hardened iron pans may have significant relief effects. They may so impede drainage as to make the highly permeable sands, on which podsols often develop, impermeable and so greatly affect drainage development. If they are sufficiently continuous they can affect relief as a local hard capping. Again, there is a tendency in the pedocal soils of semi-arid areas for calcareous concretions to develop at a depth in the subsoil depending largely on the climate, as this controls the rise back to the surface of solutions containing calcium carbonate. Such deposits occur in places like the western United States and in India, where they are known as kunkar: it has been suggested that the 'cornstones' found in the Red Marls forming the lower part of the Old Red Sandstone of the Welsh Borderland may have been produced in this way.

Two types of residual deposits are usually given most emphasis: they are the residual deposits found on limestones and the tropical residual deposits of laterite and bauxite.

The best known residual deposit on limestone is terra rossa, an unfortunate name because it merely means red earth and in some countries material called red earth, or its equivalent in the local tongue, is no limestone residual deposit, for example the terra roxa developed on the basic igneous rocks of the coffee region of Brazil. Terra rossa, as a limestone residual deposit, occurs widely in the limestones of the Adriatic areas of Italy and Yugoslavia, while many of the limestones of the interior of Provence in southern France show variable thicknesses of superficial, bright, rusty red material veining down through cracks into the underlying rocks. It seems to have been formed by the solution of considerable thicknesses of limestone and the concentration of the insoluble residue found therein, with the red colour being supplied by the oxidation of iron compounds derived from the limestone. Where rainfall is heavier it is alleged that terra rossa is washed away. Even in the dry limestone areas of southern Europe it tends to be concentrated in solution depressions, the sotchs of the Causses of southern France, where its presence is revealed after ploughing by the brilliant red of the soils and by the green of the crops grown there contrasting strongly with the surrounding desert of barren limestone and brown withered grass, herbs and perhaps stubble. Yet, although it may affect landscape in the general sense, terra rossa can hardly be called a relief former.

The same is true of the Clay-with-Flints, which may in part be an analogous deposit on the Chalk of southern England and northern France. The name itself is often used to cover a great variety of material, the thin and usually disturbed remnants of many deposits which from time to time have covered the Chalk: these include not only solution residues but also Tertiary and Quaternary beds of various ages and lithologies. Even the more stony clay aggregates are by no means certainly solution residues as was very clearly pointed out more than sixty years ago by Jukes-Browne (1906). The essence of his argument was that a calculation of the likely content of clay and flints in the Upper Chalk revealed that the residue was more likely to be a clayey gravel and that some of the thicknesses quoted for Clay-with-Flints would have involved the solution of a very great thickness of Chalk and virtually no loss of the insoluble residue by denudation. The argument was pushed further by Sherlock and Noble in dealing with part of the Chilterns, in that they showed that the maximum thickness of Chalk which could have been dissolved was insufficient to account for the observed thickness of so-called Clay-with-Flints. Furthermore, the latter contained in places such interesting material as laminated brickearth, an improbability in a purely residual deposit. They were led to suggest a glacial origin for the Chiltern Clay-with-Flints, but such an origin obviously could not apply to the Clay-with-Flints on the Chalk south of the Thames. Jukes-Browne commented on the association of this material with small outliers of Tertiary Reading Beds and with the sub-Eocene surface in general, whereas it was virtually absent from those areas where the Chalk had been eroded well below the level of the sub-Eocene surface (Fig. 5.5). As much of the content of the Clay-with-Flints could be

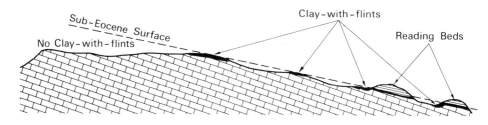

FIG. 5.5. Relation of Clay-with-flints to sub-Eocene surface

ascribed to the Reading Beds, he suggested that it was largely formed under periglacial conditions by frost action on the Reading Beds. There may be a pure solution residue form of Clay-with-Flints but it must be very much rarer. Wooldridge suggested that this was in fact so and that the material was essentially confined to the mid-Tertiary peneplain of southern England, where it affected the landscape, if not the relief, in producing an uncharacteristic (for the Chalk) amount of damp woodland and pasture. Jukes-Browne's general ideas have been on the whole supported by recent detailed research on certain soils of the South Downs (Hodgson, Catt and Weir 1967).

Laterites and bauxites, respectively residual deposits composed largely if not entirely of iron hydroxide and aluminium hydroxide, are very controversial in origin. Briefly, the problem lies in the mechanism by which these hydroxides have been left while the silica has been leached away. This is contrary to the general process in temperate latitudes, where silicates, the clay minerals, are the products of weathering and not the hydroxides. The facts requiring explanation seem to be the following. Laterites are nearly always tropical residues: bauxites usually are. Where they are not, for example in southern France and Yugoslavia, they may well be fossil deposits from a previous hotter climate. In the tropics they are not confined to silica-poor rocks, but occur on a great variety of parent material including granite, gneiss, basic igneous rocks, schist, sandstone and slate, while in Jamaica bauxite occurs on limestone. They are usually found as surface crusts of cellular, slaggy appearance.

It is difficult to find a fully satisfying explanation of these features. It was early suggested that alkaline water may have got rid of the silica, because this is attacked by strong alkalis. But this is one of that class of explanations which consist of little more than a restatement of the problem, because it is difficult to see where the necessary supply of strong alkali—and something like caustic soda would be needed—could have been derived from. Another form of explanation visualised the iron in a colloidal gel state: in seasonal rainfall climates this may have been brought to the surface by capillary action in the dry season and precipitated there by an irreversible chemical reaction. With this explanation it is difficult to understand how any thickness could have been built up, as presumably successive accretions would have to have been on the upper side of the developing crust; also laterites are not confined to areas of seasonal rainfall. Yet another explanation sees the laterites as the accumulation (B) horizons of former soils from which the overlying horizons have been stripped. This may be accepted easily in theory and visualised as happening over relatively small areas, but it is very difficult to imagine how stripping of vast areas with no trace of deposits being left could have happened.

Whatever their origin laterites can have significant relief effects. They may be the only virtually consolidated horizons in thicknesses of weathered residues. Thus they tend, where present conditions favour the destruction of the superficial weathered rocks, to form the cap rocks of minor plateaus and escarpments. Many laterites are associated with old pediplains and as these pediplains are reduced by the encroachment of newer and lower surfaces, the lateritic cappings ensure a much more perfect preservation of remnants than could have been otherwise achieved.

Laterites and bauxites are ferruginous and aluminous examples of a range of residual deposits for which the general term is duricrusts. This class in addition contains calcareous and siliceous crusts, known as calcretes and silcretes. In this terminology laterites are better referred to as ferricretes.

Fragmental deposits

The fragmental or clastic deposits are primarily differentiated on grain size, but there is no universally accepted series of limits between the different grain sizes. This is partly due to differences of opinion as to where to draw arbitrary limits in something which forms a continuous range of variation and partly due to different ways of measuring particles. This is especially true of the very small particles but may be easily demonstrated from larger particles. If one is trying to compare a spherical flint particle with a flat slate particle, does one take the mean of three diameters at right angles (obviously one diameter will not suffice), or does one abandon this and take weight which can be converted into equivalent sphere sizes? Although apparently precise, the last method would destroy the concept of shape which may have an important effect on the erodibility of fragmented material. Further, should an allowance be made for variation in specific gravity or not? With very fine particles visual selection and measurements are no longer possible. Particles are separated by sieving, provided that there is little clay or fine silt present, but sieving really measures the length of the middle axis (if one considers size in terms of three axes at right angles). Very fine material is separated by the settling velocities of the particles in water. This requires a column of water long enough for a balance to be achieved between the acceleration of the particle due to gravity and the resistance due to the viscosity of the water. Settling velocity is a function of size, shape and specific gravity and is not easily directly related to size. In fact analyses performed in this way are usually expressed in terms of settling velocities and not in terms of particle size. The larger particles cannot be easily separated in this way simply because their settling velocities are too high. Hence, a very heterogeneous sediment might have the sizes of its components determined in three different ways: direct measurement, sieving and settling velocity. The integration of the three into a common scale is somewhat arbitrary and does not give a grain size analysis of absolute precision.

Sedimentary analysis in terms of grain size, grain shape, grading etc. is a highly specialised part of sedimentary petrography. It yields information about the physical properties of sediments and also about the transporting and depositional processes they may have undergone. It involves a whole series of graphical and statistical techniques.

However, in order to give some idea as to the approximate size of the particles in the various classes of fragmental sediments, the following table gives the general range of the boundary grain sizes which have been used.

	Max. grain size (mm)	*Min. grain size* (mm)
Gravel	—	1·0–5·0 (usually 2·0)
Sand	1·0–5·0 (usually 2·0)	0·05–0·2
Silt	0·05–0·2	0·005–0·01
Clay	0·005–0·01	—

Perhaps the most astonishing limit to the layman, who is probably used to thinking in terms of concrete making or road mending, is the smallness of the lower limit of gravels, viz. 2 mm.

Some of the categories, especially the sands and gravels, have sometimes been further subdivided. In ascending size the gravels have been subdivided into gravels (or granules), pebbles, cobbles and boulders with average sizes of the order of 3 mm, 35 mm, 150 mm and over 250 mm respectively.

Sands have been divided into very coarse, coarse, medium, fine and very fine, the divisions being based on a scale of powers of 2 and not on a simple arithmetical scale.

Rudaceous deposits

By definition these are deposits of coarse material of which typical modern representatives are beach gravel, river gravel, screes, lag gravel in deserts and glacial outwash fans. It is possible to classify them on a number of different bases. Apart from the size of fragments, which is only a further refinement of the general basis of the division of clastic sediments, we may use the following:

(*a*) Shape of fragments. The basic division is into those with angular fragments, known as breccias, and those with well-rounded fragments, known as conglomerates when cemented.

(*b*) Composition of fragments. The most durable element is quartz and, if much transport and weathering has been endured, most of the weaker elements will have been eliminated: such homogeneous conglomerates may be called oligomict. With less transport and weathering, fragments of rocks or weaker minerals may be left to produce polymict conglomerates. In a sense this emphasises the basis of division (*a*) because the same processes which round fragments off will also eliminate the weaker elements.

(*c*) Proportion of rudaceous material to matrix. This criterion will separate conglomerates formed by long processes of transport and sorting from those produced by other processes which do not result in very good sorting. The former have been called orthoconglomerates: the latter, where the matrix exceeds the gravel-sized fragments, paraconglomerates.

(*d*) Mode of origin. Pettijohn distinguishes epiclastic, cataclastic and pyroclastic conglomerates. The first are produced by normal surface processes, the second by tectonic processes and the last by fire or volcanic processes. The epiclastic class especially is capable of much further subdivision into marine, fluvial, glacial, fluvioglacial and so on.

As this is only a simple account of rocks as the basis for understanding their effects on denudation, the first of these possible methods of classification will be the basic one used.

breccias   Although they are by no means confined to very resistant silica rocks, most breccias will be oligomict as they are closely related to the beds from which they have been derived almost by definition. This is apparent when the modern forms are considered. The screes of Carboniferous Limestone areas are good examples. Indeed, even when a hillside of alternating rock types, for example a limestone–shale alternation, is exposed to weathering and denudation, the smaller size of the shale fragments may well result in a crude sorting in the scree merely under the effects of gravity, so preserving an essentially oligomict quality to the breccia. Fossil Carboniferous Limestone breccias are known, for example the brockrams of the Permian of Cumberland. These are probably desert scree and detritus fans ranging from hundreds of decimetres (feet) in thickness near the Pennine fault to a few decimetres (feet) only on the Cumberland coast. Another example can be quoted from South Wales, where the breccia known as the Dolomitic Conglomerate is banked against the scarp marking the southern edge of the coalfield syncline. A further British example of a monstrous breccia is found in the Kimeridge (Upper Jurassic) of the Brora region of north-eastern Scotland. The deposit here consists of angular fragments of Old Red Sandstone, the largest of them being 45 by 27 by 9 m (150 by 90 by 30 ft) in size, embedded in Jurassic clay. These beds have been variously described as transported by ice or by torrential rivers (some torrent!), as consisting of collapsed sea stacks rather like the present Old Man of Hoy, and as a collapsed submarine escarpment affected by earthquakes. The last explanation seems the most likely.

Volcanic breccias are known as agglomerates, an unfortunate term easily confused with conglomerate. They are liable to be readily weatherable like the volcanic rocks of which they are composed.

The most common type of cataclastic breccia is the fault breccia. These shattered rocks are very often recemented with calcite or quartz. Good examples of brecciated mudstones recemented by quartz can be found in the Silurian and Ordovician of Central Wales.

CONGLOMERATES Unconsolidated gravels are usually of Tertiary or Quaternary age. Among them many of the variations classified above may be recognised. Examples of the oligomict types may be found in the pebble beds, such as the Blackheath Beds, of the lower part of the Tertiary succession of the London Basin: these consist of beautifully-rounded, ellipsoidal, black flint pebbles, which are so distinctive as to be readily recognisable even when they occur as derived pebbles in later beds. The glacial outwash Cannonshot Gravels of Norfolk, some of which are virtually spherical, are equally distinctive. Modern beach gravels afford further examples of largely oligomict gravels.

Polymict gravels are rarer as stream transport rapidly eliminates the weaker elements. They do occur, however, in the general class of glacial gravels, which were probably derived by short-distance transport from tills and contain glacial erratics the weaker of which, for example chalk and clay, may have been transported in a frozen state.

Poorly sorted gravels—paraconglomerates or conglomeratic mudstones —are very common in the Pleistocene. They include the tills or boulder clays and many of the solifluxion deposits derived from till. These are curious sedimentary rocks in that bedding is usually rudimentary or absent. In addition, they contain many angular fragments, partly because transport embedded in ice may preserve them from contacts and partly because secondary frost shattering may produce angular material.

Among the conglomerates proper, that is the consolidated rocks, a similar range is to be found.

Pure quartz conglomerates demonstrate either long continued transport and attrition, or a derivation from an earlier conglomerate which was already very mature, i.e. quartzitic in composition. Oligomict conglomerates can only be expected to be derived from materials not susceptible either to chemical weathering or to granular disintegration. Thus, rocks such as limestones and basic igneous rocks are quickly eliminated on the one hand and rocks with a two-stage breakdown into jointed boulders and individual crystal grains, such as granites and weakly-cemented sandstones, are destroyed on the other. The ideal rocks are quartzites and vein quartz in which irregular jointing produces a mass of fragments, which may be further reduced only by marginal attrition. An excellent example is the Lower Tertiary Hertfordshire Puddingstone, which consists of typical Lower Tertiary flint pebbles cemented with silica into natural concrete. It occurs as isolated cemented masses, which are usually found derived as constituents of Pleistocene deposits. Similar material derived also from Tertiary Beds produces the sarsens found in the superficial deposits of many southern England Chalk areas, for example near Marlborough: these conglomerate blocks may often be found built into old walls. The banket of the Transvaal is a gold-bearing conglomerate of vein quartz pebbles in a quartzitic matrix. A final good example of this type of rock is the red and purple quartz conglomerate occurring at the base of the Cambrian in Pembrokeshire.

Examples of polymict conglomerates may be quoted from Scotland. The oldest true conglomerate in Britain must be the basal Torridonian conglomerate, consisting largely of pebbles of the underlying Lewisian Gneiss, which can be seen in the Stoer peninsula north of Ullapool in north-west Scotland. In about the same latitude on the other side of Scotland the base of the Devonian includes a conglomerate containing, among other ingredients, large numbers of Moine Schist pebbles from the beds below. The Ingletonian Series of Yorkshire, which are probably Pre-Cambrian in age, contain conglomerate beds composed of a whole range of minerals, quartz, felspar, etc. as well as fragments of a great variety of igneous and metamorphic rocks.

As well as the orthoconglomerates, the paraconglomerates or conglomeratic mudstones, are well represented in the geological column. Beds are known which either cannot or can only with great difficulty be distinguished from Pleistocene tills in all essential characteristics. It has been suggested that they were the products of earlier Ice Ages and hence they have been called tillites. Upper Carboniferous tillites are known from South Africa and India, while Late Pre-Cambrian tillites have been reported from Africa and India, and, indeed, from the Dalradian of western Scotland and Ireland. There are, however, other probable explanations of some of these conglomeratic mudstones: these include desert streamflood deposits, mudflows, solifluxion and turbidity current deposits. Such rocks, to which the term tilloids has been given, are most likely to have been produced on a large scale by submarine mudstreams caused by turbidity currents. The efficacy of such currents in scouring submarine canyons and breaking transatlantic cables off the Grand Banks of Newfoundland has been postulated. If this belief is correct, turbidity currents may be wide-spread features of considerable power. Further, the intercalation of tilloids in bedded greywackes also points strongly to a submarine origin for those rocks. This explanation has been proposed for certain Lower Palaeozoic rocks in the west of the Southern Uplands and in Wales, where they lie in the marginal parts of an ancient geosyncline.

Although they should perhaps be classed separately, the deposits known as intraformational conglomerates are fairly close to the conglomeratic mudstones in many ways. Practically all other rudaceous rocks have individual pebbles which are older than their matrix, often considerably older. In intraformational conglomerates the pebbles and the matrix are almost contemporary. They may be formed by any process which tends to erode recently deposited material, transform it into rounded particles and embed the particles in further deposits of the same material. One can imagine a calcareous mud being deposited on the sea floor, a shallowing of the sea allowing the deposits to be disturbed by the waves and transformed into calcareous mud balls, then a further deepening of the sea with the deposition of more calcareous mud around the mud balls. The Melbourn Rock at the base of the Middle Chalk may well have been formed in some such way as this. Intraformational breccias may be produced by the desiccation of shallow water deposits, which allows a detailed polygonal crack system to develop and the edges of the polygons to curl up as they dry out. Such features are commonly to be observed on muds at the present time. In the next phase of reflooding and deposition pieces broken from the polygons may be cemented by more mud without themselves being broken down completely to their constituent fine particles. In some places curious edgewise conglomerates of platy fragments have been formed: in them the individual fragments are at right angles to the bedding. Although frost-heaving can produce such alignments, it is unlikely that edgewise con-glomerates have been formed in this way. It is much more likely that the packing has been caused by the action of the sea, which, on some beaches, produces this sort of packing where platy fragments absolutely predominate.

In considering the effects of rudaceous rocks on relief a number of facts need to be borne in mind. There is little point in separating breccias from conglomerates, for more important features are:

(*a*) whether the beds are cemented or not;
(*b*) whether they are oligomict or polymict;
(*c*) whether they are well-sorted true conglomerates or conglomeratic mudstones;
(*d*) whether the beds are continuous or very local in extent.

The degree of cementation is very important, but it is not true that uncemented conglomerates have little effect on relief, because their great permeability may give them resistance to denudation. Uncemented conglomeratic mudstones such as till will be no more resistant than the clay of their matrix because this is the weak link in the system. But uncemented well-graded gravels preserve their form often for long periods: the permeability of the gravels provides the reason why such glacial features as kames, eskers, crevasse fillings, stadial and terminal moraines are often well preserved, for example the Cromer Moraine of north Norfolk which dates from the end of the Gipping Glaciation. The high permeability of some of the Lower Tertiary sands and gravels is responsible for the local appearance of a Tertiary escarpment on the North Downs dip slope east of Croydon and also for the formation of strike valleys which join the Mole gap from either side near Mickleham. These probably developed at the foot of a small Tertiary escarpment now almost vanished and represented only by the small Tertiary outlier near Headley.

The question of the composition of the individual pebbles of a conglomerate bears upon the long term resistance to chemical weathering. The difference is really between the quartzose conglomerates on the one hand and all the rest on the other. Cementation in a rock as coarse as a conglomerate will only be effective if it is virtually complete. The cement is usually silica and hence a cemented quartz conglomerate is a highly resistant rock and can be a magnificent scarp former. The conglomeratic Devonian Plateau Beds capping the Brecon Beacons, for example, give their summits a characteristic slabby appearance. In the same general area faults crossing the headwater streams of the Neath, the Mellte for example, expose massive beds of quartz conglomerate in the Millstone Grit which give rise to waterfalls (Plate 48). It must not be thought that polymict conglomerates are quickly erodible—the Old Red Sandstone conglomerate of north-eastern Scotland quoted above is a resistant rock—but in the long run, especially in tropical climates with their intensive chemical weathering, they must tend to succumb more readily than quartz conglomerates.

Just as cemented conglomeratic mudstones such as till are only equal in resistance to the clay of their matrix, so the tillites and tilloids are virtually equal in resistance to the mudstones and greywackes with which they occur; these two rocks are discussed in more detail below. Similarly, intra-formational conglomerates are usually equal in resistance to the beds from which they have been formed. They are of more geological than geomorphological interest. A good example is the Chalk Melbourn Rock. Although it has been alleged to form terrace or scarp-like features in front of the South Downs and Cambridgeshire Chalk escarpments, more detailed investigation has in each case failed to substantiate this claim (see Chapter 12).

When the mode of formation of many conglomerates is considered it is obvious that many of them must be highly localised. Glacial gravel features are mostly restricted and disjointed apart from some outwash plains; screes are confined to the hillsides producing them; waste fans may be on a larger scale but not on a scale comparable with some marine sedimentation; turbidity currents are by their very nature localised features; stream channel deposits are highly localised and even if the stream meanders over its plain producing a thin sheet of gravel this is still confined to a valley which on the scale we are at present using is a very local feature. Some gravel deposits may be more widespread, for example till and tillite, but as has been said these are not very resistant forms of conglomerate. It is possible in mountain-grit desert basins for a confluent series of waste fans to develop and so produce a widespread deposit. Perhaps, however, the most widespread deposit is the basal conglomerate produced as the beach deposit of a transgressive sea. Yet basal conglomerates are of very limited thickness because they only form in shallow water: as soon as the water deepens a different type of sedimentation tends to take place. In practice, many basal marine conglomerates are only a few metres (yards) thick and hence affect the detail of the relief more than its major lines. Irregular cementation effectively reduces the relief-forming potential to very little, as has happened in the case of the Hertfordshire Puddingstone and the other gravel deposits from which sarsens have been derived: the individual blocks of conglomerate are very resistant but they are isolated in non-resistant beds. In short, because of their thinness and very localised occurrence conglomerates on the whole are not major relief-forming rocks. But where they are thick, as in some fold mountains, they may become much more important. Pettijohn quotes a series of conglomerates, the San Onofic, the Price River and the Wasatch, in the Western Cordillera of the United States, which reach 1500 m (5000 ft) in thickness, persist along the strike and seem to be associated with contemporaneous faulting. These and similar formations in other mountain ranges may be expected to have considerable effects on relief.

ARENACEOUS DEPOSITS

From the petrographic point of view sands and sandstones have probably been studied more closely than both coarser and finer rocks, probably because they are readily handled by mechanical sieving methods and the individual grains can be studied by normal microscope methods. As a group they are much more homogeneous than the rudaceous deposits as the range of grain size is far less. They are produced in a variety of environments, marine, fluvial, lacustrine, fluvioglacial and aeolian, very often from the breakdown of older arenaceous deposits.

Various features of the sandstones, many of them related to the derivation and formation of the rocks, bear to some degree on their potential for producing landforms.

SHAPE OF GRAINS The shape of the grains composing the sandstone reflects in part the original shapes of the rock and mineral grains and the changes that have affected them between their derivation and their deposition as a sedimentary rock. Two grain properties are commonly spoken of: they are sphericity and roundness and they need to be kept clearly separate in mind.

(*a*) Sphericity is really equivalent to being equidimensional, or, if one wishes to think of it in terms of axes, having three equal axes at right angles. Thus, cubes, cubes with their corners rounded to varying degrees, and spheres all have maximum sphericity (Fig. 5.6). The property is related

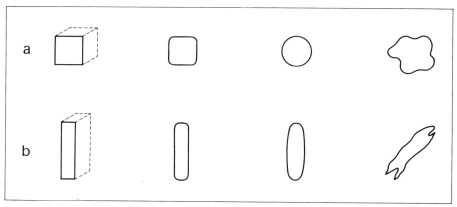

Fig. 5.6. Sphericity of particles: a. High sphericity. b. Low sphericity

mainly to the shape of the original particle or to the fragments to which the material's cleavage will allow it to be reduced during transport. Thus an equigranular igneous rock such as granite might be expected to produce a lot of quartz fragments of high sphericity, while any mineral crystallising in the cubic system and with cleavage parallel to the faces of the cube will do the same. At the other end of the scale a schist or slate will not produce many if any fragments with high sphericity, neither will mica, with its cleavage into thin flakes, ever produce detrital material with this characteristic. There is a general tendency for long distance transport to produce high sphericity, for example in some gravels, but the material has to be devoid of strong directional properties, such as those discussed above. This can be readily seen in the way in which the gravels of slaty beaches retain their low sphericity even when they become well rounded.

(*b*) Roundness is primarily related to attrition during transport, whether that attrition occurred during the formation of the rock in question or during the formation of an earlier rock from which the present particles have been derived. It is a property relating to the sharpness of corners and angles and is independent of shape, which as we have just seen has an important bearing on sphericity. If it is to be determined with precision it involves the measurement of the radii of curvature of the corners of sand grains. This is extremely time-consuming especially as large numbers of grains would have to be measured for a representative description of the sediment to be given. Accordingly a number of standard types are usually drawn and visual comparisons made of the material studied with the standard shapes. Figure 5.7 sketches three classes in such a scale. Sub-angular and round classes could be interpolated in the scale, while very

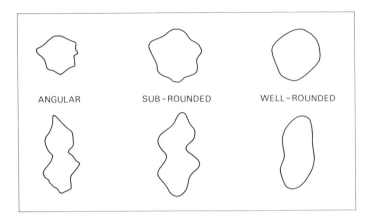

FIG. 5.7. Roundness of particles

angular and perfectly rounded types could be put at its ends. Most of these diagnostic scales are based on mathematical definitions of roundness and are usually logarithmic in character because it is much easier to recognise fine differences in roundness at the angular end of the scale, whereas it is much more difficult to do this at the rounded end. Hence the range of curvature included within the classes increases from the angular end to the rounded end of the scale.

Shape of grains is used in determining the porosity and permeability of sandstones and in discussing their possible origins. It also affects the sorting of the sediment during its deposition.

If we take the last first, it has already been shown that shape of grains is an important factor affecting settling velocity. Thus, if an area of deposition regularly receives pulses of mixed highly spherical and platy particles, the former will settle first and so give rise to a natural sorting and segregation into layers. In this way micaceous sandstones and flagstones may be formed. Flagstones tend to have segregations of mica particles at closely spaced intervals and they owe their fissility, and hence their use in the past as paving stones, to this property.

The arrangement of particles is usually described as packing. Analyses of the possible variations of packing with perfect equal-size spheres has been done mathematically. The most open arrangement of unit cells of eight spheres is a cubic one in which the centres of the spheres are centred at the corners of an imaginary cube: this produces a porosity, i.e. ratio of voids to total volume, of nearly 48 per cent. With the closest packing, which is in a rhombohedral form, the porosity reduces to almost 26 per cent. If one remembers that the volume of voids can vary between approximately one quarter and one half of the total volume no great damage will be done to the truth. With chance non-systematic packing, the condition most likely to occur naturally, the void volume is usually somewhat less than 40 per cent. These figures, relating to the total volume of voids, refer to the porosity of the rock, i.e. the total volume of liquid it could contain within its void spaces. The permeability of the rock, i.e. its ability to transmit water through itself, will vary with the cross-sectional area of voids, particularly the minimum cross-sectional area. The variation in the cross-sectional area of voids can be seen easily in the case of the cubic packing of spheres: if a section is taken through the equatorial planes of the spheres (Fig. 5.8) the area is about 30 per cent of the total, whereas if it is taken

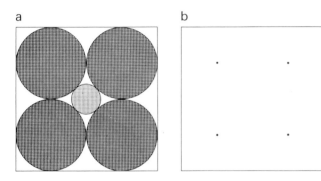

F IG. 5.8. Idealised void areas: a. Void area in equatorial plane of spheres in cubic packing. The central shaded circle is the largest particle which can be introduced after deposition. b. Void area in horizontal plane through contacts of spheres in cubic packing

through the poles of the spheres it is almost 100 per cent. In the case of rhombohedral packing the minimum void area in any cross section is about 10 per cent. Thus the total porosity of a rock composed of equal-size spheres can vary by a factor of about 2, but its permeability, related to the minimum areas of voids in horizontal cross-sections, varies with a factor of about 3 depending on packing.

In discussing porosity it is desirable to consider not only the shape but also the size of the grains. The highest porosities are displayed by equi-granular rocks. With non-uniform particles there is much more chance of the voids being filled. Again this can be seen by reference to the cross-hatched circle in the centre of Fig. 5.8a. Unfortunately there is no simple direct relationship between porosity and range of grain size because of all the other factors involved. Incidentally, it can be seen that the cross-hatched circle in Fig. 5.8a represents the largest particle that can be dropped into the cubic pattern after it has been formed. If larger subsidiary particles

179

are included they must have been deposited simultaneously. With the closest packing arrangements the largest particles included after deposition must not exceed about 0·15 times the diameter of the spheres; with the most open cubic arrangement the ratio is a little over 0·4. Assuming that close packing arrangements predominate under natural conditions any sediment in which the diameter of the fines usually exceeds about one sixth that of the coarse fragments has probably all been laid down at one time.

It has also been demonstrated that grains with high sphericity tend to pack with minimum pore space: Pettijohn quotes a void space of 38 per cent for experimentally compacted dune sand compared with 44 per cent for crushed quartz. This is only approximately one half of the void space of a recently deposited sediment of flat particles, e.g. a clay.

Another property, which is little understood and in which no direct cause–effect relation should be inferred, is that finer sands have larger void spaces than do coarse sands. One has to be careful in interpreting this because both the porosity and the grain size may depend on another property such as particle shape.

Finally, there are the pressures to which the sands have been subjected during compaction. With deep burial, movement of particles and elimination of fluids, natural sediments tend towards minimum porosity, apart from any further reductions of porosity which may be achieved by cementation. The reduction in porosity is far less in sands than in finer materials composed mainly of platy fragments: shales and clays regularly show large decreases in porosity on compaction.

In sands porosity is fairly closely related to permeability and depends on the 'useful porosity' (a term used by Pettijohn to refer to the minimum void area in the horizontal plane), the pressure difference and the viscosity of the fluid. Permeability has been shown to increase with increasing grain size and also with increasing perfection of sorting. All non-porous rocks are impermeable but not all porous rocks are permeable. The relation breaks down in the fine-grained rocks such as clays which are porous but impermeable. The clay minerals swell on wetting and so decrease permeability, while the passages are so fine as to offer large frictional resistances to flow. The function of permeability in reducing run-off and so reducing surface erosion is so well known as to make it unnecessary to stress the relations between permeability and erosion.

The shape of the grains in a sand has also been used to infer their origin (Fig. 5.9). Broadly speaking the greater the degree of angularity the closer the material is to its original source and the less severe and frequent have been the collisions between particles. This can readily be seen by a consideration of extremes.

Sands contained in tills are poorly-sorted and very angular. Fluvioglacial sands are also characteristically angular because they have the same origin, namely mechanical fracturing, and have been transported largely in the same way, in an ice sheet or glacier. The distance they have been

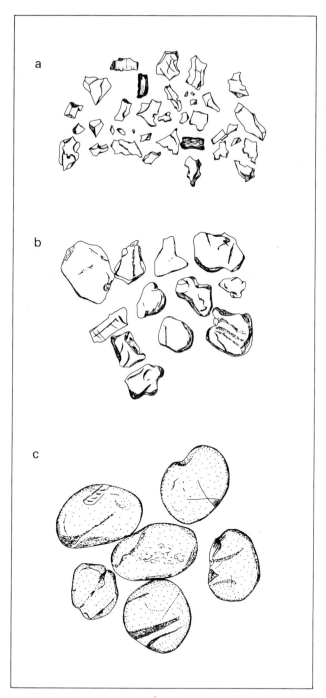

FIG. 5.9. Shapes of sand grains: a. Glacial sand. b. Beach sand. c. Desert sand *(after Hatch and Rastall)*

transported by water, in which some rounding might be expected, has been very small. In addition the grains have fresh fractures and there is a variety of weatherable and unweatherable minerals present. Such a mixture might well be termed 'immature' as it cannot survive in its present condition for a very long period.

The maximum maturity, involving the longest and severest series of collisions, occurs in wind-blown desert sands, the force of the collisions being greater in air than in water. Pure desert sands are extremely well-rounded and well-sorted; they are often called millet-seed sands. During the long process of grain bombardment readily fractured material is eliminated. Micas, for example, are split into cleavage fragments which are winnowed by the wind. A certain amount of felspar plus some resistant iron ore grains are retained: it would require more intensive chemical weathering than is normal in deserts for them to be eliminated. Very often desert sand grains have a frosted appearance and it has been said that this is imparted in the sand blast process, but there is no proof of this. Two other causes of frosting have also been suggested. It may be corrosion of the surface rather like the way in which the frosting of glass can be achieved with hydrofluoric acid: in the case of sand grains many may have been derived from calcite-cemented sandstones in which surface etching of the quartz grains may have been produced by alkalis. It may also not be a pitting effect, either chemical or physical, at all, but an incipient encrustation with secondarily-deposited silica.

Between these two extremes of fluvioglacial and desert sands lie most of the water-deposited sands, whether marine or fluvial. Perhaps on average the former are slightly more rounded than the latter, but there is such a range of variation as to make diagnosis of origin uncertain. One might think up nice theoretical examples to illustrate progressive rounding with transport from a source such as the head of a river valley or a cliffed headland. Unfortunately the natural position is usually so complex as to make such a neat reference back to an origin rarely possible. Comparatively small distances of transport are usually sufficient for the material to attain a relatively stable degree of rounding.

Origins have also been inferred from percussion markings on grains, usually in rocks coarser than sandstones: they are usually held to indicate very rapid transport or marine action—the effects are the same, namely frequent and violent collisions. Recently (Krinsley and Funnell 1965) surface features of sand grains have been examined under the very high magnifications of the electron microscope and it has been suggested that different types of sand can be diagnosed from microscopic surface markings on the grains. The technique, of course, is elaborate and slow and, while of the utmost scientific interest, not yet fast or cheap enough for anyone with an idle interest in the origin of a sediment to indulge in.

All inferences of origins are beset by derived grains, whether based on angularity of grains or on features of the surface of grains. In fact most examples which come to mind are probably from mixed sources. The north-east Trades blowing off the Saharan coast of Africa are likely to ensure that much of the 'marine' sand found there will be excessively well rounded. The sand in any river basin is as likely to be derived from the breakdown of sandstones within the basin—unconsolidated sandstones are among the most erodible of rocks—as from the primary formation of sand grains from larger mineral particles. The same is equally true of many shore sands. Where marine, fluvial, glacial, and periglacial aeolian environments have either coexisted or succeeded each other rapidly, as in many places in the Pleistocene, it is probably safer to infer origins of sediments from larger-scale textural and structural features of the beds.

All that has been discussed in relation to the shape of grains really only adds adjectives to the common word sandstone. They may be well-sorted or poorly-sorted, lacustrine or aeolian, mature or immature depending on the features discussed. The mineral composition of the grains offers the basis for a more fundamental division.

MINERAL COMPOSITION OF GRAINS Within the range of particle size of sands it is possible to get particles of different types of material, just as it is with the rudaceous rocks. If organic matter is excluded, these fragments will be either minerals or rocks. The minerals really fall into two main classes. Quartz and other forms of silica such as flint and chert are extremely resistant, while the remaining minerals are generally much less resistant. Rock fragments will obviously have to be small to fall within the possible size range. One cannot, for example, have chips of granite, other than chips of their constituent minerals, as common material in sands. Rock fragments are much more likely to form part of coarser rocks: thus they are much more common in rudaceous rocks than in coarse sands and more common in coarse sands than in fine sands.

As the proportion of minerals other than quartz is liable to decrease with chemical and mechanical action, sandstones with considerable proportions of these minerals, or of rock fragments, must be considered immature, while those from which the other minerals and rock fragments have been largely eliminated are mature. Another feature of increasing maturity is the increasing degree of perfection of the sorting involving the reduction of the clay content. These two strands form the bases of most of the classifications adopted and discussed in the works on petrography listed at the end of this chapter. Because of the necessity of distinguishing between a variety of rocks which are all within one common generic term, sandstone, a number of terms have to be adopted or invented.

Following basically the nomenclature used by Gilbert the terms are as listed below. Other authors may use different terms for some of the rocks, but the various classifications and names have a lot in common. Sandstones may be divided fundamentally into:

I. IMPURE SANDSTONES. These are called wacke and are poorly sorted with much clay matrix (Gilbert suggests over 10 per cent).

II. PURE SANDSTONES. These are called arenite. They are well sorted and have a low content of clay matrix (under 10 per cent according to Gilbert).

Pettijohn uses a similar basic division but wacke and arenite are replaced as terms by graywacke and sandstone, the dividing line being drawn at a clay matrix content of 15 per cent.

Both of the main classes above may be subdivided as follows:

A. *Wacke*
(*a*) Wacke with many unstable constituents.
    (i) Lithic wacke—containing abundant rock fragments. (It should be noted that even at this level the sediment may contain a lot of quartz: up to 75 per cent is accepted by Pettijohn in his equivalent division.)
    (ii) Felspathic wacke—containing abundant felspar fragments. (Gilbert uses this term for those with 10–25 per cent felspar content. Above the higher value he uses the term arkosic wacke.)
(*b*) Wacke with mostly stable constituents.
    (i) Quartz wacke—contains less than 10 per cent of both rock and felspar constituents. (This rock is obviously more mature than those mentioned above.)

B. *Arenite*
(*a*) Arenite with many unstable constituents.
    (i) Lithic arenite. This corresponds to the lithic wacke and may be separated by the same criterion used to define the lithic wacke.
    (ii) Felspathic arenite. This has 10–25 per cent felspar content: above the latter figure it becomes arkosic arenite. The older and more general terms for these felspathic sandstones is arkose.
(*b*) Arenite with many stable constituents.
    (i) Quartz arenite. This is the pure sandstone of earlier and vaguer terminology. The corresponding Pettijohn term is orthoquartzite, but this must have at least 95 per cent of its grains composed of quartz whereas a quartz arenite has a minimum of only 80 per cent.

In this classification the immature rocks are at the beginning and in the wackes increasing maturity is denoted by increased sorting. This is continued through the arenites by the elimination of weatherable sand grains until the quartz arenite, the most stable form is reached. Pettijohn uses the term graywacke where Gilbert uses wacke, but the former has an old and widespread usage as a term for a dark-coloured impure or immature sandstone.

The general effects of these various rocks on landforms are dealt with below in groups rather than individually.

Greywacke Most of the dark-coloured impure sandstones referred to as greywacke have been deeply buried and were characteristically formed in rapidly subsiding basins and geosynclines. They have a number of common features. The bonding material or cement is usually an indurated mixture consisting mostly of micas and clay and approximating in composition to slate. Within this the grains of felspar, quartz and rock fragments are angular and often dispersed. It is not known whether the matrix is the result of original mixed deposition or has been derived by the chemical alteration of original unstable minerals. Chemical cements (see below) are almost unknown in greywacke. Most greywackes occur in beds ranging from a few inches to a few feet in thickness and are commonly associated with mudstones and shales (Plates 5 and 23). Within the individual greywacke beds there are few signs of finer bedding and the material is tough, but it is weakened by the intervening mudstones and shales. Many greywackes show graded bedding, i.e. coarse at the bottom becoming finer towards the top. This is a feature produced by turbidity currents, or any other process which stirs up a great shower of material and then allows it to settle. Usually the rocks occur in geosynclinal tracts where turbidity currents are to be expected. But there are greywackes without graded bedding and it is probably better to call turbidity current deposits, turbidites, a genetic term, and to keep greywacke as a descriptive lithological term.

In Britain greywackes are common in the Lower Palaeozoic sediments of Wales, the Lake District and the Southern Uplands of Scotland. Although individual beds are tough, the general lack of really massive beds promotes a general uniformity of resistance to erosion of the shale–mudstone–greywacke successions of these areas. The fine matrix and the induration render the rocks impermeable and hence unlike most sandstones, which are permeable. They are, therefore, susceptible to surface erosion and owe what resistance they have to their degree of induration, which is generally high, for they are old rocks which have mostly been deeply buried. They play a large part in the detailed cliff scenery of Wales. Here the complex structures throw the beds to all sorts of angles and the pattern on both wave-cut platforms and in the cliffs themselves is often determined by the thicker sandstone horizons (Plate 42). Theoretically, the fact that they contain felspar and rock fragments makes them more susceptible to chemical weathering than quartz wacke, although the chief control on resistance is probably the strength of the cement as is so often the case. In fact many of the more unstable minerals may have been converted to matrix in the older examples.

Felspathic sandstone These have been subdivided in the classification into wackes and arenites of various types. Many of these must have been derived almost directly from the disintegration of large areas of crystalline rocks with abundant felspar content. Provided that the transport is short and the deposition rapid it seems likely that detrital felspathic rocks may be formed in a variety of climates. Hatch and Rastall quote the fact that the sandstone rocks of California of all ages from Jurassic to Recent are felspathic in spite of a wide variety of climates. Varying degrees of alteration of the felspar not conforming to distance from modern weathering surfaces suggest differing degrees of weathering before deposition. This suggests a short distance of transport rather than a desert or arctic climate in which chemical weathering was largely inhibited. Massive bedding and the inclusion of lithic fragments are other features which also point to mechanical weathering and rapid sedimentation. Many good examples are also found in the Torridonian (Pre-Cambrian) of western Scotland and its presumed Scandinavian counterpart, the sparagmite.

Theoretically, the wackes should be less resistant than the arenites because not only will their constituents be equally susceptible to chemical weathering, but they should also be less permeable and therefore an easier prey to surface erosion. In practice, few of the arenites will be uncemented and such a simple statement will require endless individual modifications.

Lithic sandstone   Examples of lithic wacke will be found in areas of grey-wacke sediments. One can really say little about their resistance to denudation that has not already been said about the felspathic sandstones.

Quartz sandstone   These include both quartz wackes and quartz arenites. Being composed almost entirely of quartz grains, which are extremely resistant to weathering, their overall resistance will depend upon other factors, such as thickness and nature of cementation. Without cement the rock is soft, permeable, unweatherable, but fairly easily erodible compared with its resistance when cemented. It may well form low upland like the Folkestone Beds of west Sussex because of its permeability. When well cemented, especially with silica cement, the rock becomes extremely resistant, for example the Stiperstones (Ordovician) quartzite, which forms a marked ridge capped by tor-like features (Plate 40) west of, and parallel to, the Longmynd in Shropshire. Rocks of intermediate degrees of cementation, such as the Penrith Sandstone (Permian) of Cumberland or the Fell Sandstone (Carboniferous) of Northumberland tend to form moorlands—uplands rather than mountains. Here, the main effects are caused by variations in cement and are discussed below.

Before proceeding to a discussion of the effects caused by the nature of the cement, a few earlier and vaguer rock terms need to be defined. A grit is a common name for a coarse sandstone in which the fragments are often angular: good examples of these rocks are found in parts of the Millstone Grit (Carboniferous). Flagstone is micaceous sandstone in which the mica layers cause the rock to be readily split into flagstones. Freestone is a masons' term applying to a building stone which can be freely worked in

PLATE 23. Weathered greywackes and shales, west of Plynlimmon in central Wales. Bent strata caused by surface downhill creep or glacial drag?

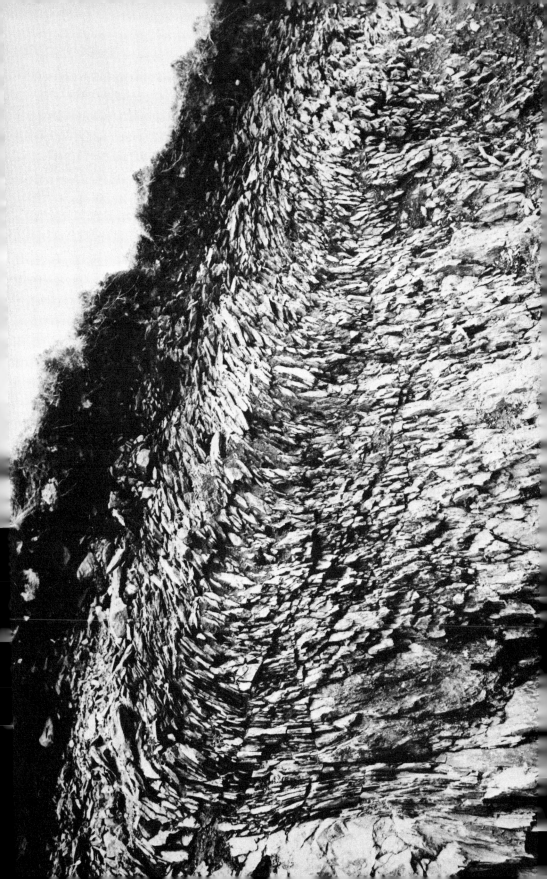

all directions. It thus applies not only to sandstones but also to some limestones such as the Bath and Portland stones. It is a word to be avoided in precise writing about rocks because of this ambiguity.

THE NATURE OF THE CEMENT   So far the nature of the cement has only been mentioned incidentally except in connection with greywackes with their hard fundamentally argillaceous cement. In addition to argillaceous cementing material, the two most common cements are quartz and calcite, with iron cements of some importance as well. Examples are also known of cementation caused by the deposition of many other minerals, e.g. barium, calcium and strontium sulphates, but these are rare enough to be of more petrographic than geomorphological interest. These chemical cements are almost confined to pure sandstones and do not occur in wacke type sediments.

Argillaceous cement   Inherently the packing of the voids with argillaceous material, whether this is due to initial mixed grain size deposition or to the alteration of chemically unstable constituents, does not give rise to a very resistant material unless the cement is partly recrystallised. Any coherence given by the cement is probably offset by increasing impermeability and hence susceptibility to surface erosion. This can be seen in the case of the Sandgate Beds, the middle division of the Lower Greensand, in west Sussex: these are silty sands and have been eroded by the strike valley of the River Rother. In the Central Lowlands of Scotland the Carboniferous contains 'rotten-rocks', which are clay-cemented sands tending to break down easily. The denudation processes which affect clays seriously, notably freeze–thaw, will equally affect clay-cemented sandstones. Greywackes are much more resistant because the deep burial and partial recrystallisation of the cement has hardened them to a greater degree.

Silica cement   Consolidation by silica may take place in two main ways. With deep burial the points of contact between the grains will be subjected to high pressures, local solution will take place and the silica will be redeposited in the pore spaces. The pressure and solution ensure an often interlocking mass of grains which is then cemented by the redeposition mentioned. The rock is very resistant indeed. These rocks are sometimes called orthoquartzites to distinguish them from the metaquartzites formed by metamorphism. In view of the ambiguity of the term quartzite, the orthoquartzites are included here as quartz-cemented sandstones.

Alternatively, the quartz grains may be cemented by deposition from percolating solutions without any alteration of the initial shape of the grains. The cement may be deposited in continuous films over the surfaces of the grains so that they gradually build outwards and fill the voids. The silica may be deposited in optical continuity with the grains, i.e. it continues the crystal structures of the grains it encloses, so that under the microscope it has the same optical properties as those grains. The grains grow until they interfere with each other. Sometimes new crystal faces are formed: sometimes they are not. The outlines of the original grains can usually be discerned under the microscope by films of impurities, for example ferric hydroxide, at their margins. This is the most common type of quartz cement, but examples are also known where the interstices (voids) are filled with minute crystals or where the grains are bonded by fringes of crystals growing between them. Usually the cement is the crystalline form of silica, quartz: but more rarely it may be amorphous, in which case the rocks become cherty quartz arenites or cherty sandstones.

Obviously, enlargement of the grains will provide the firmest bond and the toughest rock, for example the Carboniferous ganister of Britain. The rock itself may be somewhat brittle, but this is only a relative term as will be known to anyone who has tried cracking quartz-cemented sandstone. It is so resistant—in fact, one of the most durable rocks known—that the increasing impermeability with cementation does not really matter. In any case, quartzites are usually jointed and any loss of mass-permeability caused by cementation will be offset by fracture-permeability.

The resistance of such rocks will also vary with the degree of perfection of the cementation. Partially cemented sandstones are obviously weaker than completely cemented ones. The differences are similar to those between the bond of two pieces of steel spot-welded together and that of two pieces joined by a continuous weld. Impurities in the cementing material will also affect the resistance. Most quartz-cemented sandstones contain some amount of impurity, for example calcite, and, as Pettijohn asserts, the cohesiveness of the rock decreases with decreasing purity of the cement. One must also expect that, if the grains are coated with ferric hydroxide as in the Penrith Sandstone and some of the Triassic sandstones, there will be natural surfaces of weakness along these films. Cherty sandstones can possess considerable resistance even though they may not be in the same class as quartz-cemented ones: this is well illustrated by the Hythe Beds (Lower Greensand) of the western Weald. These form the highest ground in the area, even overtopping the Chalk.

Calcite cement   These and the quartz-cemented sandstones are the most common types of sandstone. As with quartz, the distribution of calcite can vary between mere spot connections at the contacts between grains, as in some barely-cemented, geologically very young beds, and complete filling of all the voids with crystalline calcite. In some examples each void is filled with a single crystal of calcite, but in others much larger crystals of calcite develop and include large numbers of sand grains within themselves. Examples of such sandstones occur in the Calciferous Sandstone of the Carboniferous Limestone series of Scotland, and in certain parts of the Hythe Beds of east Kent and the Sandgate Beds of west Surrey.

The resistance to mechanical abrasion must vary with the continuity of the cement, but probably to a smaller degree than that of quartz-cemented sandstones, because calcite has a very well developed cleavage whereas quartz has none. This must facilitate fracturing. The big difference lies in the susceptibility to chemical weathering, as calcite yields easily to solutions of carbon dioxide and organic acids. In fact in many ways these sandstones resemble limestones. Whereas most other sandstones tend to give rise to acid (calcifuge) vegetation and the development of widespread acid peat, calcite-cemented sandstones encourage calcicolous vegetation and the formation of light, readily-worked, fertile warm soils, very attractive to early settlement and nowadays to arable farming. The ensuing landscape will therefore be very different.

Ferruginous cement   Many desert sands have films of iron oxide on the grains and this may serve to bind the grains together, either with or without the addition of silica. Many examples are known from the New Red Sandstone (Permian and Trias) of Britain, while parts of the Lower Greensand of the East Midlands outcrop, e.g. in Bedfordshire and again at Hunstanton in Norfolk, are composed of iron-cemented sands known as Carstone. On a much smaller scale the iron pans of podsolised soils are impure iron-cemented sandstones or conglomerates. Other iron minerals are also known as cements but, near the surface, they mostly become altered to iron oxides.

Iron-cemented sandstones are resistant when the cement is oxide mixed with silica, though less so than pure quartz-cemented sandstones. Where other iron minerals have been altered to oxides as a result of surface weathering the rocks tend to lose coherence.

SANDSTONES IN GENERAL   It should by now be apparent that a great variety of rocks is included in the general term sandstone. They vary not only in consolidation, but also markedly in porosity and permeability as well as in hardness. Yet certain general tendencies can probably be observed in sandstone relief.

There is a natural tendency for the rock to be permeable and hence relatively resistant. In young unconsolidated sands this is a function of the interconnected pattern of voids between the grains. In older rocks, where the voids have mainly been filled with cementing material, permeability tends to be preserved by the often well developed, cuboidal jointing pattern. For example, joints are usually well developed in the Torridonian and the Old Red Sandstone. With the exception of the calcite-cemented sandstones, the richness of sandstones in silica and their poverty in bases ensure the development of acid soils and calcifuge vegetation, the developments of which are interconnected. Soil podsolisation leads to the development of iron pan in the B-horizon and thus to impeded drainage, giving rise to the curious spectacle of surface waterlogging on very permeable sands, e.g. the Bagshot Beds of the western part of the London Basin. Acid vegetation on the upland sandstone areas of Britain, e.g. the Old Red Sandstone of the Black Mountains and the Brecon Beacons, of Exmoor, and the Millstone Grit of the Pennines, includes such plants as bilberry, heather and cottongrass. It produces a sombre vegetation, which contrasts very strongly with the lighter green of limestone uplands. More important from a geomorphological point of view, the acid blanket of peat provides an impermeable cover and induces surface runoff. Thus among both the soft young sands and the harder older sandstones surface impermeability can be superimposed on a fundamentally permeable rock by soil and vegetation development.

SILTS AND ARGILLACEOUS DEPOSITS

The fine-grained sediments, including both silts and clays, are much more difficult to study than the other sedimentary rocks. This is for several reasons. The individual particles cannot be easily observed with ordinary microscopes and so the nature of the minerals forming the grains can only be determined by physical and chemical tests, for example infra-red and X-ray analyses. Such methods have been in use for a far shorter period than conventional optical analysis. Similarly, the particle size analysis of sediments of these types cannot be undertaken by ordinary sieving methods, but has to be studied by the determination of settling velocities, which have then to be translated into some meaning in terms of particle sizes. As clays are nearly all flocculated, i.e. their individual elements are clusters of particles, they have to be dispersed by chemical methods before settling velocity analysis is possible. Many of the particles are flat and at this level variations in settling velocity due to size, specific gravity and shape make simple interpretations of these velocities in terms of grain size very difficult. Further, it is unlikely that clays were originally deposited solely as dispersed particles so that deductions about environments based upon such work are not necessarily reliable.

SILTS As a group of rocks, silts and their consolidated equivalents, silt-stones, have not been widely recognised. Tyrrell was of the opinion that further study would reveal that they were far more common than had been recognised. Pettijohn, after pointing out that most shales, which are by common consent usually included in argillaceous rocks, contain a silt fraction which may exceed 50 per cent of the total rock, thought that pure silt rocks were quite rare and formed only thin beds, never very massive formations. Silt consists mainly of minute particles of minerals, such as quartz, felspar, mica and calcite, which have probably been reduced by mechanical disintegration and grinding of some sort. It does not contain large proportions of clay minerals which are largely alteration products formed by chemical weathering and other changes. The particles lack the distinctive colloidal properties of true clay particles and this affects the properties of the resulting rocks to a large extent.

The best known unconsolidated silt is loess. Characteristically, loess is extremely well-sorted and consists almost entirely of silt particles. Essentially it lacks bedding and has been widely interpreted as an aeolian deposit. It occurs in two main regions. The first of these is in north-western China where thick deposits of the material blanket the relief. It is thought that the Chinese loess has been derived from such things as old lake deposits in the Mongolian deserts and transported to its present position by north-westerly winds blowing out from the winter Siberian anticyclone. It is deposited in more humid areas probably owing to a reduction in wind velocity and an increase in humidity and precipitation. It seems to have accumulated in a steppe or prairie region, a suggestion based on the existence within the loess of many fine tubes lined with calcium carbonate thought to have been formed round the grass stems as they were submerged beneath the silt. Loess is also found in Europe and North America adjacent to glacial deposits and was probably formed under periglacial conditions by the wind winnowing the finer material from these before they were adequately protected by vegetation. In Europe it is thickest and most widespread in the Ukraine region of the south-west of the Soviet Union and tapers off westwards to northern France, with broader expanses in Silesia, the Leipzig area and the Rhine valley around Bonn and Cologne. Whether the limon of the Chalk plateaus of northern France and the brick-earth of southern Britain, e.g. that found on the terraces of the lower Thames and on the Sussex coastal plain, are identical is doubtful. Aeolian origins have been suggested for these. Their lack of bedding supports this, as do certain features of their mollusc faunas, but the occasional presence of large stones suggests some form of water action, while other aspects of their mollusc faunas suggest marshy slacks between local areas of more arid

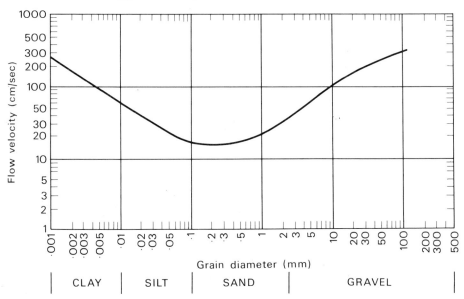

Fig. 5.10. Relation between velocity, particle size and erosion *(after Hjulström)*

micro-climate. It is sometimes alleged that these deposits were formed by rainwash: this could be true of the English brickearth but cannot apply to much of the limon of northern France which occupies a plateau-surface situation. Indeed, the similarity between the mollusc faunas of our brickearth and the Continental loess suggests a common origin.

Loess, especially the loess of China, stands well in vertical walls: to a smaller extent this can be seen in some of the European deposits. It has been suggested that this property is imparted by the number of fine tubes lined with calcium carbonate. Probably the rough vertical fissility shown by some of these deposits also helps by absorbing drainage and reducing surface erosion. But these cliffs are also capable of disastrous collapse. In north-western China cave dwellings were common in loess in the past and earthquakes in Kansu have been known to result in heavy loss of life when the loess collapsed.

Silt sediments really have no flocculation, apart from any slight degree imparted by whatever clay content they have. Hence, drying and wetting can break them down easily in a laboratory, whereas clays need chemical treatment before they can be broken down in a similar manner. This property must have some bearing on their resistance to erosion. They are certainly more erodible than clays when subject to stream action, a property which emerges clearly on Hjulström's curve (Fig. 5.10).

Among more ancient rocks it has been suggested by British authors that a number of the finer-grained Lower Palaeozoic sediments are really siltstones, while many of the Keuper marls of the Midlands (the rock is not a marl in the normal sense of that term, i.e. a transition between a limestone and a clay) are desert dusts accumulated in playa lakes, where they are interstratified with evaporites. On the other hand Pettijohn's opinion quoted above is that siltstones form only thin beds, which tend to produce resistant flaggy beds weathering out into relief.

ARGILLACEOUS DEPOSITS Clays and their equivalent consolidated rocks contain material of several different types. They include the products of chemical weathering (the clay minerals briefly described below), various amounts of detrital minerals many of which may be present as silt-sized particles for clays are not very pure rocks, new minerals that have been formed *in situ* (authigenic minerals) and varying proportions of organic matter, very often carbonaceous or calcareous.

The clay minerals usually found are kaolinite, montmorillonite and illite, but there are many others as well as 'hybrids' between these minerals. Kaolinite tends to occur under acid conditions and the other two under alkaline conditions so that relative proportions may be used in an attempt to deduce depositional environment. However, these minerals are subject to change during rock formation and great care has to be exercised in interpreting proportions present in a rock.

These minerals are responsible for many of the characteristic features of clay sediments. They possess sheet or layer structures (montmorillonite and illite resemble the micas in this respect) and a perfect cleavage, so that individual particles have a flaky habit. Many of the minerals have a lattice which is capable of expansion by the adsorption of water: this leads to the well-known powers of great expansion and contraction possessed by clays. As clays also have high porosity they can readily absorb water as well. The presence of water films between the minute flaky particles is responsible for the plasticity of clays.

Clays, which are colloidal, also have the property of base exchange: that is, various cations can be adsorbed at the surface of clay particles. The most common ions adsorbed are H^+, Na^+ and Ca^{++}. Clays with these ions adsorbed are known as hydrogen-, sodium- and calcium-clays. They have different properties due to the differing degrees of dispersal of the clay particles, a property which is well known and used in agricultural practice. The greatest flocculation occurs with calcium-clays—hence the practice of liming to improve soil texture—while sodium-clays and especially hydrogen-clays are much more dispersed and hence less permeable than the flocculated calcium-clay.

Another property possessed by some clays more than by others is that if, when firm but not consolidated, they are subjected to vibration they become mobile as liquid muds that yield to the slightest stress. This property is related to the way in which water is held in the clay. When the vibration ceases the water is readsorbed and the mud reverts to the firm clay which it had previously been. These two states are known as the gel (firm) and sol (fluid). Clays which do this readily are said to be highly thixotropic. They become very mobile as shear increases. Incidentally, the lack of bedding in many fine sediments may possibly have been produced by earth tremors affecting the material in this way. Other clays, however, actually increase in firmness with increase in shear.

The various terms used in naming clay sediments may be illustrated by considering a series of degrees of consolidation. The earliest stages might be described either as mud or as clay. The exact difference between the two varies from author to author. A non-specialist might think of mud as almost fluid and clay as plastic. Specialists vary among themselves: Gilbert uses the term, mud, to mean any unconsolidated argillaceous material and clay to be a uniformly fine-grained variant of this; with other authors there is a suggestion that muds contain organic matter. At this stage the porosity of the material exceeds 75 per cent. The sediment becomes plastic only when the porosity and hence the water content have been reduced below this level.

With increased loading the water contained in the voids of argillaceous sediments is slowly eliminated. This is a slow process as the more the voids or pores are reduced in volume the greater the resistance to flow and the slower the expulsion of the water. Finally, however, the water film breaks down and the particles come into direct contact, at which stage the porosity is of the order of 30–35 per cent. There follows reorientation of grains and a certain amount of recrystallisation, which is the main cause of the induration of the older argillaceous rocks. These tough argillaceous rocks are known as claystones or, more usually in England, as mudstones when they are compact, and as shales when they have a rough fissility.

The explanation of why mudstones occur in some deposits and shales in others is by no means clear. The natural tendencies would seem to favour shales, because the clay minerals with their flaky habit should tend to be deposited with a rough degree of parallelism. This should be accentuated by mechanical reorientation of grains produced by loading, while any recrystallisation should tend to take place at right angles to the direction of greatest pressure, in this case the loading. The problem then seems to be to explain the mudstones with their random particle orientation. It may be that the inclusion of 'impurities', i.e. material other than clay, during deposition may reduce the parallelism of clay particles. But it is also possible that random growth of minerals during diagenesis is responsible for the mudstone lithology.

Even the straight derivation of shales from clays suggested in the paragraph before the last is open to some doubt. The two are by no means identical in composition and only different in porosity. The average shale in American analyses proved to consist of about two thirds silt and one third clay and other material. This is a higher silt proportion than is typical of modern clays. Furthermore, there are differences in detail in the chemical compositions.

Apart from the terms used above further qualification of the nature of the clays, shales and mudstones is effected by the use of simple adjectives referring to specific properties.

They may indicate the composition of the material. Thus an increase in calcareous content leads to material variously referred to as argillaceous limestone, calcareous clay, marl or cementstone. The term, marl, although it means something different in some cases, e.g. the Keuper Marl, refers to a clay and calcium carbonate mixture, for example the Chalk Marl which forms the base of the Chalk in Cambridgeshire. Cementstone, a term obviously inspired by the suitability of such rocks for cement manufacture, and marlstone are indurated forms of marl. Cementstones are found at the base of the Carboniferous Limestone series in north-eastern England. The student of the Pleistocene may commonly meet the term, *Chara* marl, a freshwater deposit rich in calcareous tubelets and oospores derived from the calcareous alga of that name. Another common impurity in clays and shales is organic matter; hence such terms as bituminous clays and shales, and oil shales (found in the Carboniferous Limestone series of the Midland Valley of Scotland). Under conditions which may prevail in stagnant hollows on the sea floor, for example in some Norwegian fjords and in the Black Sea, there may be oxygen deficiency and large amounts of carbon dioxide and hydrogen sulphide. Evil-smelling black muds are liable to develop here and it is likely that the black graptolitic shales of the Lower Palaeozoic of Britain, coloured by iron sulphide and carbon, are deposits into which the remains of the graptolites have fallen from the upper layers of the oceans.

Other terms may refer to colour, sometimes a diagnostic property, sometimes not. Black deposits have already been mentioned. Red shales have been deposited in oxidising conditions, unless the colour is merely due to subaerial alteration at the surface and along joint planes. Green and blue shales indicate reducing conditions, but they are very liable to surface oxidation with the production usually of colours ranging from buff to red. This feature can commonly be seen in the Jurassic clays of the Midlands, such as the Oxford Clay, and also in the Pleistocene boulder clay derived from them.

The environment of deposition may also be indicated by words such as marine, freshwater or estuarine.

Still other terms may apply to certain physical characteristics, e.g. laminated silts, clays and shales have very fine bedding, presumably due to small-scale variations in mineral content. The term, paper shale, is sometimes used for rocks of this type with very fine laminae.

A final set of descriptive terms has been coined with regard to the economic usefulness of the clay. China-clay is a white non-plastic material, mostly formed of kaolinite, used in the manufacture of fine china and produced by the metasomatic alteration of granitic rocks. Pottery clay is plastic and aluminous and preferably free from iron, which would give rise to an unwanted reddish tinge to the finished product, although some common earthenware has this colour. Brick clay contains plenty of fluxes to promote fusion into good bricks when baked. If bituminous matter is also included, as in the Oxford Clay of the Peterborough area, so much the better because it helps to reduce the fuel bill. Fireclay has a very low flux content and is used in making refractory bricks for furnace and chimney lining, where they will have to stand high temperatures and chemical concentrations without fusion.

On the whole, clays, shales and mudstones are not resistant rocks. They are generally stable at surface temperatures as most of them are the results of surface weathering. Those that have been deposited at considerable depths are liable to undergo some form of subaerial alteration. Oxidation is one obvious example, while pyrites and gypsum contained in some varieties are also liable to lead to alteration. In terms of physical weathering they are very liable to disintegration by freeze–thaw as water forms an integral part of their composition. This is more true of the softer younger clays than of the older shales and mudstones.

The impermeability characteristic of all these rocks makes them all very liable to surface erosion, because of the high degree of runoff promoted on them, but it must not be thought that all the precipitation falling on them goes as runoff. Anyone who has watched the flow from tile drains in fields in clay areas will realise this. The very great expansion which occurs with the wetting and drying of clays means that they are generally very blocky and cracked near the surface. Water will move into and along the cracks above the level of permanent saturation. The amount so moving will depend on the season. When clays dry very thoroughly and large cracks develop, anxious and informed freeholders have been known to attempt to saturate the areas around their houses before shrinkage allows foundation movement. In these conditions the clays will absorb astonishingly large amounts of water, whereas when they have been saturated in winter even the lightest showers are sufficient to form surface puddles. Permeability will also depend on the flocculation of the clay as suggested above. Calcium-clays are the most flocculated and the most permeable. Not that clays are all that readily erodible by running water. Fine sand is the least resistant material to water erosion and resistance goes up with both increase and decrease of grain size. But as fine sands are less likely to form lowlands than clays, it seems that their lack of resistance to surface water is more than offset by their permeability. All the main Mesozoic and Tertiary clays of lowland Britain form extensive vales: these include the Lower Lias, the Oxford and Kimeridge Clays (Jurassic), the Weald and Gault Clays (Cretaceous) and the London Clay (Eocene).

Shales and mudstones are much more resistant than clays because of their induration. They share the impermeability of clays, a property known to anyone acquainted with Central Wales or many parts of the Southern Uplands of Scotland. Of the two, mudstones are on the whole more resistant, as the fissility of shales facilitates weathering especially freeze–thaw by allowing easier ingress to water. The fissility also aids stream erosion because it leads to the development of smaller flat fragments ideally suited to transport in the turbulent flow of streams. In the environment of hard rocks where they occur, the shales usually form the weakest rocks as do the clays in the areas of softer and younger rocks. Just as the Mesozoic clays tend to be dominated by ridges and escarpments of sandstone and limestone, so do the Palaeozoic shales of Wales tend to be dominated by more resistant sandstones, e.g. the Rhinog and Barmouth Grits (Cambrian) of the Harlech Dome, or the Old Red Sandstone escarpment of the Brecon Beacons, and by volcanic rocks, e.g. the Llyn Cau and Tal-y-llyn mudstones being eroded away while the adjoining lavas form the summits of Cader Idris.

Chemical and organic deposits

Although these need to be separated in the logical classification of the sedimentary rocks, they are both subdivided into classes based on the nature of the material composing them. Hence in writing of them it is easier to make the main divisions the lithological groups and subdivide each of these into chemical and organic sections where appropriate.

Furthermore, although in a discussion of the genesis of rocks it may be absolutely significant to know whether a limestone was chemically precipitated or formed from an accumulation of minute organic fragments, in the study of landforms this question is insignificant when set against the contrasts between, for example, calcareous and siliceous rocks.

CALCAREOUS ROCKS (LIMESTONES AND DOLOMITES)

Classification was simple in the case of many of the clastic sediments, but in the case of the calcareous rocks there are many possible classifications. These are discussed by Hatch and Rastall. They use as bases features such as the origin of the rocks, the grain size of the fragments and the nature of the cement. The classifications may be based on one or more of these characteristics. The one practically common feature they have is the fundamental division between limestones and dolomites. This is both a lithological and a genetic division as a rule, because the dolomite is usually regarded as a replacement of a limestone, though the argument that has arisen as to whether this need necessarily be so illustrates the drawbacks of genetic classification.

Beyond this most of the classifications distinguish between what Pettijohn calls autochthonous and allochthonous, i.e. formed *in situ* or deposited as clastic carbonate sediments. The autochthonous group includes sediments which fall into both the chemical and organic classes, for example fossil coral reefs and shell limestones are included as well as chemically precipitated deposits such as travertine (see below). Finer divisions in some classifications separate forms like coral reefs, which are strictly *in situ*, from shelly limestones, which result from post mortem sorting and accumulation due to currents as a rule. The allochthonous group are detrital fragmental limestones, derived from pre-existing limestone or shell debris. The simplest division of this group is on particle size, corresponding with gravel, sand, silt and clay and hence called respectively calcirudite, calcarenite, calcisiltite and calcilutite.

The drawbacks with such fundamentally genetic classification can be imagined from the argument which has surrounded the interpretation of reef limestones. They could be either strictly original reefs or the remnant sections of eroded banks of coral limestones: it is difficult to tell from modern exposures and in any case it is hardly relevant for our geomorphological purposes.

A further classification by Folk is much more descriptive. In this the main class is fragmental limestones, called allochemical, and subdivided on the nature of the fragments and the nature of the cement. A second class includes the microcrystalline limestones (probably the chemical precipitates of other authors) and is called orthochemical. Finally reef limestones and replacement dolomites are separated. These last two, of course, introduce genetic elements into a descriptive classification. The classification has its own terminology, for example an oolite cemented by a fine calcareous precipitate is an oomicrite, while a reef limestone is a biolithite. Such terms are necessary in a precise classification, because the more common words have vague, broad and often ill-defined meanings, but until such terminology is adopted by other than specialised limestone petrographers it would perhaps be best to stick to the more general terms.

LIMESTONES Among organic limestones those which consist fundamentally of reefs have been called biohermal. Such limestones naturally tend to be local, but are not necessarily predominantly coral, though coral would normally make up a large part of them. Associated with reefs are numerous calcareous Algae and such invertebrate animals as echinoderms and molluscs. Thus a fossil reef will be a cemented complex of this composition. Examples occur in the Wenlock Limestone (Silurian) of Shropshire, in the 'reef-knolls' of the Carboniferous of northern Lancashire, and locally in the Corallian (Jurassic), for example at Steeple Ashton in Wiltshire. There has been controversy on whether these so-called reefs are original reefs or sections of coral limestone isolated by later erosion. Algal limestones may also be formed.

Other organic limestones may be formed from accumulations of shells and shell debris. When such deposits represent a layered accumulation, as distinct from the mound-like form of the reefs, they are sometimes called biostromal limestones. If formed from broken and worn shell debris they are probably better described as bioclastic. The distinction between the two is not absolutely sharp, because a degree of transporting and sorting which would shatter fragile shells may hardly affect thick resistant ones, except perhaps to round off the hinge teeth of lamellibranchs and wear off some of the ornamentation of gastropods.

Limestones in both these classes are often named after the most significant constituent fossil or group of fossils, which may, however, only form a small proportion of the total bulk of the rock. The most general term for these rocks is shelly limestone and hundreds of such rocks could probably be named: crinoidal limestones occur in the Carboniferous Limestone and are characterised by the abundance of parts of the skeletons of these echinoderms, the sea-lilies; coral limestones are often detrital limestones associated with fossil reefs; freshwater Pleistocene *Chara* marl has already been mentioned; foraminiferal limestone is dominated by these animals and a spectacular variety is the Nummulitic limestone of the Mediterranean, so-called from the abundance of the large coin-shaped foraminifer, *Nummulites*.

Chalk is a rock which really belongs to the same group of limestones. Although it has been suggested at various times that it was either a chemical precipitate or a foraminiferal limestone originating as a deep-sea ooze, Foraminifera, even though among the most abundant of the micro-fossils, only make up a small part of the rock. It has been shown, however, that the rock is really made up of minute algal plates, known as coccoliths, and, on the evidence of various fossil and textural features, it is now regarded as a pure limestone formed in water about 200 m (650 ft) deep. It is thus an algal limestone.

Although some of the rocks included above would be classed as clastic limestones, this group contains a greater variety of rocks including those derived from the breakdown of earlier limestones and from chemically precipitated material.

The simplest chemical deposits are those occurring in caves and springs. The calcium bicarbonate carried in solution is unstable and is quickly precipitated as calcium carbonate, either through evaporation or as a result of the reduction in pressure when underground water reaches the surface again. Deposits of this type include stalactites and stalagmites attached respectively to the roofs and floors of caves, and the calcareous deposits of springs known variously as calc-tufa, calc-sinter or travertine. These materials vary from fibrous and cellular deposits, through earthy material to banded hard encrustations. With age the pores often become filled with calcareous cement. As a deposit in older rocks, and hence a relief-forming unit, tufa is relatively unimportant, although some, such as those of the Purbeck Beds (Upper Jurassic) of Dorset may form local hillside scars.

Concretionary chemical deposits may also form in semi-arid soils: this kunkar and its probable fossil equivalents, cornstones, have already been mentioned in connection with residual deposits.

Fine-grained calcareous muds may be precipitated either by chemical or by biological means. On very shallow shelves in tropical seas, such as the Great Bahama Bank, almost stagnant seas may occur, which, after evaporation caused by the heat of the tropical sun, produce a precipitation of fine aragonite crystals. Aragonite is a form of calcium carbonate produced in some parts of marine invertebrates and is far less stable than the more common crystalline form, calcite. Some of the aragonite may also be derived from algae which break up into fine needles; this cannot be distinguished from that precipitated chemically. In mangrove swamps ammonifying and sulphate-reducing bacteria are considered to be important in precipitating calcium carbonate from water rich in calcium sulphate (gypsum). Rocks produced by processes such as these are called aphanitic limestones by some authors. In the geological column they are represented by the calcite mudstones of the Carboniferous Limestone of the Bristol and South Wales districts, and by the famous lithographic limestone of Solenhofen in Germany. Lithographic limestones, so named because of their great suitability for that printing process, are dense, very fine-grained and very equal-grained, consolidated, calcareous mud.

Oolitic and pisolitic limestones are other chemically precipitated forms. They consist of spherical or ovoidal particles the size of the eggs in hard cod's roe (ooliths) and the size of peas (pisoliths). The maximum diameter of ooliths is about 1 mm. These particles display a concentric structure in their banding and, in the case of modern examples, by crystals tangential to their surface indicating how they might have been built up. But older specimens usually have this obscured by a radiating structure of fibrous calcite which is a secondary development. Most ooliths and pisoliths have a nucleus, a fragment of shell or a particle of sand or limestone, but this is not always present.

Pisoliths are being formed at the present day in certain very turbulent waters highly charged with calcium carbonate, such as hot springs and in streams in caves. The turbulence ensures an even concentric deposition and also a tendency for roundness induced by abrasion. However, most oolitic limestones are of marine origin.

At the present time ooliths of aragonite are known to be forming in a number of places. Aragonite tends to be precipitated in waters rich in sulphate ions (SO_4^{--}), but is unstable at surface temperatures and pressures, changing to calcite, of which the ooliths in older oolitic limestones are composed. The common factors in the localities where ooliths are being deposited now, the Bahamas, Florida and the Gulf of Suez, are shallow warm water and high evaporation rates. The shallowness implies continuous disturbance by waves and currents and the removal of any very fine precipitated material. Under these conditions the developing ooliths are kept in motion either on the sea-floor or in the intertidal zone of the foreshore. It would be interesting to know why they tend towards a uniformity of size. Oolitic limestones are very common in the English Jurassic, where their prevalence led to the Middle Jurassic being termed originally the Inferior and Great Oolites, but they are also found in other horizons of the Jurassic, for example the Portland Stone, and in the Carboniferous Limestone in various localities, e.g. the Gower peninsula of South Wales.

DOLOMITES Dolomite is both a mineral and a rock. The mineral is a double carbonate of calcium and magnesium; the rock is formed almost entirely of this mineral. Magnesium carbonate may also be held in small quantities in solid solution in calcite crystals without forming true dolomite.

In view of the problem of the origin of dolomites, it is best to adopt a descriptive classification. That favoured by Pettijohn has four main classes:

1. Magnesian limestone (5–10 per cent dolomite or magnesium carbonate held in solid solution).
2. Dolomitic limestone (10–40 per cent dolomite).
3. Calcitic dolomite (40–90 per cent dolomite).
4. Dolomite (over 90 per cent dolomite).

In general and rather loose usage the term, magnesian limestone, has probably been used to describe rocks with more dolomite than indicated, dolomite to describe rocks with less dolomite than indicated, and dolomitic limestone to describe the whole range of intermediates.

There are apparently no known examples of dolomite forming at the present day, nor is dolomite secreted by any living organism. But fine grained and oolitic dolomites occur and the question arises whether dolomite may have been directly precipitated in the past, formed by alteration of limestones soon after deposition, or by alteration long after deposition. Various threads of evidence point to all three as possibilities.

Dolomites occur as members of evaporite sequences, which is just the type of environment in which original deposition might have been expected. A similar original deposition has been suggested for the dolomite crystals found in the Triassic marls of parts of the Midlands.

When widespread uniform dolomites are found between limestone beds it must be concluded that they were either original or formed soon after deposition before a further layer was deposited over them. Otherwise, how could this overlying layer have escaped dolomitisation? If the fossils are dolomite then it is hardly likely to have been original deposition, because, as stated above, it is not known as an original secretion of invertebrates.

Where dolomitisation is related to cavities, bedding planes, joints and faults, either being confined to such places or being more pronounced there, then it is most likely to have occurred long after the deposition of the rocks, for the rock must have been indurated, jointed and faulted before it could have taken place. Further evidence of post-depositional dolomitisation is provided by the way in which structures such as ooliths tend to be blurred in dolomitic rocks.

Dolomitised limestones occur in most thick limestone sequences, for example in the Carboniferous Limestone series of Scotland, the Isle of Man and South Wales. The type area is, of course, the Dolomites in the South Tyrol. Magnesian limestone occurs in the Permian of north-eastern England.

Before proceeding to a general discussion of some of the effects of calcareous rocks on relief, the question of the relative resistance of limestone and dolomite deserves ventilation. It has sometimes been suggested that dolomites are less resistant than limestones. Their often carious nature, interpreted as a solution effect, has been cited as evidence, but this could just as easily have been caused by selective solution of calcite from dolomitic limestones. The presence of cavities from which fossils have been weathered has been used to suggest the superior resistance of dolomite by the following argument: fine-grained calcite is much more readily dolomitised than compact crystalline calcite; it is, therefore, possible to have a dolomitised rock in which the tougher, thicker fossils remain as calcite; with later weathering these have been removed because, it is alleged, the calcite is less resistant to weathering than the bulk of the dolomitised rock. Yet another explanation of the carious nature of dolomite is available. Replacement of calcite by dolomite molecule by molecule, a process which does not always seem to take place, would result in a volume reduction of about 12 per cent, thus giving rise to a somewhat more porous and cellular rock than its parent limestone and having no interpretation in terms of solution rates. Finally, in mixed dolomite rocks surface etching ensures that the dolomite crystals often stand out in relief. This higher resistance to acids is used in laboratory tests with dilute hydrochloric, acetic or formic acids to give differential etching on calcite and dolomite, which are not easily distinguishable by the optical properties on which the study of rocks is usually based.

RELIEF ON CALCAREOUS ROCKS A lot has been written on the relief of calcareous rocks and at the present time this subject is being actively pursued as one of the main fields of geomorphological interest: there is, for example, a special commission of the International Geographical Union at work studying karst.

The interest in calcareous rocks probably stemmed originally from the predominance of chemical weathering and the results this had in the formation of underground pot holes, caves, passages etc. Just as the glacial and arid areas early attracted attention for the challenge they put to exploration and the aesthetic appreciation they aroused, so too did the underground relief of limestone areas which challenged physical endurance and revealed wondrous sights. Scientific interest followed exploration. The concentration on underground relief has probably contributed to the identification of a so-called typical limestone relief, that of the karst regions of Yugoslavia, the Causses of France, certain areas of the calcareous Alps of Provence, and, nearer home, the Ingleborough district of Carboniferous Limestone in Yorkshire. These are all regions of thick, massive, well-cemented, well-jointed limestones, deeply-dissected so that the water table is low and the surface nearly devoid of permanent drainage. All percolation is directed along joint and bedding planes to produce there the solutional relief of typical karst. So far has the identification of limestone relief been attached to this type of rock in the eyes of some students that the reasoning seems to proceed as follows: karst occurs in some Pennine areas: these are Carboniferous Limestone; therefore, where karst occurs the rock is Carboniferous Limestone; so the karst proper and the Causses are formed on Carboniferous Limestone. In fact most of the major karst regions of Europe are formed on Jurassic limestone, some of it even on Cretaceous and Tertiary limestones. Not that age has anything to do with it except in so far as the older the limestone the harder and more crystalline it usually is. Unfortunately in Britain most of our Jurassic and Cretaceous limestones are too soft to produce this type of relief.

A cycle of karst erosion to cover essentially areas of this type of limestone was put forward by Cvijić (Saunders 1921—a brief summary is given in Sparks 1960). Briefly, it is envisaged that the drainage will begin on an impermeable cover overlying a limestone below. When the drainage pattern erodes through the impermeable cover and reaches the limestone below it becomes disintegrated by the formation of sink holes which lead it underground. After a long period of underground drainage, the development of sub-surface landforms, and a very slow, gradual lowering of the limestone surface largely by weathering, the drainage begins to reappear at the surface by cavern collapse. Presumably this must take a very long time because the surface is more in a state of arrested dissection than active weathering. Surface gorges result from the collapse and the drainage is re-established on the impermeable rocks below the limestone. The interfluvial areas are finally reduced to little hillocks known as hums. Within this general evolution it is possible, though perhaps not very profitable, to define periods in terms of youth, maturity and old age. Such a cycle can only really apply to a large area of limestone, a great thickness of limestone, a well-lithified limestone with pronounced joints and bedding, an area of considerable elevation with a water table well underground, and an area where by chance an impermeable cover was found and an impermeable substratum existed at very little above base level, though whether these last two are essential conditions is very doubtful.

In fact the range of relief developed on limestones is vast and there is much more difference between the extremes than there is between the relief developed on those extremes and that developed on other types of rock. The extremes are probably represented by the relief of Chalk, or some fairly weak Jurassic limestone on the one hand, and on the other by true karst areas, such as interior Provence, and also by high mountain limestone areas, for example the Dolomites or the French Pre-Alps, where distinctive tabular and cliffed relief, different from that of the Dolomites, is found on such massifs as Grande-Chartreuse or Vercors. Relief, but not vegetation and soils which probably contribute as much as or more than relief to general landscape effect, on Chalk is not very different from that developed on some soft and permeable sandstones, for example parts of the Lower Greensand. Again, though the relief of high glaciated limestone mountains is not the same as that on similar mountains of other rocks, it is even less like Chalk scenery.

Some authors use karst as a synonym for limestone relief but, in view of the tremendous range of relief mentioned above, this practice to my mind tends to perpetuate the mistaken concept of there being one typical kind of limestone relief. In any case it is a term derived from one particular area and there seems to be no more sense in having type areas than in having type specimens of plant and animal species unless it is to act as a norm against which others are compared. Therefore, I propose to use the general term limestone relief and to reserve the term, karst, for the areas similar in relief to the type area of northern Yugoslavia and the adjacent parts of Italy.

In the sense used here karst does not develop on any limestone but requires certain special conditions, some of which are concerned with the lithology of the rock.

In the first place, the limestone must be a hard, well-cemented, non-porous limestone, in which permeability is controlled by movement through joints and not through the mass of the rock. On the whole massive limestones are more likely to produce good karst features than are thinly-bedded ones, which collapse more easily underground and which are less likely to produce well-developed limestone pavement features.

Secondly, there must be a considerable thickness of limestone above the subjacent impermeable stratum or above the water-table for the full development of underground drainage at the expense of surface drainage.

Thirdly, there must be a considerable area of limestone. Otherwise the limestone is going to form such a minor part of the drainage basin as a whole that the incision of streams into the limestone, one of the main factors controlling the fall in the water table and leading to karstification, will be controlled largely by what happens on adjacent impermeable rocks.

Within an area of karst relief may be distinguished three main groups of relief features and each of them has given rise to theories and argument. They are:

(*a*) The features of limestone surfaces; the clints and grykes of the British limestone areas; the lapiés of the French limestone regions; and the groups of similar features classified by the Germans into various types of Karren.
(*b*) The closed depressions characteristic of limestone areas and the channels which lead from these underground. Here the main argument is whether these have been formed by solution from the surface or by underground solution and ensuing collapse.

In connection with both of these classes of form one of the main controversies is of the relative roles played by climate, with its control over the nature and rate of the operative processes, and by the lithology of the limestone.
(*c*) The underground cave and passage systems. The main questions which have arisen concern the level at which the caves were formed. Was this above the water table by water flowing under the effects of gravity or was it below the water table by water flowing under hydrostatic pressure?

To these three questions may also be added a fourth major issue, the overall rate of limestone denudation and whether this varies markedly from one type of climate to another.

Limestone solution   Both for the formation of surface limestone features and underground features it is necessary to delve into some detail of the chemical processes involved. The reaction of dissolved carbon dioxide on limestone can be generalised into the equation

$$CaCO_3 + CO_2 + H_2O \rightleftharpoons Ca(HCO_3)_2$$

but in order to have a fuller understanding of the process it is better to think of it in terms of five equations, each representing separate equilibrium states.

In the first place, calcium carbonate dissociates to a slight extent in water in which there is no carbon dioxide present, i.e.

$$CaCO_3 \rightleftharpoons Ca^{++} + CO_3^{--}$$

Secondly, that part of the carbon dioxide dissolved in water to form carbonic acid will dissociate in two stages, i.e.

$$H_2CO_3 \rightleftharpoons H^+ + HCO_3^-$$

and

$$HCO_3^- \rightleftharpoons H^+ + CO_3^{--}.$$

Thirdly, water itself dissociates to a small extent, i.e.

$$H_2O \rightleftharpoons H^+ + OH^-.$$

Finally, a certain amount of the carbon dioxide dissolved in the water combines to form carbonic acid, i.e.

$$CO_2 + H_2O \rightleftharpoons H_2CO_3.$$

In the first four cases the law of mass action states that, at a given temperature, the product of the two right hand terms divided by the left hand term will equal a constant, K, known as the dissociation constant. In the case of the last equation the amount of carbonic acid divided by the total dissolved carbon dioxide, at a given temperature, is constant. As an example of this the dissociation of carbonic acid into hydrogen and bi-carbonate ions at 25 degrees C is

$$\frac{[H^+][HCO_3^-]}{[H_2CO_3]} = K = 10^{-6\cdot4}.$$

(For a fuller account of carbonate equilibrium see Garrels and Christ, 1965, or Kern and Weisbrod, 1967, or Trombe, 1952.)

In addition to all these equilibria, which apply to the liquid phase of the system, one further condition of equilibrium needs to be taken into consideration. It is the equilibrium between the carbon dioxide of the atmosphere and that of the water, or perhaps one had better say between the gas phase and the liquid phase because we shall have to consider not only the free atmosphere but also the soil atmosphere. The total amount of carbon dioxide dissolved depends on the partial pressure of carbon dioxide in the atmosphere (i.e. the proportion of atmospheric pressure due to carbon dioxide) and on the temperature. The higher the pressure, the greater the amount of carbon dioxide dissolved, the higher the temperature the smaller the amount of gas dissolved. The decrease in solubility with rise of temperature is quite marked: at 20 degrees C the solubility is approximately one half of its value at 0 degrees C.

If any part of these various equilibria is disturbed it engenders a series of reactions which will ultimately restore equilibrium throughout. For example, if water percolating from the surface underground has remained in contact with the atmosphere and the limestone surface for a sufficient length of time it will achieve a definite calcium carbonate content in solution. If it is cooled underground, and provided that the underground atmosphere has the same partial pressure of carbon dioxide as the surface atmosphere, it will be capable of dissolving more carbon dioxide which can only be brought back to equilibrium by the solution of more limestone. Thus water, which was incapable of further surface solution, is capable of solution underground by a fall in temperature. When it returns to the surface, again provided that the partial pressures of carbon dioxide do not change, the increase of temperature will involve decreased solubility of carbon dioxide, hence a transfer of the gas from the water to the air, and an ensuing inequilibrium which can only be rectified by the deposition of calcium carbonate as travertine or tufa.

Another interesting conclusion follows from this type of argument. It concerns the use made of the dissolved calcium carbonate content of outflowing rivers as a measure of limestone denudation. If the rivers are in equilibrium with the partial pressure of carbon dioxide above them, which is purely atmospheric, the calcium carbonate content will reflect that pressure and the temperature of the water. If we assume that the partial pressure of carbon dioxide in the atmosphere is the same the whole world over, it will reflect the ambient temperature and should mean that tropical rivers have lower calcium carbonate content than those of cold regions. But in order to understand the degree of karstification the details of the whole process must be looked at a lot more closely, because the carbon dioxide content may vary widely during the passage of the water from precipitation to percolation or run-off.

For the expression of the potential action of any sample of water, i.e. solution or deposition, limestone geomorphologists have in recent years increasingly used a form of diagram first introduced by Trombe (1952). This diagram, a simplified form of which is included as Fig. 5.11, consists of a plot of the pH value of the water against its content of calcium carbonate in parts per million. The pH value is a measure of acidity: it is the logarithm of the reciprocal of the concentration of hydrogen ions in gramme ions per litre. In simpler words it is a measure of acidity, in which neutrality is expressed by pH = 7, higher figures denoting alkalinity and lower figures acidity. A change of a unit in the pH value, i.e. from pH 7 to 6 or 7 to 8, indicates a tenfold increase of acidity and alkalinity respectively. As we have noted above the solubility of carbon dioxide and hence the possible concentration of hydrogen ions varies with temperature so that temperature curves for equilibrium have to be included on the diagram as well.

If the pH is plotted against the calcium carbonate content and the point falls exactly on the line representing the prevailing temperature, for example A on Fig. 5.11, then the solution is saturated with calcium carbonate and can be expected neither to dissolve nor to precipitate. If the

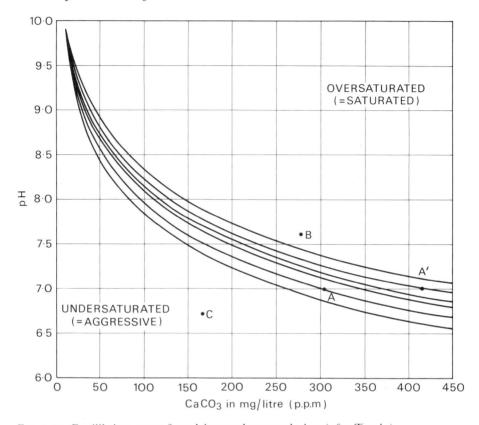

Fig. 5.11. Equilibrium curve for calcium carbonate solution *(after Trombe)*

point falls above the prevailing temperature curve, e.g. point B, then it is oversaturated temporarily and may be expected to deposit: if it falls below the temperature curve, e.g. point C, then further solution might be expected. These two states are often called saturated and aggressive respectively. The terms oversaturated and undersaturated or non-aggressive and aggressive would seem preferable, because the state of saturation is strictly indicated by a point falling on the temperature curve.

The diagram can also be used to illustrate the example quoted above of the water percolating underground, being cooled and hence becoming capable of further solution. The saturated condition at 30 degrees C is illustrated at point A with calcium carbonate content at about 305 p.p.m. With a fall of temperature to 10 degrees C the maintenance of equilibrium would require point A to move to A′, where its calcium carbonate content would be about 415 p.p.m.

The real question of limestone solution has been very much simplified in the above discussion because the solution of other carbonates, especially magnesium carbonate, and other salts, for example calcium sulphate (gypsum), has not been considered, whereas the solution of these and others as well would have to be considered in a real example.

Limestone pavements A very interesting attempt has been made to relate various aspects of the solution process to the morphology of limestone pavements (Bögli 1960), that is to the small scale solution forms of ridges and runnels known collectively in German as Karren. In this connection Bögli made use of the various equilibrium states discussed above, and divided the whole chemical process into four phases.

Phase 1 (Fig. 5.12) consists of the dissociation of a limited amount of

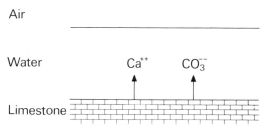

FIG. 5.12. Limestone solution, phase 1 *(after Bögli)*

limestone in the presence of pure water only. This phase takes place very quickly, but leads to only a very limited amount of limestone solution. In fact the amount of calcium carbonate in solution which can be accounted for in this way varies from 10 p.p.m. at 8·7 degrees C to 14·3 p.p.m. at 25 degrees C. The relative importance of this can be evaluated against the fact that concentrations up to 400 p.p.m. have to be accounted for in nature; for example, a series of figures given by Ingle Smith for the Cotswolds (Sweeting and others 1965) ranges from 250 to 360 p.p.m.

Phase 2 (Fig. 5.13) sees the association of hydrogen ions, formed by the dissociation of the carbonic acid into hydrogen and bicarbonate ions, with the carbonate ions released by the dissociation of the limestone. In this connection it should be noted that only a small percentage of the dissolved carbon dioxide forms carbonic acid. At 4 degrees C 0·7 per cent is

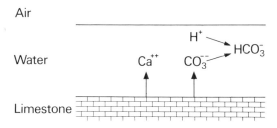

Fig. 5.13. Limestone solution, phase 2 *(after Bögli)*

'chemically' dissolved as carbonic acid: the remaining 99·3 per cent is described as 'physically' dissolved. This ratio is of course an equilibrium. The association of the hydrogen ions from the carbonic acid with the carbonate ions from the limestone results in inequilibrium at two points. First, the equilibrium between the calcium and carbonate ions derived from the limestone is upset and can only be restored by the dissociation (i.e. solution) of more limestone. Second, the equilibrium between 'physically' and 'chemically' dissolved carbon dioxide is upset and is restored by the conversion of some of the 'physically' dissolved carbon dioxide into 'chemically dissolved' carbon dioxide.

In effect, phase 3 (Fig. 5.14) consists of the transformation of some of the

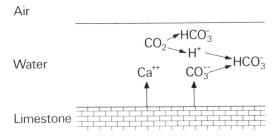

FIG. 5.14. Limestone solution, phase 3 *(after Bögli)*

'physically' dissolved carbon dioxide into ionised carbonic acid. But, as soon as this happens, imbalance occurs between the carbon dioxide dissolved in the liquid phase and the carbon dioxide content of the atmosphere. This leads to the final rectification of the equilibrium between the partial pressure of carbon dioxide in the atmosphere and the carbon dioxide content of the water (Fig. 5.15), a balance which has been seen to vary with temperature.

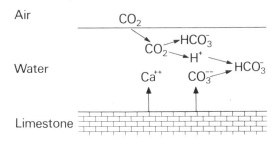

FIG. 5.15. Limestone solution, phase 4 *(after Bögli)*

The rate of reaction and hence the amount of calcium carbonate dissolved per second are high for all the first three phases, and further they rise with increases in temperature. A rise in temperature of 10 degrees C approximately doubles the rate of solution. Thus it can be said that the rate of solution caused by these three phases is roughly four times as high in the Tropics as in Arctic climates.

Compared with the first three phases Phase 4 (Fig. 5.15), which is concerned with the diffusion of atmospheric carbon dioxide into the water to restore equilibrium, takes a long time and keeps going the chain of reactions of the first three phases until equilibrium is reached. Estimates of the time taken for this phase vary considerably, as it depends on a great variety of factors. Among these may be included the thickness of the water layer; the nature of flow whether laminar or turbulent; whether the flow, in the case of underground water, is in gravity or pressure channels; the available surface area of the rocks and so on. It seems that the minimum estimate of the time taken to reach equilibrium is of the order of twenty-four hours but may be very much longer and reckoned in terms of days. The general rate of solution in this phase is low and decreases with time: it is reckoned to be vastly less than that of the previous three phases.

On the basis of these chemical phases Bögli divided the whole process up into three morphogenetic phases, which are depicted in Fig. 5.16.

Morphogenetic phase 1 comprises chemical phases 1 and 2. Reaction is complete in about one second and, although the total solution may not be very great, the rate of solution is very high indeed, some ten to one hundred times that of morphogenetic phase 2.

Morphogenetic phase 2 corresponds with chemical phase 3 and, depending upon the partial pressure of carbon dioxide in the atmosphere, involves varying amounts of solution. It takes approximately a minute for equilibrium to be realised.

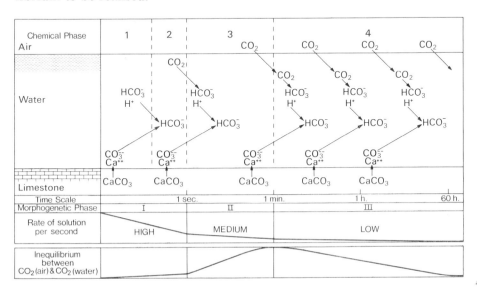

FIG. 5.16. Limestone solution. General scheme *(after Bögli)*

Morphogenetic phase 3 equals chemical phase 4. It involves the restoration of equilibrium between the carbon dioxide content of the air and of the water, which had reached a maximum of inequilibrium at the end of morphogenetic phase 2. The total amount of solution effected is very great, but the time taken is so long that the rate of solution per second is very low indeed, approximately one thousandth to one ten thousandth of that of the previous phases.

Because of the varying times taken for these three phases it is arguable that they act under different natural conditions.

Phase 1, because it is complete in such a very short time, takes place on the first contact of lime-free water with the rock. Thus forms produced by it are confined to the surface of bare limestone areas. It cannot occur beneath soils or underground because, by this time, the water has been in contact with the rock far too long for the phase to be still operating. The distinguishing forms, according to Bögli, are Rillenkarren, which are very closely spaced small runnels. Obviously, the slope of the surface will also have a considerable effect. If the surface is almost flat there will be little run-off, while the steeper the slope the greater the run-off distance covered in a second, the approximate time required for phase 1 to be completed. Again, in natural examples one cannot isolate the beginning of flow and measure the length of action of the first phase from that point for the simple reason that new precipitation is being added as the flow proceeds. This consequently extends the zone of action of phase 1 and blurs the transition to other phases.

Phase 2, according to Bögli, sees the development of a carbonate-rich lower layer of water and an aggressive upper layer. This will tend to lead to the solution of all projections and the sides of any slight depressions, while the floor of the channel or hollow is not much affected. Thus, the characteristic forms, whether they are depressions or runnels will be broad and flat-floored. The characteristic forms are said to be the Trittkarren, which are flat-floored, minute, almost cirque-like features, often retreating through headward erosion.

By the time that phase 3 comes into operation sufficient time has elapsed for most of the run-off to be concentrated into channels, so that there is a considerable contrast between the forms produced by this phase and those formed earlier. Solution taking place beneath soils and also underground must be largely referred to this phase of action, during which reaction takes place very much more slowly.

As these various morphogenetic phases are affected in different ways by temperature changes, their understanding helps to clarify the problem of the variation of karstification with climate. But in this connection another factor first needs to be considered, namely the role played by soil carbon dioxide.

The general content of carbon dioxide in the free atmosphere, and it does not vary much, is 0·03 per cent. In the soil owing to the respiration of roots and the decomposition of humus this figure may be raised up to a hundred times. Indeed, absolute maxima for soil carbon dioxide in cool humid climates of up to 25 per cent have been quoted. As one of the main factors in the solubility of carbon dioxide is its partial vapour pressure, these figures show the overwhelming importance of soil atmosphere. As humus soils are usually impermeable, the slow movement of water through them will mean that there will be enough time for at least a large part of the diffusion of carbon dioxide into the water to take place. This should result in very high measures of solution taking place immediately below the soil layers. Obviously, high concentrations of vegetation and humus are more likely to be formed in the tropics than in cold climates; hence the importance of biological carbon dioxide will be greatest there.

One might think that as soon as the water percolates below the carbon dioxide-rich soil layer it would come into contact with an atmosphere with a lower concentration and hence start precipitating calcium carbonate immediately. However, this process, the reverse of chemical phase 4 is slow and, further, if the water is flowing in a channel which it completely occupies with no room for a gas phase, it forms a closed system and neither solution nor precipitation takes place. This, incidentally, is the reason why precipitation does not take place in the pressure channels in which water often flows through limestone.

If one accepts that the main control of the degree of karstification is the rate of solution per second and not the total amount of solution irrespective of time, then the factors controlling that rate, i.e. the morphogenetic phase and the temperature, become very important indeed.

In morphogenetic phase 1 the rate of reaction rises with temperature and the total amount lost through solution rises from cold to warm climates by about 50 per cent. As the main forms produced in this zone are Rillenkarren it is not surprising that these are best developed in the humid tropics.

The effectiveness of morphogenetic phase 2, which is mostly bound up with the solubility of carbon dioxide, is reduced with temperature even though the rate of reaction may rise somewhat. The total calcium carbonate dissolved roughly halves from the cold to the hot regions.

Morphogenetic phase 3 is probably the most important for surface karstification. An important aspect of this phase is the increase in the rate of carbon dioxide diffusion with increasing temperature. This means that in hot regions more of the action of this phase is concentrated on the surface and less underground than in the colder regions. Bögli gave some statistics which showed that at 3 degrees C only 15 per cent of the possible solution due to this phase had been achieved at the end of a runnel 3·5 m long, while 26 per cent had been achieved at the end of a runnel 13 m long. These figures are probably high because they relate to a near-horizontal runnel in which flow was slow as a result of the precipitation being in the form of drizzle. Very much higher values for the solution occurring in phase 3 have been measured in the Tropics.

213

Because of the slowing down of the rate of action of phase 3, the effect of impermeable soils becomes very important as the slow percolation through these allows equilibrium to be more nearly attained between the soil atmosphere and water carbon dioxide content. This is illustrated in Swiss examples by the fact that water from moraines with much fine-grained sediment has higher calcium carbonate content than that derived directly from karst areas. Although it is possible in cold areas for water to reach a higher carbon dioxide content if the passage through humus soils is slow enough, this is not likely to occur in practice because such soils are far less likely to occur there. Thus, in the tropics there is every possibility of very high carbon dioxide content in water percolating through humus soils, every possibility that this will react quickly with the underlying limestone because of the increase in diffusion rate of carbon dioxide with temperature, and, therefore, an explanation is available for the advanced degree of superficial karstification in such regions. The other side of the picture is that, following this high rate of solution, the rate of deposition must be equally high when the water once more settles into equilibrium with the partial carbon dioxide pressure of the free atmosphere. Evaporation is another factor which tends to raise this in the tropics. The evidence for this is provided by the large amounts of sinter deposited on rock walls and in caves in the tropics. Thus, the tropics seem to be zones of intense near-surface karst development and high rates of sinter deposition in the limestone areas. The total calcium carbonate content of the effluent rivers may be less than that normally found in other climatic zones, as this depends on equilibrium with the carbon dioxide content of the free atmosphere, which does not vary much from one climatic zone to another, at a given temperature.

This review of Bögli's ideas tends to suggest that variations in process, largely depending on temperature variations, dominate the production of surface forms on limestones, but there are geological controls as well. These are very evident in the limestone pavements of England and Eire which have been discussed by Sweeting (1966) and Williams (1966). The multitude of Karren forms, discussed by Bögli, do not play a very large part in the relief of the limestone regions of the British Isles capable of forming karst, namely the Carboniferous Limestone areas principally of the Pennines and western Eire both of which have been glaciated. In these regions under suitable geological conditions limestone pavements are well-developed. These form the clint and gryke country of the Ingleborough district of Yorkshire in which the enlarged joints form the grykes which may be up to 0·5 m wide and 3 to 4 m deep, so isolating joint-bounded blocks of approximately 2 by 1 m which form the clints. Such furrowed pavements are best developed in both areas on massive limestones. Impure thin-bedded limestones tend to form much less spectacular features, because they are more erodible and also much more susceptible to shatter by freeze–thaw.

It seems to have been realised for a very long time that the stripping of surfaces to provide the limestone pavements must be attributed to glacial action. Obviously in near-horizontal rocks the opportunities for the stripping of bedding-planes is greatest, although the factor which determines the selected bedding plane remains obscure. It has been suggested that this might be a thin shaly parting between massive beds but Williams found no evidence for this in Eire. Alternatively he suggests the importance of master bedding planes separating cycles of rhythmic limestone deposition. These cycles may begin with poorly-cemented, rubbly, easily eroded limestones and change to much purer limestones upwards. Thus the weak layers at the base of the cycles might control glacial stripping. Whatever the geological control on the stripping process, it is also affected by the

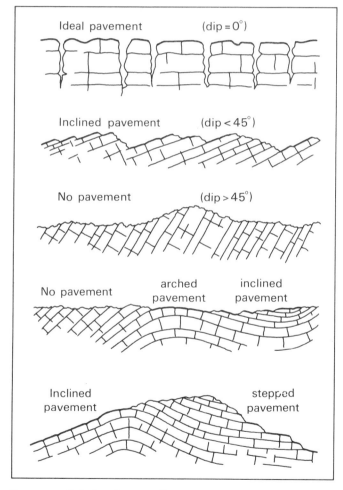

FIG. 5.17. Relations between limestone pavements and dip (*after Williams*)

relation between the area and the advance of the ice sheet. On slopes facing the ice sheets there seems to have been mainly a smooth rounding of projecting relief, while the process of stripping structural surfaces seems to have been confined to the lee slopes. Finally the formation of pavements is related to the amount of the dip of the beds. This is illustrated in Fig. 5.17, on which the ideal case is shown on horizontal limestones. With increasing dip the pavements become smaller and less perfect and finally disappear at about 45 degrees dip, according to Williams.

In some of the examples described by Sweeting and Williams the limestone pavements, with their attendant clints and grykes, occur beneath layers of glacial drift. Where this is so we can assume one of two hypotheses: either the pavements are interglacial features preserved by the drift or they are solution features formed beneath the drift. The latter case is unlikely unless the drift is devoid of calcium carbonate. Williams has described areas where calcareous glacial drift overlies the limestone; here, there seems to be no development of clints and grykes beneath the drift.

In fact he would attribute the whole of the development of limestone pavements, clints and grykes, to the Post-glacial period when the limestone surfaces were exhumed from beneath their drift cover. The suggested stages of this succession are shown in Fig. 5.18. After the drift is removed the

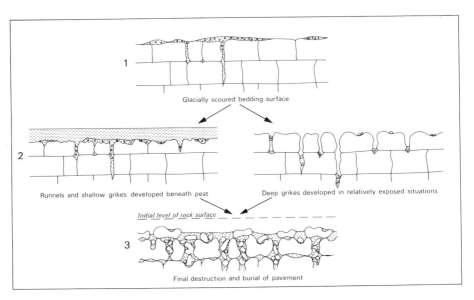

FIG. 5.18. Evolution of limestone pavements *(after Williams)*

exact nature of the pavement depends on whether a wet vegetation cover develops or not. If it does the solution beneath the vegetation is likely to lead to a comparatively shallow well-rounded clint and gryke pattern, whereas in the open a much deeper pattern is likely to develop. The end phase of both of these conditions is the destruction of the clints by the enlargement of the grykes and the development of solution hollows, so that the limestone surface is broken down into a bed of rubble on which vegetation may become ultimately established.| On this hypothesis the limestone pavement is essentially a feature developed at its best in the earlier parts of Post-glacial time and now on its way to destruction. Such a hypothesis accords also with the fact that limestone pavements are known from glaciated Carboniferous Limestone areas, for example the two that have been mentioned in this discussion, but are not known from unglaciated areas on the same rock, for example the Mendips. It is also implicit that, in addition to the climate and the nature of the rock, the past must be considered because the limestone pavements are essentially a disappearing legacy of past conditions.

Solution holes and depressions   In most karst areas the surface of the limestone is disorganised into a series of closed depressions, which may vary enormously in size and also in steepness of sides. They are obviously a distinctive feature of limestone areas and can only be paralleled in other regions by forms which usually have different origins. In glaciated regions, for example, glacial overdeepening of valleys and corrie formation may also lead to closed depressions, but these will be formed irrespective of rock type. Again, in periglacial regions the development of ice lens features, either in soft sediments or in those very susceptible to freeze–thaw, can result in a lunar landscape of very shallow depressions and intervening higher ridges, for example Walton Common, some 12 km (8 miles) east by south of King's Lynn. In desert regions closed depressions form oases: they may have been formed by irregular accumulation of sand or from tectonic basins.

Simulating forms are only likely to be confused with solution depressions where they occur on calcareous rocks. This is a possibility with ice lens features of periglacial origin which are best developed on highly cryoturbated Chalk in East Anglia. It may be necessary to cut sections through parts of such features before they can be satisfactorily differentiated from solution features.

The Great Plains depressions mentioned by Thornbury (1965) are another possible example of simulation. These pits, which occur in the Pliocene Ogallala formation, are very abundant in Texas and some are shown in Fig. 5.19. They range in size from tiny holes not a foot deep to large depressions several miles across and several dozen feet deep. Yet the rubbly limestone or 'caliche' which forms the caprock of the predominantly non-calcareous Ogallala formation is only 3–9 m (10–30 ft) thick. Various explanations have been offered for the depressions: buffalo wallows, deflation hollows, solution subsidence and differential compaction of the sediment. Of course, the first two suggestions might well represent later uses of holes formed in some other way.

217

It might vulgarly be said that the surface of many limestone areas consists of holes and hills. The nature and form of these vary from one climatic zone to another according to some authorities.

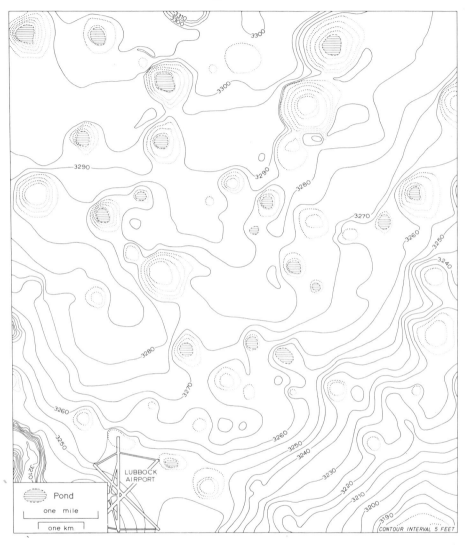

FIG. 5.19. Great Plains depressions in northern Texas. Contour interval 5 feet. Contours round depressions are dotted.

Even in the temperate areas of Britain and Europe there is a considerable variation in these forms. Two British areas where solution holes are common are in the Mendip Hills and on the northern side of the South Wales coalfield. In both of these the relief is probably best described as plateau with holes, because, although the frequency of solution holes reaches high values in South Wales, the unity of the surface is not completely destroyed. Most of the solution holes are fairly clear cut features, although gentler and steeper forms do occur.

Over much of the Causses in France there is an abundance of extremely broad, very shallow closed depressions in which the terra rossa soils accumulate. These features may run up to hundreds of metres in diameter and have side slopes usually less than 5 degrees and often much less than this.

In Yugoslavia again regional variations show. Shallow depressions of the type described above can be found in the northern coastal areas south of Rijeka. Much further south around Ploče, on the northern side of the mouth of the Neretva, depressions of a similar diameter have much steeper constant slope sides, of the order of 15–20 degrees and form very impressive features, especially where they are so near the coast and their bottoms so low that they have been inundated. Finally, the road which runs from Zagreb through Karlovac and on to Rijeka, crosses areas which are much more vegetated and cultivated than the very karstic coastal regions of the country. Here the solution holes are very much smaller, many of them of the order of 100 m (300 ft) in diameter, steep-sided and almost coalescent. The landscape in form seems to resemble what the British areas might develop into were there enough solution depressions to destroy the whole surface. The fact that most of the area is cultivated adds to its bizarre appearance.

In the tropical parts of the world two types of karst appear, which do not seem to characterise the temperate regions. These types are cockpit karst, which is described by Sweeting (1958) as 'a succession of cone-like hills with alternating enclosed conical depressions or "cockpits"' (p. 188), and tower karst, which 'is made up of steep-sided, forest-covered hills or mogotes, whose slopes vary between 60 degrees and 90 degrees. Each hill or group of hills is separated by a more or less flat alluvial plain which is often inundated' (p. 192). The German terms for these two are respectively, Kegelkarst and Turmkarst, although even these terms seem to be used differently by different authors. Jennings and Bik, for example, use Kegelkarst as a general term to include Turmkarst and what they term Kugelkarst (literally ball or spherical karst): the latter is very similar to Sweeting's cockpit karst except that the hills are not cone-like as much as ball-like.

In Jamaica, the area described by Sweeting, the relative relief of the cockpit karst and the tower karst is of the order of 100 to 160 m (300–500 ft). The diameter of the hills in cockpit karst is of the order of 300 m (1000 ft) or so and the side slopes are roughly 30 to 40 degrees. According to Sweeting tower karst may develop from cockpit karst when the floors of the cockpits reach down to water-table level. Under such conditions solution at the base of the hills—and in this connection it must be remembered that in the tropics the immediate effects of solution may attain very high values—steepens the side slopes and slowly reduces the hills. The weird landscape of a well-developed tower karst, in which extremely steep hills project from a flat floor, occurs in many tropical regions, the West Indies, Malaya, South China and adjacent regions.

Although Kegelkarst, Kugelkarst and Turmkarst seem to be largely tropical phenomena and hence emphasise the importance of climate in karst development, there is probably a measure of lithological control involved herein as well. However, it must be stressed that these tropical forms emphasise as it were the dominance of the holes over the hills whereas in the temperate regions the reverse is generally true. In Jamaica cockpit karst and tower karst are developed only in areas where hard crystalline limestones occur, while, in the case of tower karst, the presence of strong vertical jointing may be an important subsidiary factor. In areas of marly, or impure limestone and lower rainfall in Jamaica a form of doline karst not unlike that of Yugoslavia may be developed (Sweeting 1958). In these areas generally rolling country is interspersed with an irregular pattern of solution depressions.

This correlation of certain forms of tropical karst with certain lithological types does not seem to hold good everywhere, however. In a description of the altitudinal variations of karst forms in New Guinea, Jennings and Bik (1962) declare that at the lower elevations the dominant karst forms are those of hemispherical hills and depressions, i.e. a Kugelkarst irrespective of the lithology of the limestones, while at higher elevations the type of karst developed is described as doline karst. From the description it seems to be a form of tower karst.

It is apparent that there are many different types of solution depressions—and little mention has been made here of the differences between those depressions which are essentially funnel-shaped and those which are clearly-defined holes in a near horizontal surface—and that both lithological and climatic factors may play roles of varying importance. Even time may be a consideration.

Yet in the formation of these solution depressions there remains the question of whether they develop from surface solution or from underground solution and ensuing collapse. Briefly, the protagonists of surface solution reckon that water goes underground, probably at the intersection of major joints, that the sink so formed will slowly become enlarged and that as it does so the slopes leading into it will become graded or regularised. This could lead to a symmetrical hollow in due course. On the other hand, it has been maintained that most of these solution depressions are due to collapse into underground caverns.

In general, it is easier to prove the second possibility, because of the evidence from areas where the vast majority of solution forms occurs on non-calcareous rocks. This is the situation, in fact, on the northern side of the South Wales coalfield (Thomas 1954). Although there are many solution forms on the bare outcrops of Carboniferous Limestone, the best and the largest features occur on the Basal Grit of the Millstone Grit above it (Fig. 5.20). These obviously cannot be surface solution forms, because the Millstone Grit is a siliceous rock not at all susceptible to solution. The curious fact is that solution forms seem to develop at the surface where the

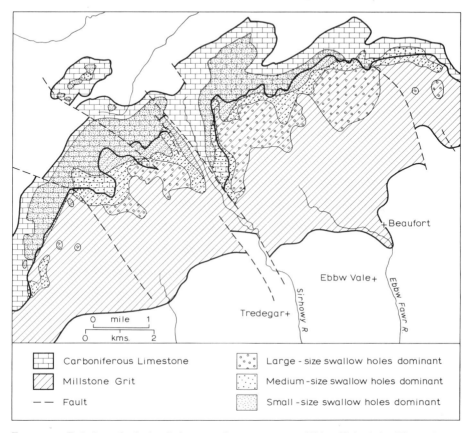

+Beaufort

Ebbw Vale+

Tredegar+

Sirhowy R.

Ebbw Fawr R.

0 mile 1

0 kms. 2

	Carboniferous Limestone
	Millstone Grit
— —	Fault

	Large - size swallow holes dominant
	Medium - size swallow holes dominant
	Small - size swallow holes dominant

FIG. 5.20. Relation of solution holes to rock outcrops near Ebbw Vale *(after Thomas)*

Millstone Grit is up to 60 m (200 ft) or so thick. This must imply subsurface solution at the Millstone Grit–Carboniferous Limestone junction on an enormous scale. In fact Thomas calculated that for each of his major solution holes a cavern the size of the largest known in Britain must have been formed below. The reason advanced for the greater size of the solution holes on the Millstone Grit than on the Carboniferous Limestone is that the Basal Grit of the former is jointed on a larger scale and that, therefore, collapse does not occur until the underground cavity reaches large dimensions. Whatever the exact mechanism, the fact remains that the presence of these forms on the Millstone Grit virtually proves sub-surface solution and collapse, as shown in Fig. 5.21.

A similar conclusion was reached by Coleman and Balchin on the

221

Mendip Hills. Here the arguments are not quite as direct except where the presence of Jurassic rocks, which are not soluble, over the Carboniferous Limestone makes the two cases comparable. One interesting point which derives from this area is that the solution pipes, which are common in

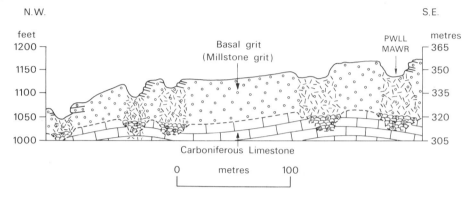

FIG. 5.21. Section through solution holes near Penderyn *(after Thomas)*

limestone areas and which contain ferruginous material, nowhere seem to coincide with the development of surface solution forms. This suggests that solution from the surface does not necessarily result in a surface form.

On the other hand the collapse hypothesis may not, almost certainly does not, apply to all areas of dolines. Even in the Mendips all the solution holes may not be collapse features. In a review of the 600 closed depressions which occur in this same region, the Mendips, Ford and Stanton (1968) have concluded that 90 per cent lie in the floors of an old dry valley system and that depression frequency varies inversely with the gradient of the long profile. The concentration clearly suggests that the agent of solution is surface water and the variation with gradient suggests that it is most effective on low gradients where its movement is least, a result which might have been expected from first principles. There is little sign of underground collapse, for collapse features rarely exceed depths of 18 m (60 ft) and seem to be due to the local collapse around pipes subject to vigorous solution action.

Again, Jennings and Bik are convinced that regularly scattered patterns of dolines up hill and down dale in New Guinea are hardly likely to be due to evenly-distributed underground drainage features. It is interesting that these authors argue that a regular pattern of dolines tends to rule out the possibility of collapse, while Ford reached the same conclusion from the non-random character of the pattern. These arguments are in a sense based upon intuition, but the case becomes very much stronger when one thinks of areas like the interior parts of northern Yugoslavia mentioned above and the areas of tropical cockpit karst, where the holes are so frequent that if they were due to collapse the whole of the interior of the limestone must be little more than a virtually continuous complex of caves. In a paper given to the International Geographical Congress in London in 1964, Aub presented strong evidence that most of the cockpits in an area he had studied in detail in Jamaica were due to solution from the surface. Wherever it was possible to penetrate into the bottom of the cockpit it could be seen that the floor of the cockpit was undisturbed, and even where no penetration was possible there was no sign of the sort of rock disturbance associated with collapse. Aub was prepared to state categorically that at least 60 per cent of the cockpits were formed by solution from the surface. One interesting fact was that the hollows received about 14 per cent more rainfall than the hills and one is tempted to speculate whether initial surface depressions, however formed, might not by concentration of both precipitation and drainage be transformed into a series of self-perpetuating cockpits.

Climate versus lithology   The possibility of both climate and lithology having important effects on the form of limestone surfaces and the pattern of solution depressions formed therein has been mentioned several times above. In a way the hypothesis of dominant climate is a more attractive one as it has an elegant simplicity, but the chaos introduced by stressing lithology probably represents a step nearer to the truth.

The dominance of climate, especially in the overall rate of the processes of karstification, was strongly stressed by Corbel (1959), whose arguments were based mainly on the calcium carbonate content of drainage waters in different regions. This, according to Corbel's figures was much higher in cold regions than in hot regions. The apparently great measure of karstification in some hot regions was to be explained by the vast amount of time that had been available under unaltered climate for this process to be completed, whereas in cold climates the late intervention of glaciation meant that, although the process was powerful, the time available was too small to allow a similar development of karst.

There has been a tendency in recent years for stress to be put once again on to lithology as of primary importance in the development of limestone landforms. This has been applied to both large and small scale landforms, to temperate and tropical forms, and to mountain and plateau limestone scenery.

The importance of the joint spacing and of the relative thinness or massiveness of the bedding in the control of the micro-relief of limestone pavements has been stressed by Sweeting (1966) for the Carboniferous Limestone districts of Yorkshire. Massive, coarsely-jointed beds of limestone need a powerful agent of erosion to form pavements on them and the pavements, once formed, need a powerful agent to destroy them. Only such massive beds can exert a dominance over the agency of erosion in the development of relief. Finer jointing and thinner bedding mean that areas can be more easily smoothed by processes of powerful erosion, and that factors such as freeze–thaw can operate readily to destroy any features that

may be formed. The variation of possible forms is as large as the variation of joint and bedding characteristics. The best limestone pavements, with broad, plateau-like clints separated by deep-furrowed grykes are on the massive, coarsely-jointed rocks.

Sweeting (1968) has recently stressed the possible significance of limestone lithology in the formation of caves. In the first place the presence of thin shale beds within the limestone may control the shape of caves especially because of their tendency to produce flat roofs to caves. Apart from this Sweeting discusses the possible effects of sparites (i.e. sparry rocks with a high content of crystalline calcite) and micrites (i.e. microcrystalline ooze—the calcite mudstones of the Bristol area come into this category) on the form of caves. She suggests that in the sparites the caves are largely confined to enlarged joints and have a predominant vertical extent, while in the less sparry rocks the caves are broader due to collapse and less confined to joint planes (Fig. 5.22). This is mainly because of the

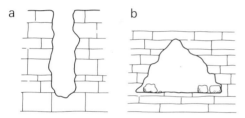

Fig. 5.22. Yorkshire cave forms: a. In sparry limestone. b. In less sparry limestone *(after Sweeting)*

greater porosity of the rocks. If these preliminary observations prove of general applicability, it may mean that geomorphologists will have to introduce a modern classification of limestones into their discussions.

The importance of both climate and lithology in the development of karst features in Puerto Rico has been emphasised by Monroe (1964). The southern part of the island is semi-arid, and, although there are important limestones there which are most likely affected by solution, there is little or no development of true karst. In the north of the island rainfall ranges up to about 3000 mm (120 inches) per year and comes in short intense falls. In this part karst features are well developed, but they vary with the type of the limestone: the true Kegelkarst with mogotes is only developed on thick, massive, pure, homogeneous limestones, while certain other features appear to be confined to other lithologic types. Monroe makes an interesting point that the greater part of the soil is washed off the mogotes into the hollows and that this causes richer vegetation and more carbon

PLATE 24. The Cristallo massif from the Tre Croci Pass, Italian Dolomites. Effect of vertical jointing on Triassic dolomite

dioxide to be present there than on the much barer hills. Thus solution is greater in the hollows and relative relief tends to increase. One is tempted to add to this Aub's observation of the increase in precipitation in the Jamaica cockpits as another factor tending in the same direction.

A final example of lithological effects may be taken from the Dolomites (Plates 24 and 25) in the South Tyrol (de Smet and Souchez 1964). Here the environment is quite different. At an elevation of just over 3000 m (10,000 ft) these mountains lie in a zone of glacial and intense periglacial action, with freeze–thaw as a dominant process. The part of the Trias succession significant here is as follows:

3. Main dolomite
2. Raibl marls and marly limestones
1. Sciliar dolomite.

The Raibl beds are generally stripped back to form a ledge separating the relief of the other two formations. The Sciliar dolomite is a reef formation and hence has no marked horizontal bedding, but it is dislocated by pronounced vertical fractures. These led in pre-glacial times to a strong karstification of the surface with the development of closed depressions, which were later modified by glaciation. The cliffs developed on it are sliced up by great chasms developed along the vertical fractures. This relief is found in the Catinaccio massif. On the other hand in the nearby Sella massif the other two formations are present above the Sciliar dolomite— there is even a small patch of younger rocks but for the present purpose they may be ignored. The main dolomite is a massive, near-horizontal rock, which gives rise to a slab-like relief bounded by vertical cliffs, which lack the tortured character of those of the Catinaccio massif. Even the Ice Age modified the relief differently, because the flatter surface of Sella encouraged an ice cap which protected the surface from the rigours of the extreme freeze–thaw to which the Catinaccio massif was subjected. Where the Sciliar dolomite appears on the northern edges of the Sella massif a development of relief similar to that of Catinaccio is to be found.

It is apparent in the description of this part of the Alps that one is dealing not only with an interaction between rock and processes controlled by the present climate, because the effects of pre-glacial karstification of the surface and of glacial action are present. The existence of palaeo-karst is mentioned fairly frequently by writers on limestone regions. Souchez (1963) attributes an area on the northern side of the Ardennes in Belgium to early Tertiary tropical karstification producing a polje with mogotes which are still present as a series of conical hills. Much of the Tertiary period in Europe was characterised by hot climates and it is to be expected that features dating from this period are likely to be preserved where one has

PLATE 25. The Cristallo massif from the Cortina side, Italian Dolomites. Effects of vertical jointing and modest amounts of modern scree

old resistant massifs that did not suffer the full effects of the Ice Age. These mogotes are formed in Devonian limestone, and it should be mentioned that they could be fossil reefs, but Souchez considered this possibility carefully and dismissed it.

Again Sweeting's (1952) description of abandoned levels of caves in the Ingleborough district of Yorkshire related to higher base-levels, which are also represented by subaerial erosion surfaces, makes this area another example of a fossil or palaeo-karst.

Finally, for an example of fossil karst the account by Gilewska (1964) of limestone scenery in Poland might be mentioned. The examples quoted above might be expected as they refer to comparatively recent (in a geological sense) karstification. Gilewska, however, recognised four phases of karstification in Polish geological history: the Permo-Trias, Upper Trias—Lias, Lower Cretaceous and Tertiary, the last being the most important. All these periods are of course times of continental sedimentation.

Underground limestone features   Theories about the origin of caves and similar underground features are closely bound up with differences in ideas about the nature of the water table, if any, in limestone regions.

It is possible to visualise the whole of the cracks and crevices in a limestone below a certain level being filled with water so that the surface of this fill represents a true water table. In such a case, presumably, gradients in the surface of the water table would exist in order to overcome the frictional resistance to flow underground. Such gradients would vary with the nature of the limestone. In these conditions underground flow would be predominantly, if irregularly, downwards to the water table and then laterally along the surface of it, the vast majority of the water constituting an underground reservoir.

On the other hand there may not be a coherent zone of saturation in limestones. Such ideas have been prompted by the finding of air-filled pockets well below 'water table' level, and by the presence of certain features in underground passages that suggest flow by water under great pressure. On this hypothesis deep flow in limestones is confined to certain channels and occurs under considerable hydrostatic pressure. There need be no continuous saturation zone within the limestone. If flow does occur in such conditions then caves and passages may be formed in this, the phreatic zone; or they may be formed at or above the water table in the vadose zone; or they may be initiated and in the phreatic zone and modified to varying degree in the vadose zone. The evidence points here in one direction and there in another. A summary of the classic papers is available in Thornbury (1954).

The general idea that caves are formed at the junction between the two zones is the older idea of the two. It is usually accompanied by the idea that much of the erosion is controlled by joints and that corrasion as well as corrosion is important. This is suggested by the quantities of under-

228

ground alluvium of varying calibre available. This is really the idea behind the development of successive cave levels caused by a falling water table, mentioned above in connection with Sweeting's work on the cave levels in the Ingleborough district of Yorkshire. Cave formation at this level would require water percolating underground to remain aggressive. One can understand that its passage downwards may be so rapid that it still has solution capacity at the time it starts moving horizontally along the water table level. A lot would depend on the ratio between the time spent in contact with the carbon dioxide-rich soil atmosphere and with the limestone below it. Other factors of course enter the picture, such as the possibility of increased aggressiveness of the water being caused by a fall in temperature underground. The reverse might hold in winter because of the more equable nature of underground climates. Obviously in some caves the great profusion of stalactites and stalagmites and dripstone in general suggests deposition rather than corrosion.

Some features of underground passages seem more explicable in terms of an origin beneath the vadose layer by streams under hydrostatic pressure. Tubular passages with blind pockets in the walls and roofs, with evidence of flow equally in ceiling and floor, point in this direction. The up and down nature of the long profiles of some caves is similar to that of some eskers, and the latter are usually attributed to subglacial streams under hydrostatic pressure. The general pattern of underground cave complexes is often a three-dimensional network and not the two dimensional one that would be expected from movement along the surface of a water table. The mechanism of corrosion is a little difficult to understand at this level. If the liquid completely fills the passages and there is no gas phase, no interchange will take place with the walls of the passage, and hence there will be neither erosion nor deposition. A solution of this problem has been pointed out by Bögli (1964). It is based upon the fact that the relationship between the amount of calcium carbonate in solution and the equivalent carbon dioxide is not arithmetical. The equilibrium curve is shown in Fig. 5.23. If we imagine two water masses in equilibrium, W_1 and W_2, to be mixed, the calcium carbonate–carbon dioxide ratio of the mixture will lie on the straight line connecting the two, its exact position being determined by the original proportions of the two sources. Because of the form of the curve these points will always be in the aggressive zone and will give the water new powers of dissolving calcium carbonate. This renewed power of corrosion through mixing can occur not only beneath the 'water table' in closed passages but also in the vadose zone and at the earth's surface.

By trying to attribute the origin of caverns entirely to phreatic or vadose conditions one may be creating a false problem. It would be nice and simple to envisage a global hypothesis in which solution took place by pressure flow in the phreatic zone, deposition took place in the vadose zone in the form of stalactites and stalagmites etc., and solution took place at the surface. Thus, the active levels of the landscape would be at great depth and at the surface with a zone of fossil caves between.

This is much too simple and the protagonists of both hypotheses have more or less allowed for modification by the other. Those who believe in

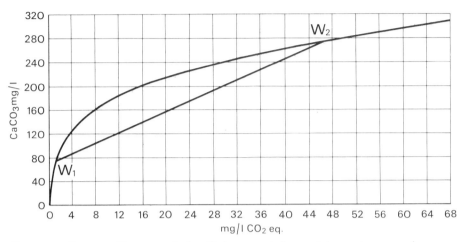

FIG. 5.23. Curves to illustrate solution likely to result from mixing water masses in limestone areas *(after Bögli)* For explanation see text.

the phreatic origin of caves cannot really disallow later modification in the vadose zone, even if only minor modification, while the adherents of a vadose origin cannot preclude the formation of phreatic passages, however vestigial, which later guided the movement of water and solution in the vadose region, in which the main part of cavern formation took place. In some areas the great amount of deposition suggests the caves are fossil and are now being filled: in others such as the Lost River region of southern Indiana, described by Thornbury (1954), the lack of deposition in caves in the vadose zone suggests continuing active development. Once again, as in many geomorphological problems, man, at least western man with his monotheistic conditioning, may have been trying to oversimplify a problem in which like, but not necessarily identical features, may have had different origins or at least origins in which different factors played roles of differing significance.

Overall rates of limestone denudation Limestone denudation, calculated largely from the amount of dissolved load in rivers, is pretty rapid everywhere. Corbel (1959) thought it to be much slower in the Tropics, but his suggestions attribute intense karstification in these areas of limestones, often as old as Cretaceous, to the length of time these limestones have been in existence, as though they have been open to karst processes for their whole span of existence. He suggests that younger Tertiary limestones often show a smaller degree of karstification, and quotes younger Tertiary limestones of Jamaica and Yucatan, Mexico, as examples. On the other hand Munroe (1964) has described well-developed tropical karst on the late Oligocene—early Miocene limestones of Puerto Rico and Jennings and Bik have described well-developed karst on the folded Pliocene limestones of New Guinea, i.e. karstification is almost certainly Pleistocene in age here. Yet, even if we avoid the heated argument about the relative effects of different climates, the figures produced for overall limestone denudation are astonishingly large.

Most investigators have relied on Corbel's formula, or some modification of it, to arrive at an overall denudation rate for limestone. The original formula (Corbel 1959) applicable to a basin wholly composed of limestones was $x = 4ET/100$ where x is the annual erosion loss in cubic metres per square kilometre (or alternatively surface erosion in millimetres per 1000 years), E is the run-off in decimetres and T is the average calcium carbonate content in milligrammes per litre. Where the basin is not wholly limestone and limestones occupy only $1/n$ of the basin the formula becomes $x = 4ETn/100$ if one is interested in the calculation of the lowering of the limestone surface in solution as a rate rather than a total amount. Corbel's formula assumes an average specific gravity for limestone of 2·5.

Various modifications have been suggested. Williams (1963) for example introduced a term for the content of magnesium carbonate, an important element in some limestone districts, and also took the average specific gravity of the limestone under consideration into account. The use of 2·5 as the mean specific gravity for the limestone introduced an error of 8 per cent into the calculations for the Carboniferous Limestone in Clare County, Eire: this has a specific gravity of 2·72 and hence the overall rate of solution will be lower than that derived from Corbel's formula. For Chalk, which has a much lower specific gravity, the overall rate of surface lowering derived from Corbel's formula would give results that are considerably too low.

In practice there are other complications, not so much in the calculation of the loss in cubic metres per square kilometre per year but in transforming this into a figure for surface lowering by solution. One has to guess at the proportion of solution occurring at the surface and the proportion occurring underground. If one is using these figures in long extrapolations one might be able to ignore this by making some such assumption as that the surface lowering over a long period will be directly related to total calcium carbonate loss because, even if the solution is underground, this will lead to occasional collapse which will temporarily accelerate the rate of surface lowering. There is also the question of how much of the calcium carbonate might be derived from rocks other than limestones; for example some British Jurassic ironstones have an appreciable content and many glacial drifts are rich in calcium carbonate. In some of the latter it is so well divided that solution is probably facilitated: Bögli (1960) has pointed to the fact that some streams issuing from calcareous moraines in Alpine areas have concentrations of calcium carbonate higher than that of those issuing from limestones.

Nevertheless, however one argues about the question, the figures derived suggest a high rate of surface lowering of limestones. Corbel's figures give something like 200 mm per 1000 years in periglacial regions, about 100 mm in mountains in our climates and 30 mm in hot, wet climates. W. M. Davis reckoned that a peneplain might take 50 or 60 million years to form. If one does a straight extrapolation of Corbel's solution rates this would give lowerings of limestone in that period of the order of 10,000, 5000 and 1500 m in the three climates cited. And, of course, one is justified in doing a straight extrapolation of limestone rates, because they are not slowed up with time, which usually involves declining denudation rates due to decreasing slope. Theoretically, they remain constant, or could even increase if slopes are so low that the water remains in contact with the rock for a longer time. For County Clare Williams (1963) arrived at a solution rate of about 50 mm per 1000 years while Sweeting (1966) in Yorkshire found a rate of about 80 mm per 1000 years. In both of these cases roughly half was probably surface solution.

Another way of arriving at rates of surface solution of limestones is in a consideration of the differences in elevation between the limestone below large glacial erratics and that adjacent to the erratics. If we can make an approximation to the date when the block was deposited and measure the difference in height between the pedestal and the adjacent limestone, the calculation of an average rate is very simple. Similarly the level of the Chalk beneath some barrows and that adjacent to the barrows (Fig. 5.24c) are often conspicuously different; all that is needed to derive a solution rate is that the barrow should be dated. It is possible to raise objections against these methods. The state of affairs with pedestal erratics is rarely the perfection of Fig. 5.24a and is much more likely to be like that shown in Fig. 5.24b, in which measurement will be much more difficult. It is possible to say that all the run-off from the glacial erratic will concentrate solution round its margins. In the case of the barrow it might be said that the land near the barrow was scraped to build the barrow, or that, even if the Chalk was not affected, all the soil was stripped thus exposing the Chalk to faster erosion. One can plead that ploughing up to the barrow has tended to shift material away from its edges, a process which occurs in some similar situations. Such objections increase the uncertainty, but even if the suggested rates of solution are halved they are still considerable. Estimates of the rates derived from boulders on the Carboniferous Limestone of Yorkshire give surface lowering rates of about 50 mm per 1000 years, a figure comparable with those obtained from dissolved calcium carbonate content. Rates may also be obtained from the obliteration of the writing on limestone tombstones and such things, though here there might be an acceleration due to urban acidic atmosphere, caused by the products of combustion, in some cases. However, the general rate arrived at for the north of England is similar to that obtained by other methods. Further, experiments of stripping limestone surfaces and allowing peaty drainage water to run over them have resulted in rapid obliteration of glacial striae and surface lowering of up to 50 mm in thirteen years. These are exceptional circumstances (Sweeting 1966).

a

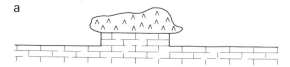

b

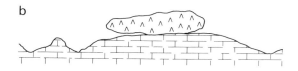

c

FIG. 5.24. Estimates of limestone solution: a. Around an erratic—idealised. b. Around an erratic—natural. c. Around a barrow

From the study of the surfaces beneath barrows Atkinson has arrived at figures of surface lowering of up to 125 mm per 1000 years. This may seem very high, but when the much lower specific gravity of Chalk is taken into account, it is quite comparable with the lowering rates calculated for the Carboniferous Limestone.

The conclusion that these rates lead to is almost inevitably that limestone is a highly weatherable rock and that its rate of lowering, being chemical, is not slowed down as slopes generally decline through the course of time. In addition Pitty (1968) has argued that the majority of limestone solution, at least in the southern Pennines area he studied, was concentrated on the surface of the limestone. In the valleys the streams already had high calcium carbonate contents and so, too, had the spring water fed into them. Cave formation probably explained little of the solution loss because it seemed that the annual solution loss accounted for more than the total volume of known caverns. The suggestion of the general enlargement of cavities in

233

the vadose layer was opposed by the fact that water seems to attain high calcium carbonate content even at shallow depth. Finally there was little evidence for the widespread widening of joints near the surface to account for the calcium carbonate in solution. Everything pointed to the concentration of the major part of the loss into surface lowering.

If one takes a loss per 1000 years of 50 mm of limestone from the surface, this means a lowering of 50 m per million years, i.e. since the early Pleistocene. As the total thickness of the Chalk is of the order of 400–500 m it should all have been dissolved since the beginning of the Pliocene. It has not. Thus, there is something suspect in very long term extrapolation. On the other hand, one should query those geomorphologists who argue that Chalk arrests dissection and is one of the best rocks for preserving old erosion terraces. Field experience suggests that the latter is a reasonable assumption to make, but it runs into direct conflict with the arguments derived from a study of solution loads and inferred surface lowering rates. Such convergence and utter conflict of conclusions derived from different arguments sets up a problem, which can only be resolved by further research. Such incompatibilities of ideas are often the sources of advances in knowledge. It is to be hoped that this one will be.

Marine denudation of limestone A word should finally be said about the denudation of limestone by the sea. Limestones in the intertidal zone are often characterised by solution forms of various types (Plates 26 and 27). The hollow is the standard form but the extent of these hollows in relation to the ridges separating them varies enormously. Variations in solution forms in relation to tidal levels and in relation to climate have been described by Guilcher (1953, 1958) and Zenkovitch (1967).

The problem of coastal limestone solution is closely bound up with the fact that all seas, and certainly those parts near coasts, are saturated with calcium carbonate. This has caused some workers, for example Wentworth, to attribute most coastal limestone solution to the action of freshwater percolating from the land, but this has not received widespread support as an idea.

Guilcher (1958) has summarised the ideas relating to the apparent anomaly of calcium carbonate solution by sea water which is usually saturated with that material. A promising way out of the difficulty seems to lie in a consideration of the diurnal variation of carbon dioxide concentration. During the daylight hours Algae absorb carbon dioxide in the process of photosynthesis. Thus carbon dioxide is abstracted from sea water and a fine precipitate of calcium carbonate is produced as the carbonate equilibrium is disturbed. This fine precipitate is easily transported away

PLATE 26. Marine solution features, Carboniferous Limestone, Southerndown, Glamorgan. Coarse solution features at or above normal high tide level affecting Carboniferous Limestone in foreground much more than unconformable Lower Lias limestone in background

from the intertidal zone by the action of waves. At night the opposite occurs and the green Algae emit carbon dioxide. As the fine precipitate formed during the day has been removed, this decreases the pH of the sea water and allows an attack to be made on any limestone occurring in the shore zone.

Another possible diurnal variation of activity could be bound up with temperature variation, but the diurnal variation of sea temperature is very small and so too is the possible effect that could be caused by an increasing of dissolving power consequent upon nocturnal cooling.

Something might be effected by biological weathering caused by limpets and certain gastropods such as *Patella* (Plate 27). These often seem to be the causes of small depressions in the rock surface, but whether they can be regarded as a major factor in the origin of solution forms is doubtful.

Biological factors in weathering Biological factors have already been introduced into the question of limestone weathering when stress was laid on the very high concentrations of carbon dioxide which arise in the soil atmosphere. These arise from the respiration of living plants and the decomposition of dead ones. Again, the attempt to explain marine corrosion forms involved the activity of Algae.

The activities of various organisms may go well beyond this but so far research into this field has little more than begun. Two Polish workers, Drzal and Smyk, have started work in this field in Central Europe and have concentrated on Bacteria, Streptomycetes and Fungi. Many of these have the ability to produce acids and to alter various rock-forming minerals including calcium carbonate. These conclusions, derived from the knowledge of the life of the organisms, have been tested by experiments involving limestone corrosion. Such organisms are most likely to live in hollows in the limestone and so to affect the surface pattern of relief. Their activity may, of course, further upset the complicated question of whether climate is dominant in limestone landform development, especially if their degree of activity rises with temperature.

Soft limestone relief Although at the beginning of the section on limestone relief, soft, comparatively-unlithified limestones were excluded from the discussion on the grounds that they did not possess all those characteristics required for the formation of true karst relief, they nevertheless respond in basically the same way to denudational processes as do hard limestones. That is, they are very susceptible to solution and they are also permeable; hence chemical denudation and waterless surfaces. But, because of the lack of other characteristics necessary for the development of karst, their relief is considerably different.

PLATE 27. Marine solution features, Carboniferous Limestone, Southerndown, Glamorgan. Fine-scale features at lower level than those shown in Plate 26. Limpets, which can be seen, may play a part in their formation

Among such limestones must be included the Chalk and most of the Jurassic limestones of Britain as well as many of those of similar ages in the scarplands of northern and eastern France. Of course, there is no clear boundary between karst limestones on the one hand and the softer limestones on the other. Various features characteristic of true karst come in progressively as the lithology changes to that needed for karst formation.

Solution rates are just as high simply because, with their great porosity and their generally much smaller-scale jointing pattern, limestones such as Chalk have a very great surface area available for solution. Again, overall surface lowering rates should be higher because of the lower specific gravities of these limestones. Perhaps because of their solubility, solution is even more concentrated at the surface than it is in the harder more openly-jointed limestones. At any rate, there is usually a much greater scarcity of underground solution features. If solution does take place underground to any large extent one must have recourse to a hypothesis such as that which suggests that rocks like Chalk are not strong enough to remain widely open below the surface, but that readjustment along the multitude of small joints closes any incipient hollows. This presumably should be followed by a developing irregularity of surface, which, however, rarely exists in Chalk.

The lack of development of pavements, clints, grykes or any types of Karren may reflect a number of features. Many of these areas were not glaciated—and it will be remembered that glacial stripping seems a prerequisite in the formation of British limestone pavements—but even where they have been glaciated their form is hardly different from where they have not. The porosity of some of the softer rocks is their worst enemy. It allows the rock to be soaked with water and so destroyed by freeze–thaw. In arid regions salt solutions may permeate such rocks and the pressure caused by crystallisation in the pores could have similar effects. Again, even where these rocks are harder and more lithified, many of them are finely-bedded and the beds traversed by very fine joint patterns. Thus their lines of weakness are far too closely spaced for the preservation of good surface features.

Solution holes also seem to be much rarer, although there are often plenty of solution pipes filled with rusty gravel, but making no impression on the surface form. These are very characteristic of English Chalk areas, where they seem to be confined to areas where thin Tertiary and Quaternary sands and gravels overlie the Chalk or areas from which such beds have been recently stripped. The descriptions of Reading Beds outliers on the Chalk nearly always contain references to the highly irregular nature of the junction with the underlying Chalk. Presumably ranker vegetation and possibly acid soils ensure a higher solution rate.

In a few places there are solution holes. These seem to occur in different

PLATE 28. Jointed limestone scarp near Castellane, southern France. The rocks are Portlandian (Jurassic), which forms a massive karst limestone in southern France

types of locality. Some of the most famous ones are those in the bed of the river Mole and near its banks where it cuts through the Chalk between Dorking and Leatherhead in Surrey (Fagg 1958). Others occur where drainage is shed from overlying Tertiary beds and glacial drift on to the Chalk: examples of these are found at North Mimms in Hertfordshire (Wooldridge and Kirkaldy 1937) and at Lane End on the Chilterns, where the drainage from a Tertiary outlier disappears into a large hole at the Chalk junction (Jukes-Browne and White 1908). Others are known in or near the Glen valley at Burton Coggles between Stamford and Grantham (Hindley 1965). The rock here is Lincolnshire Limestone and the location of the holes may be affected by a fault. On the whole, though, one is searching to name examples from these soft limestones whereas they can hardly be avoided on karst limestones.

A few caves are known, but they are very rare indeed.

On the coast the soft limestones are readily attacked by the sea, partly because of their softness and partly because of their fine-scale fracture patterns. The wave-cut benches in British Chalk and Jurassic limestone areas do not possess the solution features which the Carboniferous Limestone often has. This can be clearly seen at Southerndown in Glamorgan, where the Lias limestones rest with a gentle unconformity on the Carboniferous Limestone.

Those limestones transitional to karst limestones begin to show more karst-like features. Bare rock outcrops on the harder beds hint at the possibility of pavement and Karren formation. Occasional caves appear. Depressions become more frequent and so on.

All limestones have some problems in common. One is the origin of dry valleys, with its perennial controversy between those who believe them to be essentially river valleys desiccated by falls in the water table occasioned either by escarpment recession and spring lowering or by the action of allogenic rivers in achieving the same effect, and those who would attribute them to the catastrophic action of periglacial processes in the Pleistocene. An outline discussion of this problem is contained in an earlier book (Sparks 1960).

Hard limestone relief Many of the general features of hard limestone relief have been dealt with in previous pages in the discussion of such typical features as limestone pavements, solution holes and the origin of underground solution forms. It would be as well to summarise some of the more large-scale features of such typical limestone areas as the Causses, the southern French Alps and Pre-Alps, and the karst of Yugoslavia.

Most of the forms are related to the tendency for hard limestones to occur in massive beds, which owe their resistance to erosion largely to their

Plate 29. Cirque de Navacelles, Central Plateau, France. An abandoned meander in the river Vis gorge overlooked by scarps on the Kimeridge (Jurassic) limestones

general permeability along joints and bedding planes. It should also be noted that most of the areas mentioned above might be described as almost semi-arid. Thus the effects of rainfall are reduced to a minimum. The rocks then are to a considerable extent immune from surface erosion and hence slopes remain steep. Where precipitation is higher, as in the frontal Alpine ranges of southern Germany and Austria, limestone forms are softer and lack the gaunt character typical of southern France, though whether the differences are due solely to climate or to climate plus lithology is difficult to say.

The massive nature and the resistance are reflected in the dominance of escarpments, which are especially pronounced where the dips are low. A typical limestone scarp developed in the Portlandian, which in many parts of southern France is a massive karst limestone facies known as the *tithonique*, is shown in Plate 28. The control of the detail of the relief by vertical jointing is apparent. Again, horizontal or near horizontal beds favour the formation of stepped escarpments on the sides of river gorges. In the valley of the river Vis in the Central Plateau of France at the Cirque de Navacelles, a superb example of a cut-off meander, the Kimeridge limestones, Upper Jurassic in age, form the topmost cliffs (Plate 29). Within the Tarn gorge in the same region the harder and more resistant beds of limestone and dolomite act in the same way, though their age is slightly different, the succession in the area shown (Plate 30) ranging from the French equivalent of the upper part of the Inferior Oolite to that of the Corallian.

Gorges are quite characteristic and almost too numerous to mention. Two well-known ones have been cited above. Others in southern France include that of the Guil through the central southern part of the French Alps, mainly in Triassic limestones (Plate 31) and, most noteworthy, that of the Verdon between Castellane and Moustiers-Ste-Marie largely in the Portlandian (Plate 32). In central Yugoslavia the gorges of the Neretva and the Vrbas are well-known. But there are other spectacular if shorter and less deep gorges in other parts, for example the northern flank of the eastern Pyrénées, where the Gorge St Georges is short but slit-like.

Plateaus and gorges are best developed where the beds are near horizontal. Where the dips are high, rocky crests on varying scales occur. A minor example of this, of the dissection of a near-vertical thin sheet of limestone north of Sisteron in southern France (Plate 33), is very reminiscent of the forms that we would expect of dykes.

Most of the foregoing refers to the drier climates of southern Europe. Where limestones and dolomites reach into altitudinal zones of glacial and periglacial climate very different forms can result, though even here the nature of the rock may still exert an important influence. On soft limestones

PLATE 30. Tarn gorge, Ste Enimie, Central Plateau, France. Near-horizontal limestones and dolomites ranging in age from Inferior Oolite to Corallian (Jurassic)

freeze–thaw is a highly effective process, but the extent to which it affects hard limestones reflects the degree of absorption of water by the limestone. Although the higher parts of the Dolomites may be greatly affected by freeze–thaw, the evidence from somewhat lower elevations suggests that it may be far less active now than in the past: compare, for example, the areas of unvegetated scree on Plates 24 and 25 with the areas of forested constant-angle slope.

The effects on a different type of calcareous rock are shown in Plates 34, 35 and 36. These show the landforms developed on the eastern side of the Col d'Izoard at elevations of 2300–2600 m (7500–8500 ft) in the southern French Alps. The rocks here are Triassic limestones and *cargneules*. The latter is a term applied to spongy dolomites, the origin of which is obscure. The term spongy is not entirely a happy one, because they are not soft like a sponge but only honeycombed. Clinkery might better describe their texture and their hardness. Obviously a rock with this texture lies wide open to maximum freeze–thaw effects and this is probably reflected in the phenomenal development of constant-angle scree slopes, which in places have entirely consumed the free faces of rock. Plate 34 shows the position south-east of the top of the pass, where the scree descends in a great sheet from the cliffs above, with needle-like pinnacles, probably reflecting differing induration of the rock, projecting through it. Plate 35 shows some of these residuals in the area to the right of Plate 34, and the suspicion arises from the presence of herbaceous vegetation and conifers that even some of these slopes might be stabilised and fossil. Plate 36 shows the area north-east of the summit of the pass where scree slopes, many of them vegetated but only partly stabilised if one may judge from the look of the vegetation, have practically consumed all the original cliffs and residuals. The whole area forms a remarkable illustration of the classic view of the development of constant-angle scree slopes from free faces and the eventual submergence of the latter beneath scree slopes meeting in narrow inter-fluves.

SILICEOUS ROCKS

Both these and the other chemical rock types described in subsequent pages are not comparable with the limestones and dolomites and the various clastic sediments already described in the formation of relief.

For present purposes we may divide the siliceous rocks into two main classes: those derived directly from the accumulation of siliceous remains of invertebrates or plants, and those probably formed by replacement with silica, the source of which is not often very clear.

Three main classes of living organisms possess silica in their skeletons. They are the diatoms, which are either marine or freshwater Algae; the

PLATE 31. Guil gorge, north-east of Guillestre, southern French Alps. Mostly in Triassic limestone and dolomite

245

Radiolaria which are marine organisms, and certain classes of sponges. Accumulations of diatoms and Radiolaria form diatomaceous and radiolarian oozes in certain places on the ocean floor at the present time. When uplifted and converted into unconsolidated deposits these become diatomaceous and radiolarian earths respectively, and later with some cementation by circulating silica they are converted into diatomite and radiolarite. Some ancient radiolarian rocks form the jaspers and jaspery cherts found in parts of the Carboniferous and Devonian. These rocks have no great importance as relief formers though diatomaceous earth is important commercially for a variety of reasons, but chiefly because it is used as an absorbent in the manufacture of explosives.

Apart from these specialised rocks, silica more often occurs either as discrete nodules, often arranged in layers, in sedimentary rocks or as a replacement material, usually in limestones. The most commonly used word for either of these types is chert, incompletely replaced rocks sometimes being referred to as cherty limestones. Where nodules occur in Chalk, even though they are identical with those found in similar limestones in other geological systems, for example the Portland Beds of the Upper Jurassic, they are called flints. Chert is usually present as the amorphous form of silica, which is termed opal, or as chalcedony, the exact nature of which is obscure, but which is different from the crystalline form, quartz.

The origin of the silica in cherts is somewhat of a mystery. It is possible that silica may have been deposited contemporaneously with the sediments as blobs of silica gel precipitated directly by the action of electrolytes, but the majority opinion does not favour this view, the chemical details of which are obscure. On the other hand chert layers are sometimes involved in slump structures and intraformational conglomerates, so that in these it appears that the chert must be at least nearly contemporaneous.

But there is more evidence in favour of the hypothesis that chert is essentially a replacement material. Chert occurs along fissures in very irregular nodules sometimes enclosing limestone and partially enclosing unaltered fossils, for example in Chalk; in bands which do not necessarily always follow the exact bedding of the limestone; and as an irregular silicification within the rock. Further silicified fossils, which must have been formed by replacement, and silicified ooliths are also known. It does not follow that such silicification need have taken place very long after deposition and may well have occurred before the rock was lithified.

Where all the silica comes from is again a mystery. Many beds, for example parts of both the Upper and Lower Greensands of southern England, contain large numbers of silica sponge spicules, and these may provide sufficient silica for the formation of nodules, as may have happened in the Chalk, or of irregularly silicified beds. But where extensive beds are

PLATE 32. Verdon gorge from le Point Sublime, near Castellane, southern France. Portlandian (Jurassic) limestone and dolomite

246

thoroughly silicified some external source of the silica must be involved, though what the source might have been can hardly be guessed at.

Whatever their origin the relief effects of cherts are probably only important where they are present as continuous beds. Cherty sandstones and limestones may be likened usually to imperfectly silica-cemented rocks and thus may be fairly important relief formers, as for example the Hythe Beds of the Lower Greensand of the western Weald. Where the chert or flint occurs as layers of discrete nodules it seems unlikely that it can add materially to the resistance of the rocks in which it occurs. There seems to be no greater resistance to erosion of the flinty lower part of the Upper Chalk which can be attributed to the frequency of flint layers. Probably the chief contribution that flints and some of the earlier jaspers make is in providing a source of durable material for pebbles, which survive, derived from bed to bed, for long periods in the landscape.

FERRUGINOUS ROCKS

The precise mechanisms by which many of the sedimentary ironstones have been formed are far from clear, simply because most of these rocks are marine and would therefore be difficult to observe even were they forming at the present time, which apparently many of them are not. The great majority of the rocks appear to be chemically precipitated or to be chemical replacements of pre-existing rocks.

One exception to this rule appears to be the material known as bog iron ore. This is an earthy, unconsolidated material forming in the shallower parts of some freshwater marshes, where its precipitation seems to be brought about by various bacteria. It was earlier an important source of iron ore, because, although very earthy and impure, it contains the iron oxide mineral, limonite. Today it is little used and as a relief-former is negligible.

Iron deposition also takes place in both freshwater lagoons and in marine conditions, when the chief mineral precipitated is iron carbonate (siderite). Rocks of this type occur in association with coals in the Coal Measures. These rocks accumulated under alternations of very shallow marine and swampy, deltaic freshwater conditions. In fact a rhythmic form of deposition can often be detected indicating a recurring pattern of conditions of deposition. One of these environments was that of stagnating freshwater lagoons in which iron carbonate deposition took place. Some of the clayband ironstones so formed are continuous beds, while others are layers of nodules, the latter probably formed by segregation of iron carbonate soon after deposition. Ironstone beds of similar type are known in the Wadhurst Clay of the Weald where they provided the ore for the former Wealden iron industry. These, like the Coal Measures rocks, are

PLATE 33. Dissected vertical limestone bed, near Sisteron, southern France. The effect is similar to that of a dyke

freshwater deposits, but similar marine deposits are known at other levels of the geological column.

A rock of probably similar origin is the blackband ironstone found in the Coal Measures of the north Staffordshire coalfield. These rocks are largely free from clay impurities, but contain considerable amounts (up to 20 per cent) of coaly matter, presumably having been deposited in or near coal swamps. They were very useful ores in an earlier stage of technology, when large quantities of fuel were needed to smelt iron ore, so that their comparatively low iron content was more than compensated by their fuel content.

Because of their thinness and in many cases their nodular structure, these beds either are or have been of more economic than geomorphologic significance.

Far more important than these rocks are the Jurassic ironstones, many of which are oolitic rocks. Oolitic ironstones of this and also of different ages occur in many parts of the world. They are marine rocks and the conditions which led to the rich concentrations of iron minerals necessary for their formation are obscure. At the present time the only iron mineral which seems to be forming on the sea floor is glauconite, a greenish iron silicate occurring in the form of small pellets. It forms under suitable conditions at depths of 20 to 800 metres (10 to 400 fathoms), and also formed in the past. It is the mineral which causes greensand to be green, at least in unweathered specimens, when the glauconite concentration is high, and speckled when smaller amounts of glauconite are mixed with large amounts of quartz sand.

The Jurassic ironstones consist predominantly of two minerals: an iron silicate, known as chamosite, which usually forms the ooliths, but also occurs in the non-oolitic mudstones which are intercalated with the oolitic ironstones, and siderite which usually occurs in the iron-rich mudstones and in the matrix of the oolitic beds as well. The siderite also occurs replacing the chamosite in some oolites, where a certain degree of alteration seems to have taken place soon after deposition. On the whole, however, these ironstones are thought to have been primary deposits.

These ironstones occur at different levels in different parts of Britain. In the Cleveland Hills and in the Edge Hill area north of Banbury they are Middle Lias in age; at Frodingham in north Lincolnshire they occur in the Lower Lias; in Northamptonshire, around Corby, they are found in the Northampton Sands, which form the lower part of the Inferior Oolite of Middle Jurassic age. Compared with some of the igneous and metamorphic masses of haematite and magnetite, they are poor iron ores, the richest ores of the Northamptonshire field reaching approximately 35 per cent iron content. A number of them contain a proportion of calcium

PLATE 34. Col d'Izoard, southern French Alps. View south-east from the summit. Triassic limestone and *cargneules*. Compare with Plates 35 and 36. For explanation see text

carbonate and it is the leaching of this that leads to the phenomenon of the enrichment of the surface outcrop of some of these beds, where the iron content is distinctly higher than it is underground. In an unweathered state the rocks are often blue-green but at the surface they are weathered to the usual range of yellows, oranges and browns.

It is doubtful whether these ironstones have themselves any great relief-forming properties. The rocks are really equivalent to soft sandstones and hard mudstones. They may form parts of scarp-producing series such as the calcareous Marlstone of the Middle Lias, which forms a scarp in much of the east Midlands and again in front of the Cotswolds around Tewkesbury. But it is the Marlstone rather than the ferruginous rocks within it that is responsible primarily for the escarpment.

Strictly speaking some of the residual deposits should also be classed here, because of their high iron content. They include the laterites, the iron pans of podsol soils and some of the terra rossa.

CARBONACEOUS ROCKS

Included here are really two main groups of rocks. On the one hand is the series, peat–lignite–coal–anthracite, consisting of organic matter in various stages of lithification, with cannel and boghead coals forming a subsidiary group with an appreciable oil content. On the other hand is the group, gas–oil–asphalt, which, having been mentioned for the sake of completeness, will not be considered further because its importance is entirely economic and not geomorphological.

Plant matter is remarkably resistant to a great variety of powerful chemicals. This is well shown in the preparation of slides for pollen analysis, in which reagents as powerful as hydrofluoric acid may be used without damaging pollen grains. The chief destroyer of plant remains is oxidation, a fact well known in fenlands where water tables kept too low result in the destruction of the peat. Thus, plant debris will only accumulate under anaerobic conditions.

Modern deposits of this sort are the peats and detritus muds, the latter being the term used to describe accumulations of drifted plant debris by Quaternary botanists. Peats really fall into two classes, eutrophic and oligotrophic.

The oligotrophic peats are the acid peats of the western highlands of Britain. High rainfalls on rocks poor in soluble salts result in calcifuge vegetation, with sphagnum moss, heather, bilberry and cotton grass being important. The ensuing peat is acid. Such peats can also be formed in lowland fens if the swamp surface becomes raised out of reach of the ground water.

Eutrophic peats are fen peats. The ground water contains ample calcium

PLATE 35. Col d'Izoard, southern French Alps. Detail of area south-east of summit. See Plates 34 and 36 and text

253

carbonate in solution, the plants are calcicolous and the ensuing peat is mild.

Conditions, probably usually of the latter type, have recurred from time to time in the geological past, and have resulted in accumulations of organic matter, which have been slowly converted to lignites and coals. The conversion to coal is accomplished largely by pressure and results in a series of changes. This process of coalification involves a progressive loss of moisture, a progressive reduction in oxygen and volatile contents and a progressive increase in the proportion of carbon. Generally speaking, the older the coal the higher its rank, although the process may be hastened by earth movements. For example, in the South Wales field the high rank coals, the anthracites, are found mainly to the north-west where the Hercynian folding was most severe. Again in the Cretaceous and Tertiary fields adjacent to the Rocky Mountains, the coals are lignites well away from the mountains and bituminous coals where they are involved in the folding.

The following grades may be recognised:

(*a*) Lignites and brown coals are the youngest of the coals, usually of Tertiary or Cretaceous age. They are brown, or brownish-black, and much of the original plant material can still be recognised in them. They have a high water content, 30 per cent or so, and they crumble on exposure to the air and are also liable to spontaneous combustion in this state. They occur in thick beds at the northern foot of the Hercynian blocks of Germany and in the area between the Rocky Mountains and the Mississippi lowlands. In Britain they are rare: the lignites of the Bovey Tracey basin in Devonshire are well-known, but there are also thin and unworkable seams in the Tertiary beds of the Hampshire Basin.

(*b*) Sub-bituminous coals are also usually Mesozoic or Tertiary in age. They are black, but, like lignites, crumble easily because of a high moisture content of 20 per cent or so. They are better fuels than lignite being bright, smoky and gassy. They are rare in Britain, but were worked at Brora in Sutherland, where they are of Jurassic age. Vast reserves of coal of this type occur in western North America and some of the Australian fields are also of this type.

(*c*) Bituminous coal is household coal and coking coal, a bright streaky black material with a small scale rectangular jointing. It seems to have been formed of alternating layers of different types of material. Fusain is the dusty, friable material which gives coal its dirtiness and is probably derived from woody material dried under oxidising conditions: vitrain forms the bright layers and probably represents the alteration of wood and colloidal humus under anaerobic conditions; durain is a dull material with an irregular fracture and is probably derived from the more resistant parts of plants.

PLATE 36. Col d'Izoard, southern French Alps. View north-east from summit. Fully-developed scree slopes. For explanation see text and compare with Plates 34 and 35

(*d*) Semi-bituminous coal or steam coal formed the original basis for the prosperity of South Wales in the days, now long gone, of the coal-fired steamship. It is almost smokeless and, with its high carbon content, it has exceptionally high heating value.

(*e*) Anthracite has the highest proportion of fixed carbon and the lowest proportion of volatiles. It burns at high temperatures with a smokeless flame and is clean to handle because of the lack of fusain. It is used primarily in enclosed stoves for space or water heating.

As mentioned at the beginning of this section two types of coal stand somewhat apart from these. The first is cannel, which has a high volatile and ash content and is readily lit, burning with a bright smoky flame. It contains much spore-rich material. Boghead or torbanite is a tough, dull-coloured, oily coal and is intermediate between coal and oil shale. It was probably formed from accumulations of oil-bearing Algae. Such creatures are known to exist in shallow lakes in some semi-arid parts of Asia (Lake Balkash), Africa and Australia. In the dry seasons a scum of these Algae may become dried to a rubbery, oily substance which burns readily. Similar material is contained, mixed with argillaceous matter, in oil shales, which upon distillation will yield sometimes commercially useful quantities of oil. Oil shales occur in the Carboniferous of the Scottish Lowlands and have been used in the past. The term is a loose one, however, because there is a tendency to describe any oil-bearing rock as an oil shale whatever its lithology, which may be limestone, for example the Green River Oil Shales of the western United States many of which are dolomitic limestones.

Coal seams form members of a rapidly alternating series of rocks typical of the Coal Measures. Many of these are shales and thin sandstones. Not only are the rock types not themselves particularly resistant, but the rapid alternations also are not conducive to great resistance to denudation. Indeed, the Coal Measures form the beginnings of the lowlands on the flanks of the Pennines, while in the South Wales coalfield the northern outcrop of the Lower Coal Measures forms a marked vale between the Pennant Series and the Millstone Grit. In itself coal is too jointed and too brittle to be able to withstand much erosion.

PHOSPHATIC ROCKS

The primary source of the phosphatic compounds occurring at the earth's surface is probably the mineral, apatite, which is an accessory mineral in igneous and metamorphic rocks and occurs in workable concentrations in some pegmatites. The mineral itself is a mixture of calcium phosphate with either calcium fluoride or chloride. The phosphate which gets into solution is largely absorbed by living animals, in which it forms a main element in bone structure and by which it is also deposited as excrement. Thus, it is possible to have two types of organic phosphate deposit. One consists of the accumulation of excrement and can only survive in arid climates because of its solubility, for example the deposits of guano, which are a valuable source of fertiliser on the desert islands off Peru. They are formed by the accumulation of excrement produced by vast numbers of birds feeding off the rich fish life of the cold Humboldt Current. The other is the accumulation of bones as a bone bed whenever a catastrophic change of conditions causes a widespread and simultaneous destruction of animal life. Although such deposits are economically valuable they are geomorphologically negligible. In addition to these organic deposits it is possible to have precipitation of phosphate pellets and nodules directly from sea water. This seems to be happening at present at depths of 30–300 m (100–1000 ft) off the Californian coast under oxidising conditions.

Weathering of deposits such as these or of igneous rocks rich in phosphates may lead to the metasomatism of other rocks, such as limestones, to produce deposits of rock phosphate, again a very valuable material economically.

Beds of phosphatic nodules are known in Britain. An example is the Cambridge Greensand, which is a thin, sandy, glauconite bed, 30–50 cm (12–18 inches) thick lying between the Gault and the Chalk and containing both dark and light coloured phosphatic nodules. This was profitable to work in the nineteenth century, when several yards of overburden were removed by hand to get at the thin layer of coprolites as the nodules are termed. Its working was finished by the use of cheap, overseas, phosphate rock deposits, but not before it had revealed many interesting details of the Pleistocene beds of the county, which overlie many of the old areas worked.

EVAPORITES

Certain geological formations contain great thicknesses of salts, usually gypsum, anhydrite and rock salt, though in rarer cases including the more soluble potassium salts. In Britain the Permo-Trias basins of Cheshire and north-east England fall into this class and provide the basic raw materials for Britain's heavy chemical industry. The aggregate thicknesses of evaporites in these basins are respectively approximately 400 and 500 m (1300 and 1600 ft). Such enormous thicknesses must mean more than the mere desiccation of a relict arm of the sea: they must represent conditions where an isolated gulf is undergoing continuous evaporation and being replenished from time to time by fresh inflows of saline water. It is thought that the almost isolated gulf of Kara Boghaz on the eastern side of the Caspian may give a reasonable model picture of the operative situation. Theoretically, the evaporation of such a basin should lead to a predictable succession of salts, with the least soluble being precipitated first and the most soluble last. Although this picture is generally true in that the potash salts occur near the top of the evaporite sequences of the Permian basin in north-eastern England and the Stassfurt basin in Germany, the repeated inflows of fresh solutions cause complications in the succession.

Other accumulations of evaporites are possible, for example of sodium carbonate in the arid western parts of the United States and in the Nile

of sodium nitrate in the Atacama Desert of northern Chile.
as given rise to a lot of speculation. Vast amounts of organic
ieeded to explain the nitrate and an absolutely arid climate to
Thus the climate which would seem necessary for the preservation.   deposit would seem to be completely unfavourable to the production of the organic, presumably plant, matter for its formation.

Such evaporites have little geomorphological significance except when
they are affected by solution and give rise to a form of 'karst' topography.

6
Dating and correlating rocks

In the preceding chapters rocks as lithological units have been the objects of consideration. But, in addition to their lithological effects, the ways in which rocks were laid down, arranged and deformed by earth movements are all important in a comprehensive understanding of the essentials of relief. That part of geology which deals with the correlations of rocks is known as stratigraphy and anyone interested in regional physical landscape needs an appreciation of stratigraphy. Therefore, in the succeeding chapters, an attempt is made to provide an outline of British stratigraphy primarily in terms of lithology and relief.

At the outset two methods of dating must be considered. The first is radiometric dating based on the spontaneous decay of certain radioactive elements into others. This method gives a date in years with a definable error, which depends on the method and the age of the material in question. Most of the radioactive elements used are more common in igneous rocks than in sedimentary rocks, so that the latter have to be dated by reference to their relationships with dated igneous rocks. Sometimes this form of dating is said to give an absolute date or age to a rock, whereas correlation gives only a relative date.

The second method of dating is by correlating rocks on the basis of their fossils, or less reliably their lithology. This method forms the traditional basis of stratigraphy.

Radiometric dating

Early attempts at dating, before the existence of radioactivity had been discovered, were based on estimates of the rate at which salts were added to the oceans, assuming that none was present originally, and on estimates of the average rates of sedimentation, which were then compared with the maximum thicknesses of rocks known for each geological system to give an idea of the length of time the latter represented. The results varied wildly and are only of historical interest now that several radiometric methods are available.

The accuracy of any radiometric method depends on a number of assumptions:

1. That the rate of decay has not changed. The rate of decay is not arithmetical, i.e. a constant number of atoms does not decay in unit time. In fact the number of atoms that decays in unit time is proportional to the number of the original atoms present. Thus the decay curve is logarithmic and the point is never reached when all the original atoms have changed, though the number remaining may be in practice negligible.
2. That none of the products of decay has escaped.
3. That none of the original material has been added since decay started.
4. That no minerals indistinguishable from the decay products were originally present in the rock.

The first assumption would seem to be justified in that laboratory attempts to affect the rate of decay through temperature and pressure changes have failed.

The second assumption is certainly not true in all cases. Major tectonic events such as mountain building and metamorphism allow the escape of the products of decay, so that a radiometric date for a metamorphic rock usually gives the date of metamorphism, not of the original formation of the rock. The condition of the rock, whether glassy, finely or coarsely crystalline, can also affect the preservation of decay products and hence the apparent age. Again, in the case of helium the amount produced will affect the pressure of helium in the rock: obviously the higher the pressure the greater the possibility of escape, so that helium methods are not reliable for highly radioactive rocks. The effect of gas escape can also be illustrated from the decay series of U^{238} which decays through a series of intermediate products to Pb^{206}. One of the intermediate products is a gas, radon (Rn^{222}) which has a half-life long enough for it to escape from porous rocks. Any escape will affect the final amount of Pb^{206} and hence the date.

The third assumption may not be true if mineralisation of the rock has occurred after the date of its formation.

The fourth assumption is not necessarily true as can be illustrated from the lead methods. There are four lead isotopes, Pb^{204}, Pb^{206}, Pb^{207} and Pb^{208}. Pb^{204} is non-radiogenic lead, but original lead probably consisted of a mixture of the four and an allowance has to be made for the proportion of the other three lead isotopes present in the original lead. If it were all assumed to be radiogenic the apparent age would be too great. The correction can be made on a number of assumptions, most of them of a technical nature.

Because of the very small amounts involved and the difficulty of identifying the various isotopes, certain technical difficulties have to be faced in radiometric dating.

The first of these is concerned with the relationship between the half-life of the material concerned and the age of the rock. The half-life is simply the time taken for one half of the atoms of the original substance to decay to the final product. The length of the half-life is usually not known with absolute accuracy, and uncertainty here leads to uncertainty about the date of the rock. The half-life should also be roughly comparable in length with the date of the rock or mineral concerned. This is because any error in estimating the very small amounts of material involved will be minimised when the ratio of parent material to final product is approximately 1:1. At both extremes of the scale similar errors in determining amounts will have much greater effects on the deduced ages. This is really why the same radiometric method cannot be used for dating the Pre-Cambrian and the Pleistocene.

The second type of difficulty is purely technical. Prior to the Second World War the lead method (see below) was in use and the analysis involved weighing in a chemical balance which did not separate the various lead isotopes. These can be separated by an expensive instrument, the mass spectrometer, which measures the deflection of ions in a powerful magnetic field. The heavier the ion, i.e. the larger its mass number, the less it is deflected. Thus it is possible to measure the proportions of the four lead isotopes. Without this the old chemical method could only give an average age, which was usually too great because of the contaminating effects of original lead. Further than this, one is dealing with very small quantities, so that rigorous precautions are necessary to ensure against contamination at any stage of the proceedings.

Because of these difficulties radiometric ages are always subject to varying possible errors, but these are usually small in comparison with the age of the rock. One point that has recently received attention is the fact that the larger the igneous body the slower its rate of cooling. Above a certain temperature the products of decay may escape: below that temperature the system is closed. Rapidly cooled rocks, such as small intrusions or extrusions may give an accurate age, but larger intrusions of the same age may appear to be appreciably younger because of the slow rate of cooling. This can complicate the dating of sediments, which is done by reference to their structural relations with dated igneous rocks. It is theoretically possible for a set of sediments to be dated differently in different areas, if they are dated with respect to different sized intrusions.

A number of different radiometric methods are in use:

1. THE URANIUM–THORIUM–LEAD METHODS   There are two radioactive isotopes of uranium, U^{235} and U^{238}, and one of thorium, Th^{232}. These decay respectively through a series of intermediate products to Pb^{207}, Pb^{206} and Pb^{208}. Before the days of the mass spectrometer these three decay series were in practice averaged to give the date of the rock. Nowadays, if all three elements are present in a rock, separate determinations of the U^{235}/Pb^{207}, U^{238}/Pb^{206} and Th^{232}/Pb^{208} ratios can be made and so provide cross checks on the age of the rock. Further checks may be provided by the

ratios of the different lead isotopes, e.g. Pb^{207}/Pb^{206}, because the parent materials have different decay constants with the result that the lead ratios reflect the age. The fact that there are usually discrepancies reflects the loss of initial, intermediate or final products. The half-lives of Th^{232}, U^{235} and U^{238} are respectively 13,900 (million years), 713 m.y. and 4510 m.y. Thus, these methods are most valuable in dating comparatively old rocks.

2. THE URANIUM–THORIUM–HELIUM METHODS   Helium is the other end product of the decay of thorium and uranium, so that it is theoretically possible to determine the age of rocks by the U/He and Th/He ratios. Helium dates were used in the early days, when dating was only possible on rocks of high uranium content. Such rocks had high helium pressures and it was almost certain that false dates were obtained owing to the escape of helium. The method fell into disrepute but it may well be revived by the use of techniques capable of dealing with rocks of low uranium content, in which less helium pressure would occur and in which, accordingly, the risk of helium escape is much less.

3. THE POTASSIUM–ARGON METHOD K^{40} decays to Ar^{40} and Ca^{40}. The K/Ca ratio cannot be used for radiogenic dating because the Ca^{40} of radiogenic origin cannot be distinguished from the Ca^{40} of other origin which forms the bulk of the calcium in rocks. Even with Ar^{40} a correction has to be made for atmospheric Ar^{40} similar to the correction for common lead in the uranium–thorium–lead methods. The most suitable minerals for analysis are micas, which have a high K content and under the right conditions a high argon retentivity. Amphiboles and pyroxenes have low K contents but tend to retain argon well even under conditions of slight metamorphism. Felspars on the whole are less satisfactory as their argon retentivity is less good and they give apparently young dates.

This method is useful as it is possible to work on minerals in beds of volcanic ash intercalated in ordinary sediments. It can also be applied to the mineral, glauconite, which is a primary mineral in sedimentary rocks produced in a marine environment, as long as the glauconite is fresh. If it has been subject to temperature or pressure increases, its argon retentivity deteriorates and hence dates become less accurate.

The half-life of K^{40} is 1300 m.y. and dates of less than 1 m.y. can be obtained, thus bringing the early part of the Pleistocene within the scope of the method, though the reliability of dates at this level is not fully known.

4. THE RUBIDIUM–STRONTIUM METHOD Rb^{87} decays to Sr^{87} with a half-life of about 47,000 m.y. Thus it is a method for older rocks and cannot approach the possibilities of the K/Ar method for the younger rocks. Sr^{87} seems less liable to escape than Ar^{40} and hence the method may usually be more accurate for the older rocks than the K/Ar method. There is, however, still disagreement about the half-life of Rb^{87}, which leads to uncertainty concerning the accuracy of the dates obtained. The minerals

used are similar to those used in the K/Ar method, the micas being particularly suitable.

5. THE RADIOCARBON METHOD   This method uses the fact that C^{14} decays to N^{14} with a half-life of 5570 years. C^{14} is produced by the bombardment of nitrogen by cosmic rays and the amount in the atmosphere, and hence in plants and animals at any one time, is pretty well constant. When a plant or animal dies the C^{14} commences to decay and it is possible to date Late Pleistocene organic sediments by this method. Of course, the method is complicated by any inert carbon introduced into the fossils by groundwater. Most workers regard 20,000 years as the maximum for dating, but some workers with refined techniques have obtained dates as old as 70,000 B.P. (before present). Thus, this method dates virtually Postglacial time and with decreasing reliability the episodes of the last (Weichsel) ice age.

A number of general points emerge from this description of the elements of radiometric dating.

(*a*) Most datable rocks are igneous rocks. Sediments containing uranium minerals may be datable by the U/Pb method, those containing glauconite by the K/Ar method, and Postglacial organic sediments by the C^{14} method.
(*b*) Sedimentary rocks are usually dated by considering their relations with datable igneous rocks. This is similar in principle to the method used to date folds and faults and may give a wide or very narrow range of possible ages in different examples.
(*c*) The yawning gap in absolute dating is between the youngest K/Ar dates in the early Pleistocene and the C^{14} dates in the latest part of the Pleistocene. Thus one of the most controversial geological periods is not yet susceptible to reliable radiometric dating.

The dates assigned to the beginnings of the various geological systems by the Geological Society of London in 1964 are as follows:

	m.y. B.P.
Pleistocene	1·5–2
Pliocene	7
Miocene	26
Oligocene	37–38
Eocene	53–54
Palaeocene	65
Cretaceous	136
Jurassic	190–195
Trias	225
Permian	280
Carboniferous	345
Devonian	395
Silurian	430–440
Ordovician	500
Cambrian	570

This is, of course, a short spell in comparison with the age of the earth. The oldest rocks of the earth have ages of the order of 3200 m.y., while meteorites with ages of 4500 m.y. are known. These may be the debris from an exploded planet from which the earth derived.

The correlation of rocks

The essential stratigraphical problem can be readily envisaged in terms of the sediments at present being deposited on the earth's surface. Let us imagine it.

In the seas and oceans present sedimentation ranges from the pebble beaches of Dungeness to the abyssal oozes of the deepest parts of the oceans. In between are various sands, coral deposits, black muds of such places as the Black Sea and so on. What link is there between all these deposits to witness that they are of the same age?

On land the deposits range from aeolian to glacial, from temperate to tropical and arctic: in the tropics laterites; in the polar regions tills and moraines; in the Alps waste fans; in Britain river alluvium and gravels; in the deserts sand dunes and evaporites. Ideally a link has to be found between all these, and, more important, a link between all these on the one hand and all the marine sediments on the other.

The task is, of course, impossible. There are two main criteria for correlation, the lithology of the sediment and its fossil constituents. Let us look at these in turn.

Correlations based on lithology are suspect for two main reasons.

The first of these is that the same lithology may be produced at different dates. Thus, one must question whether in 100 m.y. time it will be possible to distinguish between a fossil laterite of our present age and one formed 50 m.y. earlier in the Eocene. Can one easily differentiate between a boulder clay or till of the last (Weichsel) glaciation and one of a much earlier glaciation?—or, indeed, can one distinguish on lithological grounds between a Permo-Carboniferous tillite and a Pre-Cambrian one? The examples could easily be multiplied.

In addition one has the problem of transgressive deposits. The best example is probably the marine conglomerate which represents the beach of an encroaching sea. The beach varies in age from place to place, so that the upper and lower surfaces of the deposit would cut horizontal time planes at a slight angle giving a typical diachronous deposit, c.f. the Tertiary deposits of the London and Hampshire Basins described in Chapter 13.

The reverse of the picture is that, at any one time, the lithology varies widely from place to place and so do the fossils with the different environments of deposition. This is virtually the picture revealed by present day deposition throughout the world. The products, sedimentary and biological, of different environments, are known as facies. This is a useful term which can be applied on a wide variety of scales. One can use it on the broadest scale to talk of marine and continental facies, for example in discussing the Devonian: it can be applied to lithological variations, for example sandstone facies or pebbly facies: it can be used for predominantly biological variations, as in coral-reef facies or, more vaguely, shelly facies.

One turns away from lithology to fossils for more precise correlations. Even here enormous practical difficulties arise, as can again be appreciated from a general consideration of the current world picture. What links can one hope to find between a dead elephant in an African swamp and a dead whale stranded in a Polar fjord? This may seem an extreme example, but even in marine deposits it would be virtually impossible to correlate a Pacific coral reef with the sands around the Wash on the basis of their fossil contents.

Certain classes of organisms are useless as fossils because they are of such a soft nature that they are not readily preserved, unless under exceptional conditions. Creatures with neither internal nor external skeletons, e.g. jellyfish and many plants, fall into this category. Other creatures are so rare that the chances of finding them fossil are practically nil, e.g. at the present time the giant panda. Others are so large that the geologist investigating a small temporary exposure millions of years in the future is unlikely to meet them, e.g. whales and elephants.

Obviously as an index or zone fossil we need to have a small creature with readily preservable hard parts, with a wide distribution and occurring in profusion. The field geologist might add that it should be readily recognisable, but the finer divisions of geological strata are often based on the subtle distinctions between closely-related species. These distinctions are real but only easily discerned by the specialised palaeontologist who is constantly handling the species concerned. Easy identification may not, therefore, be practicable.

The greater part of the earth's surface is marine and so the ideal fossil must be marine. Because of the necessity for wide distribution it must either be pelagic or able to float and so be widely distributed after death. Sessile creatures, for example corals or, in our latitudes, mussels, which occur in great profusion in places, are not of much use.

So far only the spatial or horizontal aspects of the problem have been considered, but the time or vertical aspect is just as important. Rocks need to be divided as finely as possible and this involves rapidly-evolving organisms, which can be distinguished from each other. Some creatures, for example the brachiopod, *Lingulella*, remain unchanged from the Cambrian to the present day and so are useless as zone fossils.

It has sometimes been argued that if evolution is rapid and dispersal of the species slow even fossil zones could well be diachronous. In a strict sense they must be diachronous because every organism needs time to occupy its full habitat. However, even the finest fossil zones are rarely less than $\frac{1}{2}$ m.y. long and the rate of dispersion of even sedentary creatures is much faster than this. Even the humble land snails, slow moving creatures if ever there were, spread with surprising rapidity after the last glaciation in Europe. In fact they probably occupied their full environmental range in about 10,000 years, i.e. one-fiftieth of the length of the shortest fossil zone. Thus, the diachronous element of fossil zones may for practical purposes be ignored.

Very rarely are the ideal characteristics of zone fossils realised. In fact, in the geological column in Britain only the Ordovician and Silurian and the Jurassic have ideal zone fossils. In the first two systems graptolites are used: these were freely-floating pelagic organisms. The Jurassic is zoned on ammonites, which are usually plane-coiled shells consisting of a number of sealed chambers with the animal occupying the outermost. The shells, as a result, floated after death.

In other systems assemblages of shells are used, while index fossils often come from far from perfect groups. In the Carboniferous Limestone, for example, corals are used: such sessile animals, confined to one particular environment, are theoretically bad. Inferior zone fossils and groups usually lead to coarser and less universal zonal divisions.

In any system, whether the zone fossils are ideal or not, the correlation of other facies is always difficult. It depends largely on finding the overlap between two facies. The modern equivalent would be finding the skeleton of a North American bison in the Mississippi delta intercalated in beds which contained typical Gulf of Mexico Foraminifera. Chance links of this sort are indispensable in building up a complete stratigraphy, but they are not always available. Then one may have to revert to less accurate procedures such as the arbitrary integration of two sets of beds known to be of the same general age because they occur between horizons which can be correlated. In an extreme case, this could lead to the correlation of a series of tropical laterites with a series of periglacial varved clays by proportional division.

On such accurate and inaccurate bases as these rests the generally detective art of stratigraphy. The skill and experience of field geologists probably mean that few gross errors are made, but the basis of correlations must always be borne in mind especially by the non-geologist. In many cases there is room for difference of opinion. Thus work remains to be done—and undone.

7

The Lower Palaeozoic systems

The Lower Palaeozoic rocks, belonging to the Cambrian, Ordovician and Silurian systems, are not the oldest sedimentary rocks: they were preceded by the Pre-Cambrian Torridonian sandstones in Scotland and comparable rocks in several other countries. They are not even the oldest sedimentary rocks with fossil evidence of life in them, for fossils have been found in Pre-Cambrian rocks. But they are the oldest sediments which can be reasonably well correlated over Britain and they provide the only good example in British stratigraphy of geosynclinal sedimentation, its associated facies, and the varying relief effects developed on such rocks.

Although it is not easy to define a geosyncline precisely, it is broadly an area of large extent in which, owing to continuous subsidence, a considerable thickness of sediments accumulates. It may be a simple trough or it may possess several subsidiary troughs. Let us examine a simple model example (Fig. 7.1).

When it is fully developed one might expect variable sedimentation including limestones on the continental shelves at the margins of the geosyncline. On the continental slopes there would be zones of coarser, greywacke type of sedimentation, which could well include slumped beds

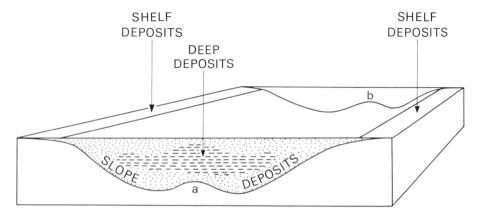

Fig. 7.1. An idealised geosyncline

(turbidites). In the deepest water fine-grained muds might be deposited. Whether these would occur in the deepest bottoms, which might be filled on the contrary with slumped sediments, or on slight rises in the deep areas (ab on Fig. 7.1) is uncertain.

If one then combines a time element with the idealised localisation of sedimentation one may imagine that, in the early stages of the geosyncline before much deepening had happened, the whole of the sediments would be fairly coarse. Later on, with maximum geosynclinal development, sedimentation such as that described in the previous paragraph might prevail. Towards the end of the geosyncline's existence, when it had ceased to deepen, there would be a phase of coarser sedimentation representing the final filling of the trough. Ultimately, of course, in the ideal pattern the sediments would be converted to mountains by the moving inwards of the stable blocks on either side.

Let us complicate the picture one stage further. It is well known that folding occurs at intervals throughout the life of a geosyncline. This is shown by unconformities in the sediments. It happened not only in this ancient Caledonian geosyncline, but also in the much later Alpine geosyncline, when some of the nappes started to form early in the Mesozoic era although the climax of the folding was not reached until the middle of the Tertiary area. The instability affects all parts of the geosyncline. At the margins it results in the continental shelf becoming land and vice versa— hence unconformities and breaks in sedimentation. It may also cause a change in the character of sedimentation through alterations in relief and the products of denudation. In the deeper parts rising sections of the sea floor may control sedimentation. Finally, in any unstable part of the earth's crust, which is after all what a geosyncline is, vulcanicity may break out and produce masses of extrusive rocks which ultimately will form a great variety of relief.

A further complication must be made to approach reality. The geosyncline may well be composed of a number of subsidiary troughs, perhaps undergoing sedimentation for different periods and being converted into mountain ranges at different times.

One final point needs to be stressed. When orogenesis succeeds geosynclinal sedimentation the main mountain ranges will be produced where the sediments were thickest, and where they are thickest they are usually most homogeneous, i.e. in the centres of the former troughs. Such areas with complex structures but lithological uniformity, whether the rocks are metamorphosed or not, will lead to relief monotony, apart from the possible presence of igneous rocks, either volcanics extruded in unstable phases of the geosyncline's history or intrusives emplaced during orogenesis. On the contrary, the shelves of the geosyncline will be comparatively rigid areas where the varied sediments will not have been subjected to such intense structural processes. Here the full effects of varied lithology may be brought into relief.

With this necessarily oversimple picture in mind the general trends of the Lower Palaeozoic rocks and their effects on relief may be more readily appreciated. Figure 7.2 attempts a composite picture of Lower Palaeozoic geography, but at no single phase did such a situation exist.

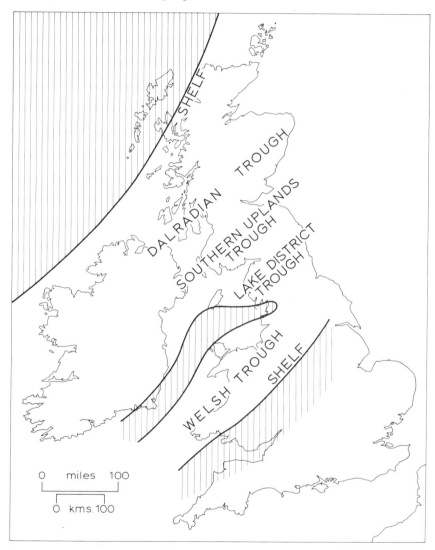

Fig. 7.2. Composite diagram of Lower Palaeozoic geosyncline

In the far north-west of Scotland shelf deposits are represented by quartzite-carbonate deposits of Cambrian and Lower Ordovician age. These have greater fossil affinities with the northern side of the Lower Palaeozoic geosyncline in North America than with the southern margin of the geosyncline in Britain. These facts have been used as evidence for the drifting apart of Europe and North America. To the south the Dalradian trough was probably developed in the later Pre-Cambrian and Lower Cambrian, and was soon thereafter converted to mountains, the rocks undergoing a comparatively low grade of metamorphism. The Southern Uplands trough was a much later development, the sedimentation of which was predominantly Ordovician and Silurian. The Lake District trough was probably very similar in age, although it may have started a little earlier. Finally, the Welsh trough started perhaps early in the Cambrian and did not become finally extinct until late in the Silurian: the uncertainty about its date of origin derives from the unfossiliferous nature of the lowest sediments found in it. In fact, practically the whole of the Harlech Series (see below) is devoid of fossils. In the south-east the shelf deposits of the southern margin of the geosyncline are best represented in Silurian times—rocks of this age exert the most profound effects on the relief of the area. The fluctuating nature of the Lower Palaeozoic geosyncline should be apparent from the above summary of conditions, which is based essentially on Rayner (1967).

Cambrian rocks

I propose to omit the Dalradian rocks, which are not entirely of Palaeozoic age, and to concentrate on the rest. Reference to the model of a geosyncline which was described earlier should prepare one for an early phase of shallow water sedimentation gradually becoming deeper water in character. By and large this is true.

The main present day outcrops of Cambrian rocks in Wales are in the Harlech Dome and north of Snowdonia, where they reappear from beneath the Snowdon syncline, around the anticlinal Padarn ridge (Fig. 7.3). The general succession recognised in North Wales is as follows, the older terminology being given on the right and the more modern terminology in the other two columns.

?Cambrian	Tremadoc Slates	
Upper Cambrian	Dolgelly Beds Ffestiniog Beds Maentwrog Beds	Lingula Flags
Middle Cambrian	Clogau Shales Gamlan Shales Barmouth Grits Manganese Shales	Menevian
Lower Cambrian	Rhinog Grits Llanbedr Slates Dolwen Grits	Harlech Grits

The most interesting of these beds geomorphologically are the Harlech Grits, which comprise alternating shales and grits warped up into an anticlinal structure and hence, in spite of the detailed complications of the outcrops, giving rise to inward-facing escarpments. In fact the outermost

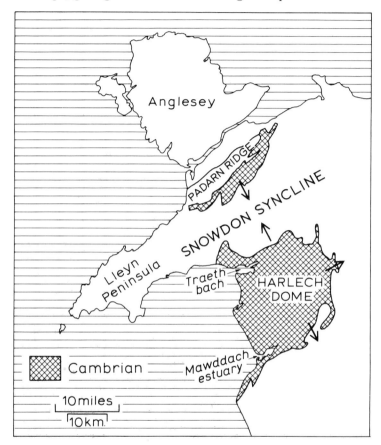

Fig. 7.3. Cambrian outcrops in North Wales

escarpment around the Harlech Dome is not formed of Cambrian rocks, but of the succeeding Ordovician which is heavily reinforced with igneous rocks. As will be seen in the next section, this consists from south round to north of the scarps of Cader Idris, the Arans, the Arenigs and the Moelwyns.

There seem to have been two sources of sediment for the Harlech Grits. On the one hand the predominance of coarser beds in the eastern part of the structure suggests land in that direction, but there are also correspondences between some of the constituent elements of the beds and the rocks of the Pre-Cambrian Mona Complex in Anglesey. A derivation from Anglesey does not accord well with two facts. One is the increasing importance of shales on the western side of the Harlech Dome: the other is the

271

generally fine grain of the sediments of the Cambrian slate belt north-west of Snowdonia. Both these facts tend to suggest deepening water to the west. But, however obscure the palaeogeography, the significant fact in relief is the grit–shale alternation especially well-developed on the west side of the dome.

Among the grits the Rhinog Grits are by far the most important. They reach a thickness of about 750 m (2500 ft) in the centre of the dome and give rise to some of the most spectacular relief in Wales outside the areas of volcanic rocks. The Dolwen Grits are only exposed in the centre of the dome south of Trawsfynydd and do not form marked relief. The Barmouth Grits are lithologically similar to the Rhinog Grits although they are much thinner. They form relief of a similar rocky type (Plate 37), for example Y Garn. Relief on Rhinog Grits is well illustrated in Rhinog Fach and Rhinog Fawr. Their absolute height is not great, some 700 m (2300 ft), but their bareness, which exposes all the detailed effects of their westward-dipping beds, sliced by many faults (Plate 38), forms a barren landscape of a type more often seen in Scotland than in Wales. In some ways, their shallow angles of dip and the serrated skyline seen from the Trawsfynydd-Dolgelly road recall the Torridonian hills north of Ullapool in north-western Scotland.

In detail, the Rhinog Grits and the very similar Barmouth Grits have a well-developed system of joints, which exerts an important control on the detail of relief in the confused area north of the Rhinogs themselves. Foel Penolau on the Barmouth Grits, provides a very good example of such joint control (Plates 9, 10 and 11).

All in all the Harlech Series add diversity to the general plateau landscape of Wales, which is often classed as monotonous.

The beds above, the Menevian, the Lingula Flags and the Tremadoc (the last are often classed as Ordovician as they contain Ordovician fossils, although they are conformable with the Cambrian but not with the overlying Ordovician) are finer-grained, deeper-water sediments. Their effects on relief are less pronounced and the outcrops are followed by the estuaries of Traeth Bach to the north and the Mawddach to the south of the Harlech Dome.

When the Cambrian beds reappear from beneath the Snowdon syncline they are on the whole much finer-grained rocks and have been converted during the Caledonian folding to some of the finest slates in Wales, quarried formerly at Bethesda, Llanberis and Nantlle. Here there is less effect on relief than in the Harlech Dome. Whether this is due to the inferior resistance of the slates or to the juxtaposition of these slates against the Pre-Cambrian volcanics of the Padarn Ridge and the Ordovician volcanics of Snowdonia is difficult to say.

PLATE 37. Barmouth Grits, Craig-y-Penmaen, Harlech Dome. Joint-controlled west-facing scarp overlooking Trawsfynydd-Dolgelly road

Although some outcrops of Cambrian rocks are associated with the exposure of Pre-Cambrian rocks in Pembrokeshire and the Pre-Cambrian inliers of the Midlands, they are primarily of geological interest. From the point of view of the geomorphologist the other interesting area of Cambrian rocks—and with them one would include the lower parts of the Ordovician which represent a continuation of the same continental shelf facies—is in the far north-west of Scotland. In this area the lower beds, which are mainly Lower Cambrian in age, form the quartzites and pipe rock of the older literature. Upwards the calcium carbonate content increases and these rocks are known as the Durness Limestone group, most of which is probably Middle Ordovician in age. There might well be an important unconformity in these beds. Lithologically the contrast is between quartzite below and limestone above, most of the exposures consisting of gently-tilted beds. The quartzites are sometimes found as unconformable cappings on Torridonian Sandstone mountains, for example Cul Mhor. They also occur as a series of very resistant, well-jointed, eastward-dipping beds on hills such as Breabag, south of Conival in western Sutherland. The Durness Limestone, much of which is dolomite, has such normal karst features as solution holes and disappearing streams. It forms a very well marked escarpment near Inchnadamph (Plate 39). The outcrop of the Cambrian rocks is narrow, as large masses of Moine metamorphics have been thrust over them from the east. This Scottish shelf facies is quite different from the Cambrian rocks of Wales. In fact it is not until much later in the Silurian that a comparable facies occurs in the Palaeozoic rocks of Wales.

Ordovician rocks

If one can briefly characterise the Cambrian rocks it might be said that, because of their generally shallow-water origins, they include a variety of types, whose varying lithology has a marked effect on relief. This holds good in different ways for both Wales and Scotland. On the whole the Ordovician, whether in Wales, the Lake District or the Southern Uplands of Scotland, is a period of greater uniformity of sedimentation, which is, however, greatly diversified by outbreaks of volcanic and intrusive igneous activity at different times in different places. The effect of these igneous rocks on relief can hardly be overstressed.

The general succession of Ordovician rocks is as follows:

$$\left.\begin{array}{l}\text{Ashgill}\\\text{Caradoc}\end{array}\right\} = \text{Bala}$$

Llandeilo
Llanvirn
Arenig

Tremadoc (? Cambrian)

PLATE 38. Rhinog Grits, Rhinog Fawr near Harlech. Detailed effects of bedding and faulting on southern side facing Rhinog Fach

Within this succession vulcanicity was widespread, though the times and places varied, as shown in Fig. 7.4. The important volcanic horizons from the point of view of relief formation are the mainly Llanvirn outbursts in the Dolgelly–Arenig district—this was responsible for such features as the Cader Idris escarpment (Fig. 3.24)—the Caradoc volcanic activity in the Snowdon district and the Llanvirn-Llandeilo Borrowdale Volcanic series of the Lake District.

	DOLGELLY ARENIG	SNOWDON	SHROPSHIRE	LAKE DISTRICT
ASHGILL				
CARADOC		■	■	■
LLANDEILO	■			■
LLANVIRN	■		■	■
ARENIG	■			

FIG. 7.4. Horizons of Ordovician volcanic activity *(after Wells and Kirkaldy)*

The volcanic rocks represented include a great variety of types. Many of them are lavas and ashes, some of which may have been deposited in the form of welded tuff. Some of the eruptions were submarine, a conclusion pointed to by the occurrence of pillow lavas. Such lavas were probably erupted on to the sea floor, became a low density, honeycomb-like rock by the retention of steam within the vesicles of the rock, and were rolled over and over on the sea floor finally to come to rest as a mass of 'pillows', the spaces between which were filled up by secondary deposition. The rock type usually involved is spilite (see Fig. 3.27). On the other hand the predominant rock type in the Lake District is andesite, while the Bala vulcanicity in Snowdonia produced predominantly rhyolitic rocks.

The over-simplified picture of a geosyncline presented at the beginning of this chapter was destroyed during these phases of Ordovician vulcanicity. The whole area was very unstable and, although some of the volcanic centres may have been submarine, others were definitely subaerial. Certain of these centres may still be recognised in Wales. The earliest was at Rhobell Fawr, north of Dolgelly, where vulcanicity occurred before the deposition of the earliest Arenig marine sediments. It was, in fact, a sub-aerial centre, which spread lava and pyroclasts on to an eroded surface of older beds, before the whole lot became submerged beneath marine

PLATE 39. Durness Limestone escarpment near Inchnadamph, Sutherland

sediments. Later a volcanic centre developed in Llanvirn times north of Builth Wells in central Wales: it apparently started as a submarine feature which built up above sea-level, became dissected and was gradually submerged by marine sediments, which filled in the pattern of cliffs and stacks developed on the volcanic rocks. Snowdon may be taken as a third example of a volcanic centre. Although its present pyramidal form is due mostly to corrie dissection, the mountain coincides with the maximum thickness (exceeding 1100 m; 3500 ft) of acid, rhyolitic lavas and tuffs. These thin to the north and the south, thus indicating the probability that the original volcanic centre was in the present position of Snowdon.

To the geomorphologist the fact of greatest importance may not be the nature and origin of the igneous rocks, but the sheer proportion of rocks in a given locality which are igneous. Outside the igneous areas the general types of Ordovician sediments are usually either the great thicknesses of turbidites, with their mudstone/sandstone alternation, or the restricted deposits of slowly-accumulating, black, graptolic shales. Neither of these facies usually compares in resistance with the igneous combination of intrusions, lavas and pyroclasts.

The greatest density of igneous rocks (Fig. 7.5) is reached in the Snowdon area and in the broken escarpment surrounding the Harlech Dome. A much lower density is present in the Berwyn Dome and in the inliers around Plynlimmon, the latter being almost entirely greywacke in type. Although glacial landforms occur in the Berwyns and again at Plynlimmon, just as they do in Snowdonia, they are not really comparable. There is a corrie on the northern side of Plynlimmon: it even has rocky cliffs fringing it on the mountain side, but it would be fair to describe it as pathetic compared with the developments on the southern side of the Nant Ffrancon valley, with the corries that bite into Snowdon, or with Llyn Cau between the northern and southern ridges of the Cader Idris block. Not only that: Plynlimmon is generally soft and rounded in relief with well-developed peat bogs slightly below the maximum elevations. Cader Idris, Snowdon, Tryfan and the Glyders are mountains of jagged rocks—in fact, mountains and not hills. Whether these differences are due primarily to glaciation or primarily to rocks is difficult to argue, for both process and rocks have contributed. It might be argued that glacial forms develop best on igneous rocks. It could also be argued that glacial forms are best preserved on igneous rocks. On the other hand, it could be alleged that Snowdonia was the centre of an ice sheet and that the Berwyn Dome was not. But even this explanation throws the ultimate responsibility back on to the rock type: if Snowdonia was the centre of an ice sheet, it was because it was higher; if it was higher, it was because the rocks were more resistant.

In Wales outside the Snowdon area and the Harlech Dome igneous rocks are still important in the relief, though usually less so than in that area. On the Welsh Borderland the Corndon Hill (Plate 15) phacolite in the Shelve area is prominent in the relief, and there are other examples in the same area of the effect of igneous rocks on relief. In Pembrokeshire the Prescelly Hills probably owe something to igneous rocks in their prominence in the relief, though this prominence is relative and in North Wales these hills would pass unnoticed. On the north Pembrokeshire coast

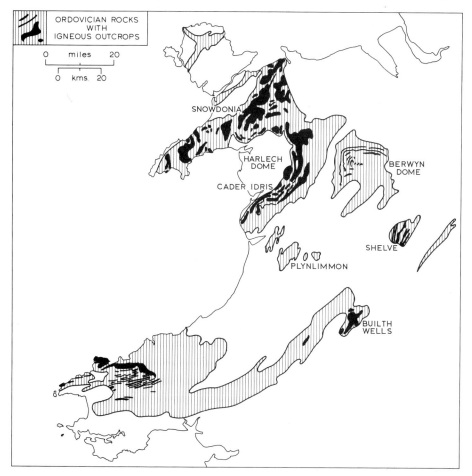

ORDOVICIAN ROCKS
WITH
IGNEOUS OUTCROPS

0   miles   20

0   kms.   20

SNOWDONIA

HARLECH
DOME

BERWYN
DOME

CADER IDRIS

SHELVE

PLYNLIMMON

BUILTH
WELLS

FIG. 7.5. Ordovician outcrops in Wales

the igneous rocks tend to form the headlands with Strumble Head providing a pronounced example.

The same type of contrast holds in the Lake District where the lower Ordovician beds (of Arenig and Lower Llanvirn age) are the Skiddaw Slates, while above them lie the Borrowdale Volcanics of Llanvirn and possible Llandeilo age. The Skiddaw Slates include a variety of rock types, greywacke-type turbidites as well as true slates. They are resistant rocks, just as comparable rocks in Wales are, and they form high relief in Skiddaw and Saddleback in the north of the Lake District, but they lack the crags, the rock outcrops and the accentuated glacial features of the Welsh Ordovician volcanic rocks. These latter features, however, are found in force on the succeeding Borrowdale Volcanic Series. These are a tremendous pile of rocks, some 3500 m (12,000 ft) in thickness, of mainly andesitic lavas, tuffs and agglomerates. They are, in fact, much thicker than the volcanic accumulations in Snowdonia. Variations in resistance between lavas and less-indurated beds together with joints and faults have produced, under the influence of glaciation, a true mountain landscape of crags and cliffs which occupies the whole central part of the Lake District (Fig. 7.6) and includes most of the famous mountains and rocks of the area.

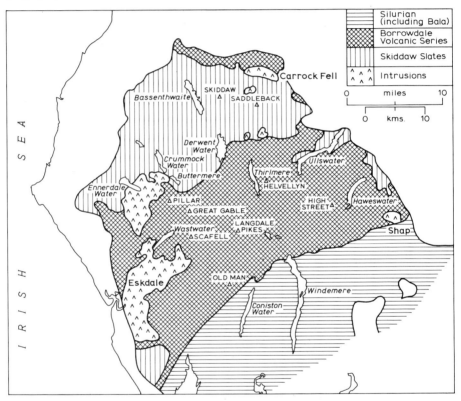

FIG. 7.6. Geology of the Lake District. Post-Silurian rocks unshaded

Ordovician rocks make up approximately the northern third of the Southern Uplands of Scotland. Volcanic rocks are much more limited in extent than in Wales and the Lake District, though the Ballantrae Volcanic Series, of Arenig age, is exposed on the west coast between the Southern Upland fault and Girvan. Most of the remaining Ordovician rocks are of

PLATE 40. Stiperstones Quartzite between Montgomery and Church Stretton, Shropshire. Shattered quartzite in degraded tor-like feature

Bala age and there is the usual facies contrast between the greywacke-turbidite sediments and the thin, black, graptolitic shales indicating slow deposition possibly under poorly-aerated conditions. The contrast between the 30–40 m (100–130 ft) of black graptolitic shales in the Moffat district and the 5000 m or more (16,000 ft) of coarse sediments, including greywackes and conglomerates, representing the same (Bala) group of rocks in the Girvan district is a classic one. Before the greywackes were so generally attributed to turbidity currents, it used to be argued that the coarse sediments represented marginal geosynclinal rocks and that the graptolic shales were caused by slow sedimentation in the deepest parts of the geosyncline. Today, as Rayner emphasises, it is thought likely that the greywacke-turbidite sediments would have slumped to the depths of the basin, and that the black graptolitic shales are more likely to have been deposited on swells in the floor of the geosyncline above the basins filled in by the turbidites.

The black shales are not, of course, resistant rocks, and the greywackes are too variable to equal the volcanic rocks in resistance. The general landscape found on the latter is the same as that found on Silurian rocks of similar type (see below).

Finally, before leaving the Ordovician with its unique volcanic contribution to the development of landforms on the Lower Palaeozoic rocks, we must mention one of the shelf sediments found on the margins of the geosyncline. In the Welsh Borderland the best examples of shelf sediments are Silurian (see below), but the Ordovician is also a shallow water series full of unconformities. At its base is the Stiperstones Quartzite, some 120 m (400 ft) of hard resistant rock, which forms a pronounced ridge, capped by tor-like crags, west of the Longmynd (Plate 40).

Silurian rocks

Compared with the Ordovician, with its fine glacial landscapes developed on and preserved by volcanic rocks in North Wales and the Lake District, the Silurian system is characterised by a virtual absence of volcanic rocks, by the development of great thicknesses of the greywacke-turbidite facies in Wales, the Lake District and the Southern Uplands of Scotland, and by the best development of a marginal shelf facies in the whole of the Lower Palaeozoic.

The main divisions of the Silurian system are as follows:

Ludlow Series
Wenlock Series
Llandovery or Valentian Series

In the main geosynclinal tracts the rocks generally become coarser

PLATE 41. Rejuvenated valley in Llandoverian greywackes and shales, near Dylife, central Wales

upwards denoting the final filling of the trough. For example in North Wales the Llandovery sequence seems to be complete and consists of 90 m (300 ft) of black, graptolitic shales, while the succeeding Wenlock and Ludlow series reach a thickness of about 3000 m (10,000 ft) in the Denbighshire Moors between the Vale of Conway and the Clwyd Valley. They are primarily turbidites by origin, the rock types being greywackes, siltstones and mudstones. A similar contrast occurs in the Lake District where the Llandovery consists of 75 m (250 ft) of graptolitic Stockdale Shales and the Wenlock and Ludlow combined embrace some 4000 m (13,000 ft) of beds similar to those found in Denbighshire. In the Southern Uplands of Scotland only the Middle and Lower Llandovery were periods of quiet sedimentation which produced the Birkhill Shales. The Upper Llandovery, Wenlock and Ludlow combined gave rise to some 6000 m (20,000 ft) of predominantly coarser sediments.

Thus over the whole of the geosynclinal areas the terminal beds of the Silurian, and therefore of the geosynclinal sediments, are typically grey-wackes and mudstones. When the great area of Silurian exposures is noted from the map, it is obvious that these types of sediments must be dominant in landscape formation over wide areas. In Wales the Silurian outcrops cover the whole of central Wales between the main Ordovician outcrops (Fig. 7.5) and even the Ordovician inliers around Plynlimmon are of similar types of sediments. Further, the Silurian fills in the north-eastern corner left by the Ordovician outcrops between the Conway and the Clwyd. In the Lake District the Silurian outcrops over virtually the southern half (Fig. 7.6). In the Southern Uplands of Scotland rocks of similar age occupy the southern two-thirds of the area. On the whole these are all areas of smooth hilly relief, probably because there are few beds massive enough to affect the relief profoundly. The rocks are not lacking in resist-ance and all form dissected plateaus often reaching heights of almost 600 m (2000 ft) (Plate 43). The greater resistance of some of the sandstone beds is probably betrayed by the ridged and grooved character of many of the grass-covered hills of central Wales, where the harder beds have probably been brought into relief by the passage of an ice sheet. Where intense rejuvenation has affected these rocks and the topmost Ordovician below them, as for example in the plateau area round Dylife north and west of Plynlimmon, they retain rubbly constant angle slopes reminiscent of semi-arid features (Plate 41). But the constant slope on rubble, rather than crags on rock, is a sure indication that they will quickly become vegetated as soon as they are stabilised. In coastal locations the harder sandstones exert considerable detailed effects on relief depending on their attitude in rela-tion to the coastline (Plate 42). These features are well-exemplified by the Aberystwyth Grits, some 1500 m (5000 ft) of Llandovery greywackes and mudstones, which are exposed in the cliffs between Borth and Aberystwyth (Plate 5).

PLATE 42. Greywacke sheet controlling cliff profile, near Borth, Cardiganshire. Aberystwyth Grits (Llandoverian)

The areas of really individual Silurian relief are in the shelf sediments of the Welsh Borderland. The beds involved are primarily the Wenlock and the Ludlow, which are unconformable on pre-Silurian rocks, the Llandovery being largely missing. The rocks involved are listed below, the older lithological nomenclature being used as it is more appropriate in geomorphological study.

Upper Ludlow Shales ⎫
Aymestry Limestone ⎬ Ludlow
Lower Ludlow Shales ⎭
Wenlock Limestone ⎫
Wenlock Shales ⎪
   (with Woolhope Limestone ⎬ Wenlock
   locally at the base) ⎭

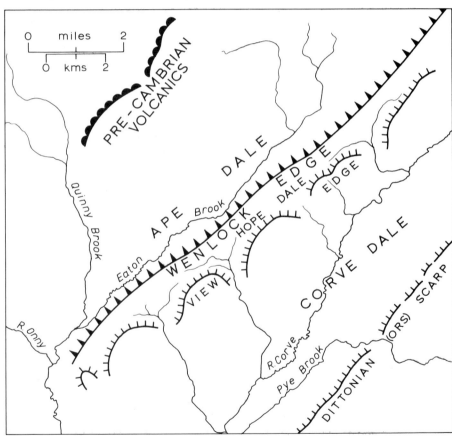

FIG. 7.7. Morphological diagram of Silurian rocks in Welsh Borderland

PLATE 43. The Welsh Plateau from Plynlimmon. Monotonous relief on Silurian greywackes, mudstones and shales

The facies are local ones. There is, for example, a rapid change towards Knighton, some 10–15 km (6–10 miles) south-west of the area shown in Fig. 7.7. Here the Wenlock and the Ludlow consist of about 2000 m (6500 ft) or so of mudstones, siltstones and shales, so that the typically Borderland Silurian relief is absent. The limestones themselves vary along the strike and the Aymestry Limestone in particular seems to be dia-chronous.

The general elements of the relief developed on these rocks may be illustrated from Wenlock Edge, almost the type area. The morphological sketch map (Fig. 7.7) and the sketch section (Fig. 7.8) should speak for

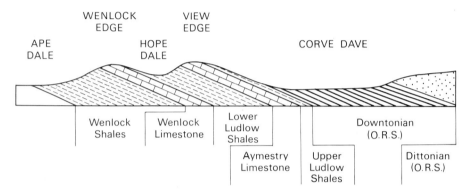

FIG. 7.8. Section of relief on Silurian rocks in Welsh Borderland

themselves. The whole area is a splendid example of scarpland relief, which must be appreciated in terms of a dip of about 10 degrees compared with the 1–2 degrees characteristic of the Mesozoic scarplands of England. The rocks themselves explain most of the relief, but not why the Aymestry Limestone forms a dissected escarpment sitting on the back, as it were, of an undissected Wenlock escarpment. The problem is not unique to the area and recurs in the South Downs and North York Moors in different guises.

The relief effects of the Silurian limestones are not confined to this area. They can be traced on the western side of the Malvern Hills and thence northwards in a little-known area that separates the Triassic Midlands from the Devonian Hereford Plain. The same limestones, plus the Wool-hope Limestone, are found in the Woolhope Dome south-east of Hereford, where the simple annular relief pattern that might be expected is broken by faulting. However, on the eastern side of the dome the scarp and vale landscape developed on the Wenlock and Ludlow rocks is quite clear, though, because of the high angles of dip, it is more a ridge and valley landscape than a cuesta landscape.

The Caledonian folding

The end of the Lower Palaeozoic geosyncline was its conversion into the Caledonian fold mountains, a process which varied both in time and in intensity from place to place. The folding was most pronounced in the north and a distinction can be made between the metamorphic Caledonian belt, i.e. the Highlands of Scotland, and the less altered regions of the Southern Uplands, the Lake District and Wales, where the highest grade of metamorphism merely produced slates. The rocks involved in the Highlands of Scotland are largely the Moine Series, which are most likely Pre-Cambrian in age and the metamorphic equivalent of the Torridonian. There were probably several phases of folding and metamorphism starting in the Pre-Cambrian and extending to the Middle Silurian. The Dalradian Series, which occupy much of the area between the Highland Boundary Fault and the Great Glen, are probably mostly Cambrian and are less metamorphosed than the Moine Series, possibly because they were involved in fewer folding episodes being younger rocks. Elsewhere the culmination of the folding was much younger, Siluro-Devonian in age.

These Caledonian areas are characterised by extensive intrusions. The Highlands of Scotland have both older and newer granites: the former are mainly deep-seated 'metamorphic' granites, while the latter, which include the Cairngorm and Aberdeen granites, are truer igneous rocks. Younger granites occur both in the Southern Uplands and the Lake District but not in Wales. Thus igneous rocks contribute to the fundamental differences between the Highlands of Scotland (metamorphics and frequent granites), the Southern Uplands (greywackes and some granites), the Lake District (greywackes, volcanic rocks and a few granites) and Wales (greywackes, shelf sediments and important volcanics).

The intensity of the folding (the details in many areas remain obscure) and the rapidly alternating sediments are allied to produce the monotonous relief dominated by erosion plateaus in many areas. Where the folding is less severe, i.e. in the shelf areas, the effects of lithology are most emphasised. In fact the areas of greatest lithological contrast are precisely where gentler structures allow those contrasts to appear most fully in the relief.

The Caledonian north-east to south-west trend is visible in the relief of many areas, for example the major faults of Scotland and the structures of Wales, but these are secondary effects developed by erosion. Fuller details really lie beyond the scope of a simple stratigraphical account.

8
The Devonian system

The geography of Britain possessed great contrasts during the Devonian. This should not be surprising, because it was, after all, a period of some 50 m.y. Accordingly the elements of Devonian geography sketched in Fig. 8.1 are highly generalised: they represent recurring units, but at any one time they would not all have been present.

FIG. 8.1. Elements of Devonian geography

The old Lower Palaeozoic geosyncline had finally been filled and converted into mountains. The new geosyncline, which was to be the site of the later Hercynian mountains, lay to the south. Its northern boundary, which fluctuated, lay roughly through North Devon, so that much of South Devon and Cornwall lay within the geosyncline, like the Ardennes massif which formed the eastward continuation of these areas. North Devon was in a marginal position, at one time sharing in the geosynclinal sedimentation and at others being subject to the continental influences from the north.

The area described in Fig. 8.1 as the Welsh Continental Margin was an extensive desert or semi-desert plain extending from the mountains to the sea. It seems to have been deltaic in general form with some marine incursions especially in the lower levels of the rocks accumulated there, but dominantly it was formed of terrestrial sediments. It existed primarily in the Lower Old Red Sandstone period, disappeared in the Middle Old Red Sandstone when earth movements occurred, and reappeared in restricted form in Upper Old Red Sandstone times when rivers deposited coarse sediment in the area, except for one brief period when it was invaded by the sea.

To the north the continuity of the Caledonian fold mountain ranges was interrupted by two principal intermontane basins, the Caledonian and Orcadian Basins. The former existed only in Lower and Upper Old Red Sandstone times, rather like the Welsh Continental Margin, but the type of sedimentation in this intermontane basin was very different from that of the Welsh area. The Orcadian Basin, on the other hand, existed in Middle and Upper Old Red Sandstone times, again primarily as an intermontane basin with continental sedimentation. At various times there were connections round north-eastern Scotland between the Caledonian and Orcadian Basins, and also in a different direction between the Caledonian Basin and areas of Devonian deposition in Ireland to the south-west. A Cheviot area of deposition ran southwards from the Caledonian Basin and may at times have connected with the Welsh shelf area.

Thus, there were three main elements of deposition in the Devonian:

1. The southern geosynclinal facies.
2. The Welsh shelf facies.
3. The Scottish intermontane basin facies.

They tend to have different effects on the relief, partly as a result of their inherent differences and partly because of the different nature of the rocks now found in juxtaposition with them.

South-west England

Although the south-west peninsula was heavily folded in Hercynian times and many of the Devonian rocks of the area metamorphosed to slates and, locally near Tintagel, to phyllites, the broad outline of the structure is quite simple. Essentially Devon consists of a synclinorium with a tongue of Permian rocks in the centre, flanked by Carboniferous Culm Measures, which in turn are flanked by Devonian to north and south. Most of Cornwall lies on the southern side of this great downwarp (Fig. 8.2). In South

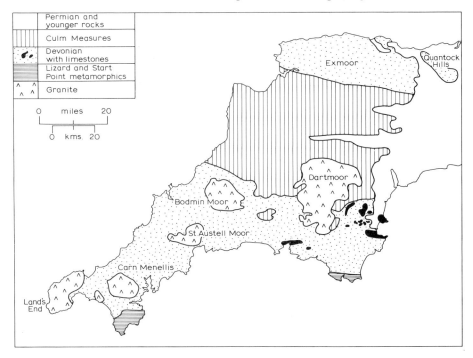

FIG. 8.2. Simplified geology of Devon and Cornwall

Devon the rocks are predominantly marine sediments; in North Devon they are mixed marine and continental. This probably represents the oscillation of the actual shoreline over a low-gradient continental plain and continental shelf stretching from Wales across the site of the present Bristol Channel into Devon.

South Devon and Cornwall

The lowest Devonian beds in South Devon, the Dartmouth Beds, are of continental origin and are some 900 m (3000 ft) thick, predominantly sandstones and siltstones. Above them, marine shales, slates and limestones, as well as lavas and volcanic ashes farther to the west, make up the rest of the sequence. Obviously we are not dealing here with an extensive geosyncline with deep-water sedimentation, or even with a great succession of turbidites as in the Lower Palaeozoic in Wales. The limestones, which occur as lenticular masses in the Middle Devonian, are shallow-water deposits: they include colonial corals and other reef-building animals and are associated with beds of detrital limestone containing broken coral and representing the wastage of the reefs. Westwards, these limestones, which occur mainly around Tor Bay (Fig. 8.2), become intercalated with submarine lavas and ashes near Totnes. Igneous activity is also associated with the Upper Devonian in Cornwall and such rocks occur around the northern side of Bodmin Moor. They are usually spilitic, a good example being provided by the pillow lavas of Pentire Point on the eastern side of the Camel estuary. Over much of Cornwall the Devonian rocks have been converted to slates. Further evidence of the proximity of land is provided by the presence in zones of thrusting, which affect the rocks near Falmouth, of blocks of quartzite and limestone, which may have been derived from land to the south. If this is so, the geosyncline must have been a fairly narrow feature, even allowing for the possibly severe crustal shortening which took place during the folding of the Devon synclinorium.

Much of this area, like Wales, is planed off by coastal plateaus. Indeed there are similarities with Wales: rejuvenated valleys, truncated rocks, and the slate-grit lithology, but differences are provided by the lower levels and the importance of granite batholiths and bosses intruded later. The limestones as usual stand out, but here mostly in the coastal scenery. Devonian limestones form the Torquay promontory and Berry Head, north and south of Tor Bay, which is itself eroded in much softer Permian sediments between. In more detail, the Torquay promontory, which is much broken by faults, shows the importance of limestones in producing headlands separated by more erodible sections of coast either on Devonian slates or Permian deposits.

North Devon

The Devonian rocks reappear from beneath the synclinorium in Exmoor. No longer are the rocks mainly marine, because two or three massive beds of sandstones and grits, each some 1200 m (4000 ft) thick, show the intervention of terrestrial Old Red Sandstone conditions from the north. The strike of the rocks (Fig. 8.2) is slightly south of east: the north coast of Devon trends due east. Hence, progressively older beds are exposed from west to east along the coast. In spite of these coastal exposures the exact succession is not clear. A useful simplification of the relations between the rocks of Devon and Wales is shown in Fig. 8.3, which indicates the massive continental sandstone beds. On the other hand, Rayner regards the Foreland Grits as the repetition of the Hangman Grits on the north side of an east–west anticlinal axis which cuts the coast at a shallow angle near Lynmouth. They have also been regarded as faulted repetitions of each other. Thus Fig. 8.3 may need to be modified to show only two invasions of continental conditions.

Although the Hangman Grits appear to be instrumental in the formation of the Great and Little Hangman headlands near Combe Martin, it is not true that these continental beds are always superior in resistance to erosion. The Pickwell Down Sandstone, for example, occupies Woolacombe Bay between the slaty headlands of Morte Point and Baggy Point. In other

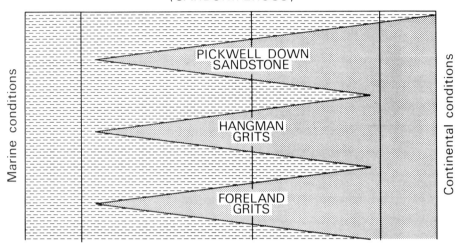

SOUTH
DEVON

NORTH
DEVON

SOUTH
WALES

(CARBONIFEROUS)

Marine conditions

Continental conditions

PICKWELL DOWN SANDSTONE

HANGMAN GRITS

FORELAND GRITS

F ɪɢ. 8.3. Relation between marine and continental Devonian in Devon and Wales (*after Wells and Kirkaldy*)

places, slates seem to be more readily erodible, as at Woody Bay a few miles west of Lynton. Where the sandstones are massive and well-jointed they form rectangular mural patterns in the cliffs in contrast to the ragged appearance of features developed on slates. Inland the Exmoor plateau, much of it between 300 and 420 m (1000–1400 ft) in elevation, is an erosion surface or a series of erosion surfaces, comparable with those of Wales, truncating a great variety of Devonian rocks. It is possible, however, that some of the explanation of the much greater height of North Devon when compared with South Devon is provided by the massive continental beds of the north.

East of Exmoor lie the Quantock Hills, a Devonian inlier, bounded on the west by a fault-line escarpment prominent in the relief.

Wales and the Welsh Borderland

A number of purely geological problems are involved in the interpretation of the rocks of this region. In the Welsh Borderland the base is conformable with the underlying Ludlow rocks so that the division between the two is to some extent arbitrary: this is shown by the different views about the base of the system adopted in the British Regional Geology handbooks. The lowest series, the Downtonian, has, as a result, been placed at different times and by different authors in both the Silurian and the Devonian. The series starts with the Ludlow Bone Bed, which is succeeded by the grey Downtonian, mainly siltstones, and the red Downtonian, formerly called the Red Marls, also mainly siltstones. The Silurian fauna became gradually impoverished as marine conditions changed to freshwater conditions. Thus, there is room for disagreement about the base of the Downtonian. Away from this region, especially to the south-west, an increasingly important unconformity develops between the Devonian and the underlying rocks.

In this account the Downtonian will be included in the Devonian. Most of the Devonian consists of the Lower Old Red Sandstone. The Middle is missing and the Upper, which rests with a slight unconformity on the Lower, is so restricted as to lead to the suspicion that it is not fully represented. A generalised succession, which however varies considerably over the area is:

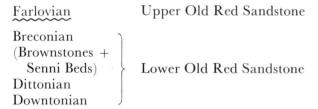

Farlovian   Upper Old Red Sandstone

Breconian
(Brownstones +
   Senni Beds)   Lower Old Red Sandstone
Dittonian
Downtonian

Lithologically there are important differences between the beds. Apart from the Downton Castle Sandstone, a relatively thin bed near the base, the Downtonian is mainly siltstone, the Red Marls of the older accounts. Although the Dittonian also contains siltstones, there are more sandstones and thin limestones in it. The latter, known as cornstones, are probably evaporites and occur either as false-bedded, lenticular deposits or as conglomeratic deposits, presumably formed by the breaking up of sheets of chemically-deposited limestone. Some of the thinner limestone beds may have originated as concretionary deposits in phases of semi-arid soil development. The Senni Beds and Brownstones are mainly arenaceous and the Upper Old Red Sandstone, at least in the outcrops north of the South Wales coalfield, is the Quartz Conglomerate. The division on the map (Fig. 8.4) shows the distribution of the two primary lithological divisions of the Old Red Sandstone.

Apart from the lithology, one other factor is needed to understand the general relief developed on the Old Red Sandstone of the area. As the area lay on the shelf at the edge of the Hercynian geosyncline, it was not involved in the intense folding that affected the Devonian rocks of Devon and Cornwall, but only in gentle tilting. It was thus comparable with the

Welsh Borderland at the end of the Silurian or the area of Torridonian rocks in north-western Scotland at an earlier date. The result is the development here of what is probably the finest erosion scarp in Britain, i.e. the Black Mountains, Fforest Fawr, Brecon Beacons, Black Mountains line

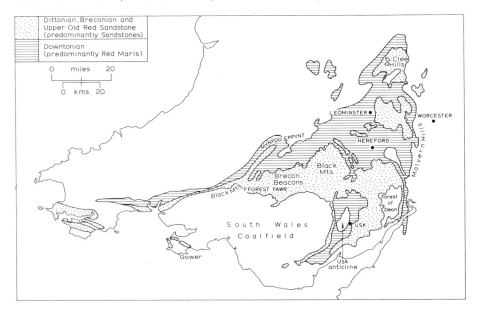

FIG. 8.4. Old Red Sandstone outcrops in Wales and the Borderland

(Fig. 8.4). There are other escarpments in Britain comparable in height, for example the Cross Fell scarp in the north-west Pennines, but they owe at least part of their being to faulting.

In the north the break of slope between Corve Dale and the Clee Hills (Figs 7.7 and 7.8) coincides with the Psammosteus Limestone at the Downtonian-Dittonian junction, while the area of undulating upland reaching to about 240 m (800 ft) between Worcester and Leominster is formed on the large expanse of post-Downtonian rocks to the south of the Clee Hills (Fig. 8.4).

South of this area the Plain of Hereford, a fertile agricultural zone with fruit and hop cultivation, lies on the Red Marls between the area mentioned above and the Black Mountains. Across this lowland the isolated patches of post-Downtonian rocks that continue the line of the Black Mountains escarpment are marked by a prominent but discontinuous hill range, including Burton Hill and Dinmore Hill, that rises to about the same elevation as the upland between Worcester and Leominster. To the south, a second range of hills, parallel to and south of the Black Mountains scarp, separates the Plain of Hereford from the Wye Valley on the northern side of the Forest of Dean. This runs from Craig Serrerthin, through Garway

Hill, Orcop Hill, Aconbury Hill and Dinedor Hill and is continued in the Woolhope Dome, which is, however, structurally quite different (p. 288). It is a higher range of hills than the northern one, falling from 415 m (1370 ft) in the west to about 270 m (890 ft) in the east.

A similar contrast between the marls and the sandstones is seen further south. A Red Marl lowland surrounds the Silurian core of the Usk anticline and is terminated eastwards by the higher ground which lies between the Usk lowland and the lower Wye valley. Here the main escarpment, which runs north–south through Llanishen, is on Upper Old Red Sandstone.

Finally, there remains the Old Red Sandstone area on the northern side of the Brecon Beacons escarpment. The Red Marl lowland is followed by the Wye through Hay and thence stretches south-westwards towards Brecon. North of it lies the Mynydd Eppynt, an upland reaching about 470 m (1560 ft), monotonous in character and with a relatively small thickness of post-Downtonian Old Red Sandstone remaining on it.

To the south lies the Black Mountains–Brecon Beacons escarpment, developed mostly in the Senni Beds and Brownstones with cappings of Upper Old Red Sandstone especially to the west in the Carmarthenshire Fans of the Fforest Fawr. Farther west still, in the western Black Mountains, cappings of Carboniferous Limestone and Millstone Grit occur.

In the Brecon Beacons the escarpment reaches about 900 m (2907 ft) at Pen y Fan; 705–800 m (2350–2630 ft) in the Carmarthenshire Fans; and about 705 m (2338 ft) in the eastern Black Mountains, though higher elevations to 800 m or so (2660 ft) occur back from the main line of this escarpment. Shelves fronting the escarpment reflect the presence of important sandstone horizons lower in the succession. There is a prominent one at 300–360 m (1000–1200 ft) in front of the eastern Black Mountains. No ground higher than the Old Red Sandstone escarpment occurs to the north until Cader Idris is reached.

On the whole these are smooth peat-covered hills, with cotton grass as a characteristic element in the vegetation, but slight variations in slope, causing presumably slight variations in drainage, are reflected in the grass species and in the coloration of the landscape. Rock outcrops occur mainly where the escarpment has been eaten into by corrie development, especially around Pen y Fan in the Brecon Beacons and around Carmarthen Van to the west. The eastern Black Mountains escarpment is much smoother.

The valleys that trench the dip-slope are characteristically U-shaped with the more massive beds of sandstone often breaking the turf cover on their shoulders. The U-shape may have nothing to do with glaciation as horizontal or slightly-dipping rocks with hard beds near the crest are conducive to this type of valley cross-profile. Many of the valleys have aprons of slumped glacial drift in them, which the present streams are slowly dissecting. The ridges between the valleys are typically stepped where beds of different resistance come in, for example at Pen Allt-mawr in the south-western corner of the eastern Black Mountains, where Upper Old Red Sandstone occurs.

Westwards into Carmarthenshire and Pembrokeshire, although the beds are very thick, the lithology changes and the dip steepens. The outcrop is thus narrow and conditions are not conducive to major escarpment formation. South of the coalfield the Old Red Sandstone appears in the Carboniferous area of the Gower peninsula in the cores of anticlines. It reaches about 180–190 m (600–630 ft) and is the highest ground of the district.

Scotland

In Scotland Devonian rocks are largely but not entirely confined to two major basins of deposition: the Caledonian Basin coincides approximately with the present Midland Valley, while the Orcadian Basin probably occupied the site of the present Moray Firth and extended north-eastwards. These basins were probably fundamentally intermontane desert basins and the rocks are to be regarded as the results of the mosaic of environments that occur under such physical conditions: mountain-front fans, piedmont deposits, playa lake deposits and so on, but may also have included conditions which do not seem to have recurred later (see the account of the Orcadian Basin below). The combined total thickness of rocks amounts to some 12,000 m (40,000 ft) compared with 1200 m (4000 ft) in the Welsh region. Two reasons are advanced for this great difference in total thickness. First, because of the more mountainous relief there are great thicknesses of conglomerate, and, second, the Caledonian Basin, an area of tectonic instability, was the site of important outbreaks of vulcanicity, the ensuing beds of lavas and ashes adding considerably to the total thickness of the rocks.

The Caledonian Basin

Broadly the basin coincided with the present Midland Valley, the basic structure of which, if one excludes the multitude of faults and minor folds, is synclinal. As a result, the major Old Red Sandstone outcrops occur to the north and to the south along the Highland Boundary and Southern Upland faults (Fig. 8.5). Sedimentation at one stage extended beyond the Highland Boundary fault, as Old Red Sandstone occurs locally north of it, and it also extended south of the Southern Upland fault, primarily in Upper Old Red Sandstone times, in the Cheviot area. The beds are usually regarded as Lower and Upper Old Red Sandstone, and are separated by an unconformity, in which warping, including movement along the marginal faults, and erosion took place. The erosion may have removed any Middle Old Red Sandstone which had been deposited. It may be,

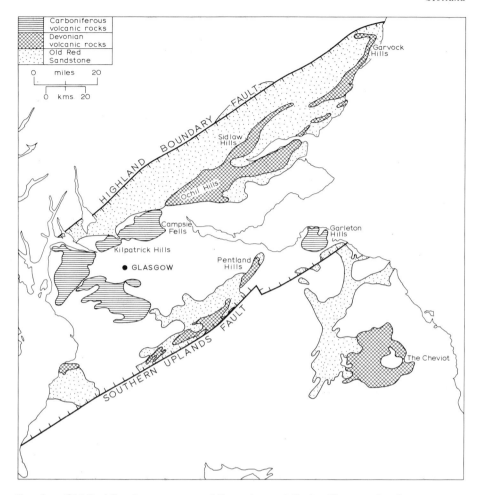

FIG. 8.5. Old Red Sandstone areas and Devonian and Carboniferous volcanic outcrops in the Caledonian Basin

however, that the Middle Old Red Sandstone is represented by the unfossiliferous upper part of the Lower Old Red Sandstone.

The main outcrop of the Old Red Sandstone is at present in the north of the area, where the Lower Old Red Sandstone reaches nearly 6000 m (20,000 ft) in thickness. It is thought that severe denudation of the Caledonian fold mountains to the north flung vast quantities of coarse detritus into the Caledonian Basin. Near the base the Dunnotar Group, almost entirely conglomeratic, is about 1800 m (6000 ft) thick, i.e. thicker than the whole Old Red Sandstone of the Welsh area. Locally these conglomerates give rise to hill masses, for example Uamh Beag (650 m; 2181 ft) between Callander and Crieff south of the Highland Boundary fault. But, on the whole, the Old Red Sandstone outcrop is a zone of low-lands, including Strathmore, Strathallan and the upper Forth valley, lying between the Highlands and the discontinuous line of volcanic hills to the south. The sediments are at least as coarse as the equivalent rocks in Wales, so that their general outcrop as lowlands is puzzling. The lowland position may be partly explained by downfaulting, but probably the major cause is that the neighbouring rocks, Dalradian metamorphics to the north and Devonian and Carboniferous volcanics to the south, are even more resistant, whereas in Wales the Old Red Sandstone is the most resistant of the beds in the area.

The Lower Old Red Sandstone period was also one of extensive volcanic activity. These rocks are responsible for the Garvock, Sidlaw and Ochil Hills to the north, where the last, bounded to the south by a fault–line escarpment, reach about 710 m (2363 ft) and are the most impressive. On the southern side of the Midland Valley the Pentland Hills, anticlinal in general form and with a core of Silurian rocks, show great thicknesses of Old Red Sandstone lavas and conglomerates—they reach about 570 m (1898 ft) in elevation. South-west of the Pentlands lies an upland area of diverse geology, where the highest points are not all on Devonian rocks, although conglomerates and intrusions of this age are responsible for many of the summits which range from 400 to 700 m (1300 to 2300 ft) and cul-minate in Tinto Hill, a felsite laccolith of Lower Old Red Sandstone age.

Outside the Midland Valley two areas deserve comment. An extensive area of lavas, known as the Lorne Plateau, occurs along the southern side of the Great Glen fault around Oban. In the context of the general resis-tance of the Highland metamorphics it is not an upstanding area like the Ochils and the Sidlaws, although the terraced appearance of parts of the landscape reflects the variable resistance of the beds.

South of the Southern Upland fault the extensive area of Upper Old Red Sandstone west of the Cheviots is of no great total resistance, although individual beds vary widely in resistance. Differential erosion on a minia-ture scale is, in fact, superbly illustrated in some of the ancient buildings made of these rocks, e.g. Jedburgh Abbey. Even the great Cheviot volcano, cased in andesitic rocks and with a granite core and radiating dykes, has generally a smooth relief with few rocky crags in spite of its overall height of 800 m (2676 ft) (Plate 44).

The Orcadian Basin

The extent of rocks preserved in this area is shown in Fig. 8.6. Generally speaking the rocks are Middle and Upper Old Red Sandstone, though, as the lowest part of the former is unfossiliferous, the Lower Old Red Sand-stone may be represented. On the whole the rocks are finer-grained than

PLATE 44. The Cheviot from the west. A smooth landscape on a dissected volcanic and intrusive mass. The fence is the Border

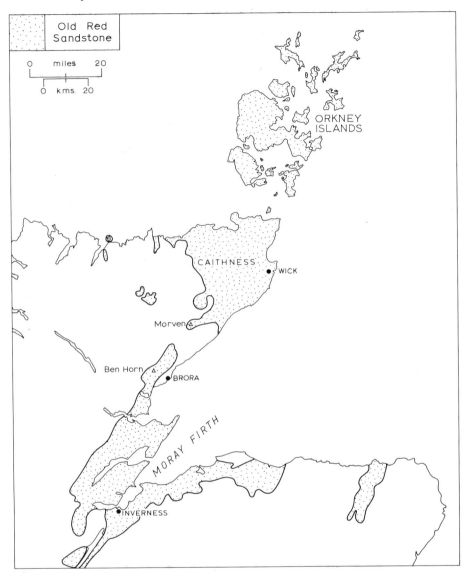

FIG. 8.6. Old Red Sandstone areas in the Orcadian Basin

those of the Caledonian Basin and volcanic rocks are virtually absent. Hence there are differences in relief.

The lowest part of the Middle Old Red Sandstone contains breccia and conglomerates, many of the pebbles being derived from the underlying Moine metamorphics. These rocks, known as the Barren Red Measures, form the whole of the surviving Old Red Sandstone south of Caithness, where much of the present exposure is preserved in a generally synclinal position on the western side of the Moray Firth. Some of this forms a syn-clinal upland, for example the Ben Horn (510 m: 1706 ft) area inland from Brora, where an Old Red Sandstone escarpment, including breccias and conglomerates, faces westwards over the Moine rocks. The general relief is typically sombre, peat-covered, smoothly sloping Old Red Sand-stone.

In Caithness there occurs in the upper part of the Middle Old Red Sandstone the Caithness Flagstone Series, a formation unique in Britain of alternating sandstones, mudstones, bituminous flagstones and lime-stones. There are some 4500 m (15,000 ft) of deposits which probably accumulated in something like a vast playa lake in a subsiding desert basin. The whole sequence of rocks seems to consist of cyclothems, though what length of time is to be associated with each is impossible to say. To the west marginal red conglomerates are known (though some of these outliers shown on Fig. 8.6 may be Torridonian) and these may represent the land-ward margin of the basin, but eastwards the rocks are primarily grey. On the conglomerates occasional high hills are found, e.g. Morven (690 m: 2313 ft), which is synclinal in form. But the majority of Caithness is a monotonous low plateau bounded by magnificent cliffs. Inland there is hardly any relief greater than the vertically set flagstones which provide the ancient form of walling between the fields. On the coast the cliffs of near-horizontal finely-bedded rocks are broken by inlets, known locally as geos, worn out along faults and joints, and possess magnificent stacks and arches.

The Upper Old Red Sandstone, the Hoy Sandstone of the Orkney Islands, is some 1050 m (3500 ft) thick and consists of false-bedded red and yellow sandstones. Although the flagstone facies is no longer present, these faulted and jointed horizontal sandstones are responsible for a similar type of coastal scenery including the celebrated stack, the Old Man of Hoy.

9
The Carboniferous system

The Carboniferous system provides probably the finest example in Britain, though the Jurassic rivals it closely, of facies of deposition and, hence, the necessity for the geomorphologist to understand the palaeogeography of the period, which is reflected in regional variations of relief development.

By the end of the Devonian the Caledonian desert mountains of Wales and the highlands of Scotland must have been worn down to areas of gentle relief—one is tempted to say pediplains, but there is no real evidence of their form. The deep water area of sedimentation still lay to the south and is represented in the British Isles by the Culm facies of Devon and Cornwall and the Carboniferous Slate of the Cork region of south-western Eire. This facies, at least in the lower part of the Carboniferous, extended across the middle of Europe and is found in Brittany, the Central Plateau of France, the Vosges, the Harz Mountains and the confines of the Bohemian massif. North of this the Old Red Sandstone plains were invaded by the sea to produce a repetition of shallow marine, deltaic and coastal swamp environments responsible for the general Carboniferous succession of Coal Measures, Millstone Grit, Carboniferous Limestone.

These are primarily facies, but in many parts of Britain they also represent a time sequence of three environments: epicontinental or shelf seas, vast deltaic developments, and coastal swamps, in which humic matter accumulated. Because they normally occur as a time sequence the facies names have acquired a time connotation, even though the rocks included in any one division may be lithologically more typical of another division. Hence the geomorphologist must beware, for example, of assuming that all the rocks described as Carboniferous Limestone are limestones: in certain areas they might include grits or even coal seams. Conversely, deltaic grits, indistinguishable from 'true' Millstone Grit, occur in all three divisions.

The detailed palaeogeography of Carboniferous times was subject to much variation, as might be expected in an area of shallow seas, but there were certain recurring elements (Fig. 9.1). The Highlands of Scotland and areas to the north-east and south-west of them north of the Highland Boundary Fault were land throughout the period. The Midland Valley and parts of the Southern Uplands was mostly land early in the period but

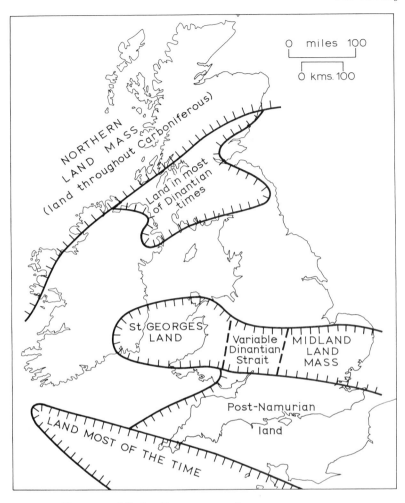

Fig. 9.1. Elements of Carboniferous geography

became an area of deposition later. A land barrier extended east to west through the Midlands and Wales, where it is called St George's Land, to eastern Eire. It separated two main areas of deposition: north Wales together with central and northern England on the one side and south Wales and the Mendips on the other. On the whole this land barrier persisted throughout, though breaks between the Midlands and Wales may have occurred in Carboniferous Limestone times. Another area of land lay south of Devon and Cornwall, and this whole area became land after the Lower Carboniferous. Within the zone of deposition certain areas tended to be areas of uplift and restricted sedimentation: the most important of these were the Askrigg and Alston blocks of the northern Pennines, which separated the basins of the Craven Lowlands to the south from the rather unusual conditions existing in the Northumberland trough to the north-east.

The divisions of the Carboniferous are as follows:

$$
\text{Upper Carboniferous}
\begin{cases}
\text{Coal Measures—Westphalian}
\begin{cases}
\text{Stephanian} \\
\text{Morganian} \\
\text{Ammanian}
\end{cases} \\
\text{Millstone Grit—Namurian}
\end{cases}
$$

$$
\text{Lower Carboniferous} \quad \text{Carboniferous—Dinantian}
\begin{cases}
\text{Viséan} \\
\text{Tournaisian}
\end{cases}
$$
Limestone

In Britain the Stephanian is mainly missing.

Because the following account is mainly in terms of contemporaneous deposits in order to illustrate facies effects on relief, the stage names on the right hand side of the table above will be used. In addition to these terms there are two sets of index letters, referring to the fossil zones of the Carboniferous, used in the literature. One set, based on the coral-brachiopod faunas of the British Carboniferous Limestone, is confined to that division. The other, based on goniatites (earlier relations of the ammonites used for zoning the Jurassic), applies mainly to the Millstone Grit and the upper part of the Carboniferous Limestone, but the rarity of the animals in the lower part of the Carboniferous Limestone and in the Coal Measures makes them less useful there.

The Culm facies of the south-west

The central parts of the Devon synclinorium are largely filled with the Culm Measures (Fig. 8.2). These were deposited in the main Hercynian trough and are quite different from the rest of the Carboniferous rocks of Britain. The Lower Culm, of Dinantian age and hence the equivalent of the Carboniferous Limestone elsewhere, lies in two strips separated by a central outcrop of Upper Culm, which is almost all Namurian in age. Generally, the Lower Culm was a period of quiet sedimentation with black shales below and thin-bedded dark limestones and shales above. The beds are intricately folded and thrust, but are fairly uniform in resistance to erosion. An exception may be made of certain beds of radiolarian chert, which are sufficiently resistant to form series of hog-back ridges across both outcrops. The Upper Culm is generally coarser and consists of greywacke–shale alternations with occasional beds of soft powdery carbon, on which the name Culm was originally based. Most of the succession is marine but near the top of the Upper Culm there is one phase of near Coal Measures conditions represented.

On the whole the Culm Measures form areas of monotonous relief, lowland and low dissected upland. There is little reflection in the relief of the intricacy of folding and thrusting affecting the beds. The folding is well displayed in the coastal sections between Clovelly and Boscastle. The sections at Hartland Quay on the western side of Hartland Point are classic examples of the truncation of structure by an erosion surface, comparable with the examples on the coast of Central Wales, where the lithologically similar Silurian is also abruptly truncated by the coastal plateaus. Apart from the cherts mentioned above, the only other note of variety is introduced by volcanic activity in the Lower Culm, which was confined to the southern outcrop: Brent Tor, west of Dartmoor, is formed of brecciated spilite of this age.

The Dinantian (Carboniferous Limestone) in Britain

Because of varying periods and types of deposition, it is best to subdivide this account regionally, and to begin in the south-west, which has the oldest marine rocks and the thickest known succession of Carboniferous Limestone.

The south-west

Apart from the Culm Measures of the Devon synclinorium, Dinantian rocks occur on the south side of St George's Land in:

(*a*) the Mendips and Bristol area east of the Severn estuary;

(*b*) the Forest of Dean, from which an outcrop runs parallel to and a little inland from the coast south-westwards towards Newport;

(*c*) the northern and southern sides of the South Wales coalfield, both the main coalfield and the Pembrokeshire section west of Carmarthen Bay.

The first series of exposures is predominantly anticlinal, the Carboniferous Limestone appearing as inliers in areas of mainly Triassic rocks. The second and third series are synclinal, so that the Carboniferous Limestone usually rests on and is surrounded by the Old Red Sandstone.

In these areas the Dinantian is complete and the succession is entirely one of carbonate rocks. The Avon Gorge was the type section on which the original zonation of the Carboniferous Limestone was based. There are variations in rock type. The Tournaisian is represented by the Lower Limestone Shales, consisting of calcareous shales and mudstones with subordinate limestones. Upwards these are succeeded by purer bioclastic limestones, i.e. ones composed mainly of accumulations of animals not in the position of growth. These rocks are commonly referred to as standard limestone. In the Viséan an inshore facies of fine, precipitated limestones, the calcite mudstones, and algal limestones was formed in lagoons on the seaward side of which the standard bioclastic limestone continued to accumulate. Locally oolites occur and the rocks are dolomitized in some areas. Finally, at the top of the Viséan there is a reversion to muddier sedimentation and the period ends with the deposition of the Upper Limestone Shales. On the whole the beds are much thinner towards St George's Land, due partly to condensed deposition and partly to erosion.

There is one important lithological variant on this succession. The Viséan in the Forest of Dean is represented by the Drybrook Sandstone, some 210 m (700 ft) of sandstones, sandy limestones and shales. The sandstone beds thin southwards with increasing thicknesses of intercalated limestone and very little is left of them by the time the Mendips are reached. Neither do they extend over into the South Wales coalfield from which the Forest of Dean is separated by the Usk anticline (Fig. 8.4). Obviously the Drybrook Sandstone is a very early example of a deltaic facies but it is difficult to understand why it occurred here, as St George's Land generally fed very little detritus into the Carboniferous Limestone sea, a fact demonstrated by the general high degree of purity of the carbonate rocks.

Where the Carboniferous Limestone is anticlinal, especially in the Mendips where it is brought up by four en échelon periclines with cores of Old Red Sandstone, it is flanked by much softer Triassic sediments and generally stands out as upland. The Mendips reach 320 m (1068 ft), but the maximum heights are usually on the Old Red Sandstone and their general summit level is about 270 m (900 ft). Again, west of Bristol the Carboniferous Limestone forms a plateau 155 m (520 ft) at its maximum. A slightly lower mass of Carboniferous Limestone flanks the coast between Clevedon and Portishead. Discontinuous outcrops are also responsible for the coastal promontories of Middle Hope, Worle Hill and Brean Down at Weston-super-Mare. Within the Mendips karst features are well developed, for example Cheddar Gorge, Burrington Combe and Wookey Hole, while the surface is scarred by closed depressions the origin of which is still controversial (see pp. 221–2).

In the Forest of Dean the Carboniferous Limestone with its overlying Upper Coal Measures, and the Upper Old Red Sandstone west of the lower Wye, forms a distinctive plateau with a strong marginal scarp, for example at Symond's Yat on the north side.

In South Wales, however, the Carboniferous Limestone on the northern side of the coalfield is comparatively thin and forms a minor discontinuous escarpment on the dip slope of the massive cuesta of the Brecon Beacons. It is also dwarfed to the south by the 600 m (almost 2000 ft) escarpment of the Pennant Sandstone. Where the Usk slices across the Old Red Sandstone escarpment between Crickhowell and Abergavenny the Carboniferous Limestone is prominent in the north-eastern rim of the coalfield as the Llangattock and Llangynidyr mountains. Yet karst features are very well developed on this restricted outcrop: it has probably the finest concentration of solution holes in Britain, though many of the best are on the Basal Grit of the overlying Millstone Grit (see pp. 220–1), and the river Mellte disappears underground near Ystradfellte (Plate 45) for 250 m or so (250 yards). On the southern side of the coalfield the Carboniferous Limestone forms upland where it appears through the Trias and Lias of the Vale of

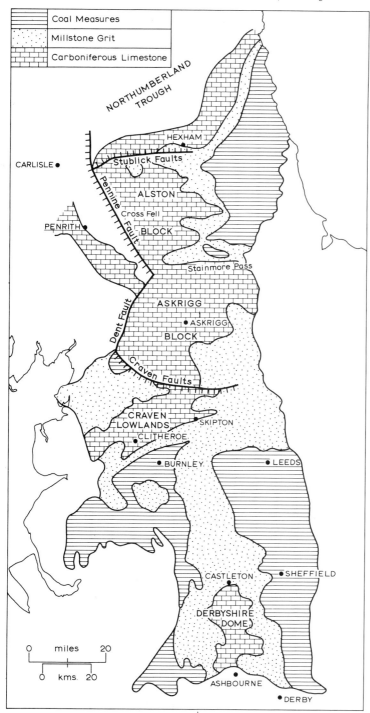

FIG. 9.2. Simplified Carboniferous outcrops and structural units

Glamorgan (for example the ridge followed by the Cardiff-Corbridge road): the situation is thus analogous to that around Bristol. Where, however, the Old Red Sandstone appears through the Carboniferous Limestone, as in the Gower peninsula, the former rock proves to be superior in resistance, exactly as it is in the Mendips and, less obviously, on the northern side of the coalfield.

In two areas, the Gower peninsula and St Govan's Head, the Carboniferous Limestone forms important coastal exposures. In these cliffs the rock is usually massive and well-jointed, especially near St Govan's Head, and the cliffs, although not spectacularly high, are very impressive and show all the orthodox relationships between direction of bedding and jointing and cliff profiles. In detail, there is well-developed intertidal solution-fretting of the limestones both in Gower and at Southerndown (Plates 26 and 27) near Porthcawl in Glamorgan. In addition, faults transverse to the coast in Gower have been responsible for the formation of small rocky valleys which are now dry (Plate 19b).

Central and northern England

The area of deposition which originally stretched from St George's Land to the Southern Uplands of Scotland was mostly invaded by the sea spreading from the south around the western end of St George's Land. As might be expected with a marine invasion, the Carboniferous Limestone transgresses a variety of outcrops and on the whole only the upper part, the Viséan, is represented, though earlier rocks may be present deep down in the trough area of the Craven Lowlands.

Deposition was largely controlled by the distribution of blocks and basins (Fig. 9.2). The former, which include the Derbyshire dome and the Askrigg and Alston blocks, have rigid basements of pre-Carboniferous rocks on which the shallow sea deposited mainly standard and lagoonal limestones, i.e. these are the areas where the limestone type most closely approximates to that of the south-western area already described. In the basins much greater thicknesses of dark argillaceous limestones and shales were deposited. The chief basin lies in the Craven Lowlands and this may well extend under the Millstone Grit of the central Pennines to the Derbyshire dome. In addition to the block facies and the basin facies, a reef facies, which includes according to some authorities, reef structures of varying types, occurs at the margins of blocks and basins. Conditions in the Northumberland basin were somewhat different.

Carboniferous Limestone is preserved in Anglesey, along the north coast of Wales around Colwyn Bay and Llandudno, and as a strip broken by the Bryneglwys–Llanelidan fault system on the western side of the Flint and

PLATE 45. Disappearing stream. Porth-yr-Ogof, Ystradfellte, South Wales. The river Mellte leaves the Old Red Sandstone outcrop through a 250 m long tunnel in the Carboniferous Limestone

Denbigh coalfields. In the north, where it lies on the dip-slope of the faulted Silurian Clwydian range it forms a broken scarp at the foot of which lies the middle course of the river Alyn, but it makes less effect here than it does farther south in Eglwyseg Mountain north of Llangollen. The limestone is thick (900 m: 3000 ft) and pure so that the jointed escarpment of Eglwyseg Mountain is a very impressive feature (Plate 46). On the coast, massive, well-jointed limestone is exposed in the cliffs of Great Ormes Head, the equivalent in north Wales of the Gower and St Govan's Head cliffs. In north-east Anglesey the northward-dipping Carboniferous Limestone forms a low escarpment north of Llangoed; Puffin Island is a detached part of it.

Although the Carboniferous Limestone of the Derbyshire dome is an anticlinal structure, it probably corresponds roughly with a true block, because marginal reef facies are known on both the western and the northern sides. Elsewhere, the area is one primarily of standard limestone, which is probably about 720 m (2400 ft) thick. In this limestone many karst features are well-developed, e.g. gorges on the river Derwent west of Matlock and in Dovedale (Plate 47) north of Ashbourne, and caverns around Castleton. On the whole the surface elevations are lower than in the north Yorkshire limestone districts, limestone pavements are less well developed, and the country in general, especially the valleys, is much more wooded. Thin lavas, known as toadstones, and volcanic pipes are present, but do not greatly influence the relief, although ancient volcanic pipes form low hills in various places, e.g. at Grange Mill 8 km (5 miles) west of Matlock.

In the northern Pennines the variations in Carboniferous Limestone facies are very well developed. These are illustrated diagrammatically in Fig. 9.3. A thick basin facies is developed in the Craven trough, where some 2000 m (6600 ft) of rocks are known to exist. There tends to be more limestone towards the base, though much of this is an impure grey limestone and not the standard limestone of the block areas, while the beds become much more shaly towards the top. Reefs are well-developed, both at the margins of the Askrigg block near Malham and some distance away from it near Clitheroe, where they form conspicuous, low isolated hills some 60 to 90 m (200 to 300 ft) high. The Craven trough facies does not produce high ground or good karst, probably because of the unusually high argillaceous content of the rocks apart from the reefs. It forms the broad lowland drained by the Ribble in which most of the hills are formed of outliers of Millstone Grit.

On the Askrigg and Alston blocks there are two main rock groups, the Great Scar Limestone below and the Yoredale Beds above. The former, which is essentially standard limestone, thins northwards. Its thickness is of the order of 210 m (700 ft) and it forms some of the best karst in Britain

PLATE 46. Carboniferous Limestone escarpment, Eglwyseg Mountain, near Llangollen

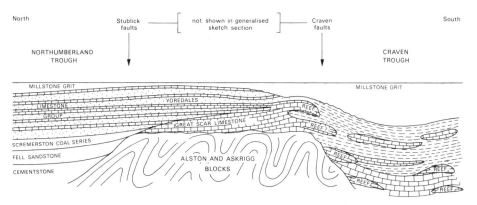

North

Stublick ———— [not shown in generalised] ———— Craven   South
faults   sketch section   faults

NORTHUMBERLAND   CRAVEN
TROUGH   TROUGH

MILLSTONE GRIT   MILLSTONE GRIT
YOREDALES
LIMESTONE
GROUP
GREAT SCAR LIMESTONE   REEF
SCREMERSTON COAL SERIES   REEF?
FELL SANDSTONE
CEMENTSTONE   ALSTON AND ASKRIGG   REEF
                              BLOCKS   REEF
                                                              REEF

FIG. 9.3. Sketch section of Carboniferous Limestone facies in northern England

especially on the Askrigg block. Bounded to the south by several fault-line
escarpments—Giggleswick Scar is by far the best of them—it forms the
high plateau round Pen-y-Ghent, Whernside and Ingleborough. In it are
developed caves and passages, disappearing streams as at Malham Tarn,
gorges as in Gordale, and the best limestone pavements in Britain (see
pp. 214–17). It possesses all the characteristics of a good karst limestone in
that it is massive, thick, well-jointed and high above base-level.

The Yoredale facies above it represents a change. It is essentially a
rhythmic sequence of limestones, shales, sandstones and seat-earths with
occasional vestigial coals. The limestones indicate marine conditions,
which then became muddy (marine shales) until the area was built up
above sea-level by terrigenous muds and sands. On the surface of the sands
soils developed and vegetation accumulated though rarely in sufficient
thickness to form coal. Six or seven of these major cyclothems occur in north
Yorkshire and they are complicated by minor rhythms as well. The exact
cause of the rhythms, which pulse through the Millstone Grit and Coal
Measures as well, is unknown.

Figure 9.4 shows how the standard limestone facies becomes Yoredale
facies with the intercalation of sediment from the land. In this area the land
lay to the north and the sandstone elements thicken in that direction. The
same diagram also shows how the Millstone Grit and the Coal Measures
represent further modification of the Yoredale pattern. The Millstone Grit
lacks the pure marine invasions which produced the limestones, the
dominant elements of the Yoredale facies, but these are compensated by
great outpourings of deltaic sediments which form the main sandstone and
grit horizons. In turn Millstone Grit may become modified to Coal
Measures by the development of large thicknesses of lagoonal mud (the
non-marine shales of today), the thinning of the sandstones, which become
also finer-grained thus denoting less violent erosion and decreased deltaic

PLATE 47. Carboniferous Limestone valley, southern Dovedale, Derbyshire

development, and finally and chiefly by the formation of coastal swamps in which accumulating thicknesses of plant debris were to form the coals of today. Although this simplified diagram probably outrages the complexity of nature, it emphasises the way in which the various facies are transitional to each other, so that we may expect all sorts of intermediate developments.

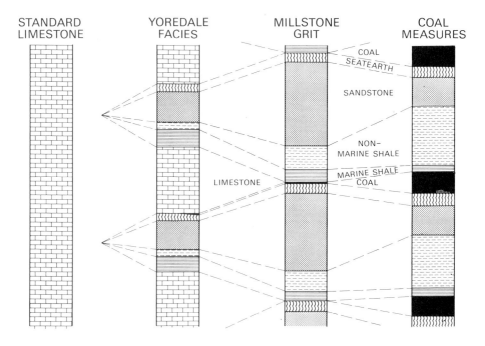

FIG. 9.4. Rhythmic sedimentation in the Carboniferous. For explanation see text

Because of the unconformity between the Millstone Grit and the Carboniferous Limestone in the Craven Fault area, the correlation of beds to the south and north is not certain. It seems, however, that the upper part of the Yoredale Series is really Namurian in age.

These Yoredale Beds, which do not extend south of the Craven faults, have an important effect on relief. The limestones are both harder and more permeable than the other beds and this, coupled with the near-horizontal attitude of the beds, leads to a terraced, stepped relief in which the 'treads' are the limestone surfaces and the 'risers' their scarps. The Yoredales as a whole are less permeable than the Great Scar Limestone below, so that water tends to be shed from the Yoredales to disappear into a series of well-developed solution holes at the Yoredale/Great Scar junction.

Although the Northumberland trough accumulated a great thickness (1800 m: 6000 ft) of sediments like the Craven Lowlands trough, the nature of the rocks, which are usually classed as the Carboniferous Limestone Series, is different. The succession here is:

Upper Limestone Group	Namurian
Middle Limestone Group	
Lower Limestone Group	
Scremerston Coal Series	Dinantian (probably Viséan)
Fell Sandstone	
Cementstones	

The lowest three beds which come in north of the Stublick faults are lithological units of non-marine origin. The Cementstones are dominantly argillaceous with thin, impure limestones and form the low-lying ground between the volcanic rocks of the Cheviot area and the Fell Sandstone escarpment. The Fell Sandstone is a series of massive, deltaic, false-bedded sandstones: the source seems to have been from the north-east for the beds thin south-westwards. This is succeeded by the Scremerston Coal Series, in some ways comparable with Coal Measures type of deposition, but con-

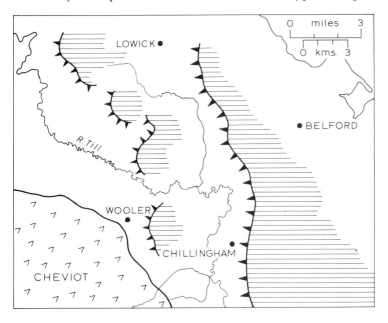

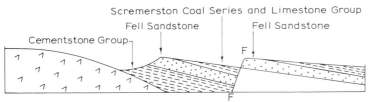

FIG. 9.5. Fell Sandstone escarpments in north Northumberland

317

taining impure limestones and far fewer coal seams. Thus, in a sense, the Fell Sandstone–Scremerston Coal Series succession represents an early parallel of the main Millstone Grit–Coal Measures succession. The limestone groups above are Yoredale in type and, like the latter, straddle the Dinantian–Namurian boundary.

None of these rocks is very resistant except the Fell Sandstone (and the series of lenticular intrusions which form the Whin Sill). The Fell Sandstone escarpment is best developed around Rothbury, where it is crossed in a deep and attractive valley by the river Coquet. In the Simonside Hills, south-west of Rothbury, it reaches heights of 430 m (1447 ft) and is sufficiently resistant to form low north-facing crags. Farther north, it is split into a double escarpment by strike faulting (Fig. 9.5), one section running north from Wooler and the other lying between Chillingham and Belford. The low ground between is formed on downfaulted rocks of the Scremerston group. The elevation of these ridges falls from about 300 m (1000 ft) in the south to 180 m (600 ft) in the north.

The Yoredale type lithology of the limestone groups above is responsible for a pronounced graining along the strike in some places, for example on either side of the Whin Sill where it is followed by the Roman Wall between Hexham and Haltwhistle.

Dinantian or Carboniferous Limestone rocks, resembling those of the Northumberland trough in general type, are also found in the Midland Valley of Scotland, but there have been difficulties in correlating the Scottish and the English rocks. Further, vulcanicity is very important in the Scottish Carboniferous, so that these rocks will be dealt with as a whole after the Upper Carboniferous of England and Wales.

The Upper Carboniferous in Britain

The Millstone Grit (Namurian)

The Millstone Grit represents a great expansion of the deltaic facies of deposition. The main outcrop is at present in the Pennines as shown in Fig. 9.2. This area might be considered as the last of a series of overlapping deltas, with material derived from the north. The earliest gave rise to the Fell Sandstone of the Northumberland trough in early Viséan times, the second occurred in the Upper Viséan in Fife, the third was the Yoredale facies described above, and the last and greatest and best-developed gave rise to the Millstone Grit.

The main area of deposition was in the Craven Lowlands trough, where 1800 m (600 ft) of sediments are known between Bradford and Skipton. Northwards the Millstone Grit thins rapidly over the Askrigg block and in the north-east of England not much more than 100 m (330 ft) of transitional beds separate the Namurian Limestone Group below from the Coal Measures above. In the main area of deposition the Millstone Grit consists of grit (often sandstone) and shale alternations. The earliest grits are found in the north and are overlapped southwards by a succession of later and higher grits. To the south the beds become not only thinner but also finer-grained and end against the St George's Land–Midland rise. The shales and thin poor coals of the south represent more the modern idea of deltaic sediments than the grits themselves. The thick, often coarse, false-bedded sandstones are more explicable as a series of confluent fluvial sediments, deposited on and channelled into an emerged delta surface. The emergence of the delta is not sufficient to explain the facies, for such great thicknesses of coarse, mechanically-weathered debris would have required a great rejuvenation of erosion in the source regions as well. The nature of the debris suggests either cold or dry conditions: the latter seems the more likely as arid conditions intervened before the end of the Carboniferous especially in the north.

The Millstone Grit of the Pennines gives rise to a distinctive type of relief. North of the Craven Faults, Ingleborough, Pen-y-Ghent and Whernside (680–720 m: 2270–2414 ft), formed of Yoredale Beds capped with Millstone Grit, are terraced hills overlooking limestone plateaus. Southwards, in the main grit area of the Pennines between the Lancashire and Yorkshire industrial regions, and also in the Rossendale Forest, there are all the variations on cuesta form that can stem from variable dips and faulting. These sandstone–shale alternations are ideal for the formation of scarp and vale landscapes, and the individual grit beds are sufficiently massive to give rise to many rocky scars or crags. Where the beds are almost horizontal, as in the central parts of the Rossendale anticline, a landscape of plateaus with stepped scarps is developed. Where the dips are slight, as in the eastern part of the central Pennines, the cuestas are broad and the west-facing escarpments are high. Where the dips are steeper, as in the western part of the central Pennines, the east-facing scarps are lower and more closely spaced. Blackstone Edge, developed on the Kinderscout Grit, is the highest of the west-facing scarps. Faulting will cause repetition of the scarps in the case of strike-faulting and offsetting of the scarps in the case of dip-faulting. The faults also provide potential lines of weakness for valley development.

These areas are mainly heather and bilberry moors where they are dry, but where peat develops and drainage is impeded cotton grass is characteristic, as it is on the Old Red Sandstone in South Wales. The rocks are generally devoid of bases so that the vegetation is calcifuge. The whole area is a source of soft water, so that many reservoirs have been built to supply the industrial areas of Lancashire and Yorkshire.

The juxtaposition of fracture-permeable grits and impermeable shales leads to spring-lines, to spring-sapping, to saturation of the underlying shales, which are then subject to mechanical failure. For these reasons landslips are common. These are assisted by the well-developed jointing of the grits, which not only increases the permeability but also provides planes of weakness along which sections can break away. The area around Mam Tor west of Castleton in Derbyshire is noted for landslips and valleys have occasionally been blocked by them. The presence of hard competent grits above shales in the valley floors is also conducive to the phenomenon of valley bulging, the weight of the massive grits on the shales tending to cause the latter to bulge up in an anticlinal fashion along the lines of the valleys. Such conditions are very liable to lead to fracturing of dams impounding reservoirs in the valleys.

Many of the landslips may be Pleistocene in age. The same is true of similar features on the Hythe Beds escarpment of the western Weald. Deep freeze–thaw may have considerably weakened the shales under cold conditions, while the vast amounts of meltwater produced by the summer thaw would have caused a high degree of saturation of the clays and hence again tended to weaken them.

Within the valleys falls may be caused, either where a stream flows over the edge of a grit bed onto the underlying shale, or where a fault, transverse or oblique to the valley, throws grit on the upstream side against shale on the downstream side. The latter condition incidentally is repeated in the Millstone Grit on the northern side of the South Wales coalfield.

Differential weathering, notably undercutting, which was probably wrongly attributed to wind erosion in earlier accounts, acts on isolated rock outcrops of variable resistance to produce such tor-like features as the Brimham Rocks north-west of Harrogate.

Millstone Grit also occurs in the North and South Wales coalfields. In North Wales it is of no great importance, but is coarser towards St George's Land in the south where the Cefn-y-Fedw Sandstone forms a feature between the Carboniferous Limestone and the coalfield.

Although it is significant in South Wales, the Millstone Grit is missing in the Forest of Dean and may or may not be present in the Bristol area, where comparatively thin sandstones occur between the Carboniferous Limestone, the top of which is missing, and the Coal Measures. They may represent a grit facies but whether it is Namurian in age is not known, for a similar facies occurs as the Drybrook Sandstone in the Dinantian of the Forest of Dean and as the Pennant Sandstone in the Morganian of South Wales.

In South Wales the Millstone Grit starts on the north side of the coalfield with the quartz conglomerates of the Basal Grits, which are succeeded by shales and at the top by the Farewell Rock. The Basal Grit is important: it forms minor features and its outcrop has the greatest concentration of large solution holes in the area (see pp. 220–1). Where the quartz conglomerates are faulted against shales in the valleys spectacular waterfalls, comparable with the Pennine ones, may be formed: there are two such in the Mellte valley (Plate 48) below the limestone tunnel of Porth yr Ogof. The Mill-

PLATE 48. Falls on Millstone Grit, Mellte valley, South Wales. The second of two sets of falls where quartz conglomerates are faulted across the valley

stone Grit, which is thin in this northern outcrop (60–120 m: 200–400 ft), disappears under the Coal Measures in the syncline and reappears south of the coalfield in the Gower peninsula, where it is 600 m (2000 ft) thick. But the whole thickness here, far from the source of the sediment in St George's Land, is much more shaly and it has no great significance as a relief former.

The Coal Measures (Ammanian and Morganian)

Coal was the basis of Britain's industrial revolution. Hence, the coalfields were Britain's greatest asset and the amount of geological knowledge accumulated from mining operations and trial borings is immense. It is probably true that more detail of structure and stratigraphy is available for these beds than for any others. Yet, geomorphologically they possess few features of great interest and on the whole are a thick series of non-resistant deposits.

It may be assumed in reconstructing the palaeogeography of the period that the Caledonian mountains had been completely worn down and that an environment of vast deltaic swamps prevailed at about sea-level. The worn-down character of the land is witnessed by the finer grade of the sediments, mostly shales, and by the greater fineness of the sandstone beds. Coal Measure swamps extended right across the centre and north of England, with the possible exception of the Lake District, and even over-lapped the Highland Boundary Fault in Scotland. A general uniformity of conditions is indicated, at least in central and northern England, by the widespread nature of certain coal seams and the way in which marine bands occur over wide areas, representing slight changes in base-level. Lancashire was probably the centre of subsidence as the rocks are thickest there. They tend to thin southwards against the Midland land mass. Of course, in this type of sedimentation the original thickness must have been very much greater, because the peats have compacted enormously to form coals and the lagoonal muds to form shales, although the degree of com-paction in the latter case was less than in the former.

The general sedimentation is rhythmic (Fig. 9.4): compared with the Millstone Grit the marine episodes and the sandstones are both reduced, although occasional sandstones in channels, known as wash-outs, show where streams trenched the surface of the deltas. Generally the rocks are most varied where they are thickest. A greater number of coal seams is found in such localities, whereas towards the land masses more continuous coal conditions are indicated by very thick seams. For example, the Thick Coal of the Warwickshire and South Staffordshire coalfields splits into a number of thinner seams to the north (Fig. 9.6).

PLATE 49. Pennant Sandstone, near Aberdare, South Wales. Funnel-shaped entrance of the Cynon valley into the Welsh coalfield plateau. The dip is downvalley. Mountain Ash in the middle distance.

The Coal Measures form lower ground than the other Carboniferous rocks, varied landforms only occurring where there is considerable lithological variation. This is particularly the case in South Wales, where the succession is as follows:

Upper Coal Measures { Upper Coal Series / Pennant Sandstone

Middle and Lower Coal Measures   Lower Coal Series

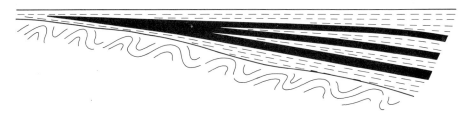

FIG. 9.6. Splitting of thick coal seam away from land

The Pennant Sandstone is a series of about 1200 m (4000 ft) dominantly of sandstone, thicker in the west than in the east. It represents a change in the proportions of the elements of the Coal Measures cyclothems rather than an absolute break, because it contains some thin coals and also because lateral variation to more normal Coal Measures occurs. It consists typically of coarse, current-bedded sandstone, the general effect on the relief being similar to that of the Old Red Sandstone. On the northern side of the coalfield the Lower Coal Series is mostly eroded into a broad vale, especially west of Merthyr Tydfil and north of the Rhondda valleys. The streams which mainly flow down-dip have narrowing funnel-shaped valleys as they enter the Pennant outcrop (Plate 49), because the dip lowers that rock down the valley sides until the whole valley becomes a narrow gorge cut solely in Pennant and not floored by erodible Lower Coal Series. The north-facing scarp (Fig. 9.7) is an impressive feature reaching 600 m (nearly

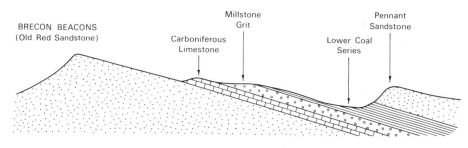

FIG. 9.7. Section of Devonian and Carboniferous relief north of South Wales coalfield

PLATE 50. Pennant Sandstone, near Hirwaun, South Wales. The highest part of the north-facing scarp looking over the Lower Coal Series

324

2000 ft) (Plate 50), while the gently-dipping Pennant Sandstone to the south forms the extensive, almost uninhabited plateaus between the crowded mining valleys. Like the Old Red Sandstone some beds are resistant enough to form local rock scars, for example around Cwm Parc west of the Rhondda Fawr, and on the north-facing scarp above the head of the latter. The Pennant is a Millstone Grit facies, though it is not Namurian in age, and must represent some upheaval and renewal of erosion in St George's Land. Similar, but thinner and less important beds of about the same age are known in the Forest of Dean and Bristol coalfields.

The Carboniferous in Scotland

The Carboniferous in Scotland is complex and difficult to correlate with the equivalent rocks in England. The lowest beds deposited on the diversified basement in the Midland Valley are subject to large facies changes. Even higher in the succession contemporaneous movements controlled the sedimentation to a large extent, so that there are still innumerable facies changes. There is a complete absence, as in Northumberland, of anything approaching standard limestone. In addition, Coal Measure conditions became fully established in the Namurian. Finally, vulcanicity was very important especially near the base of the system. Furthermore, there has been a reclassification of the rocks and those which were formerly classed as Carboniferous Limestone, i.e. Dinantian, are now known to cover the Dinantian and much of the Namurian.

The following table attempts to clarify the stratigraphical position.

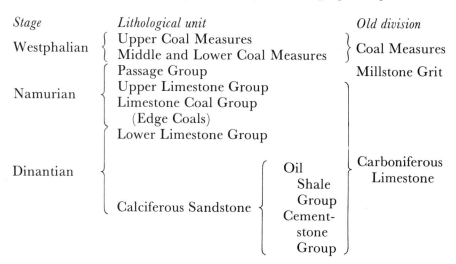

Stage	*Lithological unit*			*Old division*
Westphalian	Upper Coal Measures			Coal Measures
	Middle and Lower Coal Measures			
Namurian	Passage Group			Millstone Grit
	Upper Limestone Group			
	Limestone Coal Group (Edge Coals)			
	Lower Limestone Group			
Dinantian	Calciferous Sandstone	Oil Shale Group		Carboniferous Limestone
		Cement-stone Group		

The divisions of the Calciferous Sandstone cannot be followed everywhere: the Cementstones are well-developed in the south-west of the Midland Valley, the oil shales in a deep basin west of Edinburgh. In Fife the oil shales are replaced by deltaic, Millstone Grit facies sandstones, Upper Viséan in age and comparable in origin with the Fell Sandstone of Northumberland. The Limestone groups are Yoredale in type, while the intervening Limestone Coal Group is dominantly terrestrial and contains good, workable coal seams. It represents a very early advent of the Coal Measures facies. The Millstone Grit is a thin transitional series as in Northumberland. The Middle and Lower Coal Measures are the productive ones: the Upper are barren and often reddened. It is therefore obviously of little use to attempt to transfer any picture of stratigraphy-relief relationships from the Carboniferous of England to the Carboniferous of Scotland.

These rocks, originally deposited on an unstable basement, have been considerably warped and faulted. They are mainly not resistant rocks, but relief is provided by the diverse igneous activity that accompanied their deposition. The most important lavas are the Clyde Plateau lavas—mostly olivine basalt—which form a great west-pointing V-shaped outcrop surrounding Glasgow (Fig. 8.5). They mostly occur above the Cementstones and form terraced upland areas ranging from 360 to 510 m (1200–1700 ft). Lavas again crop out in the Garleton Hills east of Edinburgh. Deep borings suggest that these may be continuous with the Clyde Plateau lavas beneath the synclinal centre of the Midland Valley. In addition, basic sills, dykes and volcanic vents form such features as Stirling Castle rock (a sill) and Arthur's Seat at Edinburgh (a volcano). Whereas the igneous rocks tend to form higher relief inland, on the coast they are usually truncated by the sea so that they can be studied in plan.

In Scotland (as also in England, notably the Midland coalfields) the uppermost Carboniferous rocks are red. It is perhaps a little naive to see in this a foretaste of the desert conditions prevailing in the succeeding New Red Sandstone. Some of this reddening is secondary, as can be seen in the Cumberland coalfield where its base is irregular. In other places, some of the beds were probably originally red, but red beds can also be lateritic and thus need not necessarily denote an early onset of desert conditions.

10
The Permo-Triassic system

The Coal Measures—and hence the Carboniferous—ended with the Her-cynian folding. Once again, however, the main geosynclinal area to the south, like the Caledonian geosyncline and the later Alpine geosyncline, had been considerably folded before this. The main east–west Hercynian structures affected only part of the British Isles, south-western Eire, the southern part of Pembrokeshire, Devon and Cornwall, and, underground, the area roughly south of a line through Bristol and along the North Downs. In southern Pembroke the violence of the folding is attested by the crumpled and thrust characteristics of the narrow outcrop of incompetent Coal Measures. The Devon synclinorium is equally a mass of tightly-packed folds with an estimated crustal shortening of 40 per cent near Ilfracombe. Low grade metamorphism has produced great areas of slate which are locally good enough for commercial use and were quarried for hundreds of years at Delabole in Cornwall. Major thrusting carried the metamorphic rocks of Start Point and the Lizard over Devonian sediments north of them.

The rest of Britain really lay on the foreland zone north of the geosyncline and so escaped the most violent folding. The warping was comparatively broad and simple, thus allowing Devonian and Carboniferous lithology to exert a more dominant effect on the relief than in the areas of more complex folding. The main east–west structure in this area was the South Wales coalfield, an asymmetric syncline with gentler dips and better-developed relief features on its northern side. It is complicated by minor east–west folds and, as noted above, in southern Pembrokeshire by much more intense folding.

North of South Wales the predominant east–west trend is no longer in evidence. A north–south trend is apparent in the main Pennine upwarp in Derbyshire, in the Malvern and Abberly Hills of the west Midlands, and presumably in the Pennine–Dent fault system.

Elsewhere Caledonian trends occur. Renewed movement took place along the Bala fault line, and the folding of the Rossendale Forest and the area to the north is Caledonian in trend. In Scotland movement again took place along the marginal faults of the Midland Valley and also on the Pentland Hills anticline.

Igneous activity was important especially in the south-west, where all the granites from Dartmoor to the Scilly Islands are of this age. In the north-east the sill, or series of sills known collectively as the Whin Sill, is an important influence on relief in places though not throughout its entire length.

Just as the Caledonian folding was succeeded by a desert period, so also was the Hercynian folding by the deserts of the Permo-Trias. There were differences, however. Britain was in the main region of the Caledonian folding, and the subsequent desert deposits are often coarse, thick and well-consolidated beds which have a very important effect on relief. In the Hercynian folding Britain was well north of the main mountain zone, and the deposits are thinner, finer, and less-consolidated. In the Devonian the only rocks approaching evaporites are the cornstones: in the Permo-Trias limestones, gypsum, halite (rock salt), polyhalite (hydrated calcium-magnesium-potassium sulphate) and even the highly soluble potassium chloride are known. Can all this be interpreted in broad outline by assuming that the lower Hercynian uplands caused less rainfall than the Caledonian mountains, that less rain and gentler slopes resulted in finer sediment, and that the greater aridity was responsible for the formation of great evaporite sequences in cut-off arms of the sea and in inland lakes? Like most broad generalisations it is a seductive but probably incomplete hypothesis.

The Permo-Trias in Britain

The succession of both Permian and Triassic deposits in Britain is atypical. The type area from which Murchison first described the Permian system was adjacent to the Urals in Russia, while the Trias derived its name from its threefold nature in southern Germany. In Britain the rocks are more continental than in the type areas and correlation is very difficult and unsatisfactory. Some geologists regard the rocks as a whole as pure facies deposits. Others see the rudiments of a succession much complicated by facies deposits. Finally, there is a series of passage beds, very thin in Britain, known as the Rhaetic, which some authorities would put in the Trias while others place them in the Jurassic above. The complete succession in Europe is as follows:

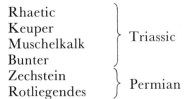

Rhaetic	
Keuper	
Muschelkalk	Triassic
Bunter	
Zechstein	Permian
Rotliegendes	

Within this succession there are two important marine horizons, albeit ones which show impoverishment of the fauna. The first of these is in the

Permian, where increasingly saline conditions, reflected in great evaporite sequences, occur in the Zechstein. The second is in the Muschelkalk division of the Trias, a marine invasion spreading from the Alpine geosyncline, which succeeded the now-defunct Hercynian geosyncline. There was probably some barrier between the geosyncline and the Muschelkalk sea because typically Triassic ammonites are missing from the latter. The Muschelkalk, a limestone formation, very important in southern Germany where its presence means that the whole of the Trias is not poor farming land, thins westwards. There is little of it left in eastern France and in boreholes in the Low Countries, though traces of it have been found in drilling for gas in the North Sea. In Britain it is missing, though marine fossils are recorded at one level in the lower Keuper Sandstone in Nottinghamshire. These may indicate a brief westward transgression of the Muschelkalk sea without the formation of typical deposits. Elsewhere the Bunter runs conformably up into the Keuper, so that some part of the succession, either Upper Bunter or Lower Keuper or both, must be the local British facies of the Continental Muschelkalk. Thus, whereas on the Continent beds above the Muschelkalk must be Keuper and beds below it Bunter, no such easy division is possible in Britain. Similarly in the Zechstein: Britain was very near the limit of evaporite deposition and the continental facies of the Zechstein evaporites, which would occur towards the west of Britain, might be indistinguishable from the Bunter.

It is curious that the inability to separate the two systems satisfactorily should affect a pair, one of which belongs to the Palaeozoic and the other to the Mesozoic. This is because the fundamental difference between the two eras is faunal and not environmental. One might expect major mountain building episodes to separate eras, but in fact none of them does. If one wants a general name for this pair of geological systems it is probably better to call them the Permo-Trias rather than to revive the old term, New Red Sandstone, for although it is new by comparison with the old, it is neither all red nor all sandstone.

The Permian in Britain is best developed in the north-east of England where the beds are:

Upper Permian Marls
Magnesian Limestone
Marl Slate
Yellow Sands

The beds are the equivalents of the Continental Zechstein, a fact demonstrated by the correlation of the fish fauna in the Marl Slate with that of the Kupferschiefer in Germany, as well as by the general similarity of the succession.

The Yellow Sands are probably marginal dune deposits formed on the shores of the encroaching Zechstein Sea. They are uncomfortable on the Carboniferous and fill hollows in the surface of that formation: this unconformity probably represents the Continental Rotliegendes period. They are not geomorphologically important but being patchy and aquifers they had an importance in early mining, because shafts on the concealed part of the Durham coalfield might become flooded if they happened to pass through a patch of Yellow Sands. Freezing overcame this difficulty. In spite of its great geological importance the overlying Marl Slate is a thin bed of no importance as a landforming horizon.

On the other hand the Magnesian Limestone, which forms part of a complex evaporite sequence, has an important influence on landforms. Near the shore of the Zechstein Sea, which for considerable periods at least was not far west of the present outcrop, the evaporite sequence of limestone or dolomite, gypsum, halite and polyhalite is incomplete and the predominant form of deposition was the least soluble part of the sequence, i.e. limestones, later largely dolomitised, and anhydrite. In County Durham there are some 500 m (1700 ft) of Magnesian Limestone, which form an important escarpment inland, and give rise to vertical cliffs with stack development on a coast which is mostly of soft rocks and subdued in relief. In the upper part of the Magnesian Limestone occur the curious cannon-ball concretions, spheres of varying size, the precise mechanism of whose formation is obscure.

Evaporites become much more important underground. Boreholes at Whitby and Scarborough have revealed several evaporite cycles which run up to the deposition of potash salts. It seems that by the time the shore of the dwindling Zechstein sea had retreated this far, the saline concentration was high enough for very soluble salts to be deposited. Repeated marine readvances disturbed the perfect sequence and complicated the stratification of evaporites found there. The great thickness of evaporites, approaching 900 m (3000 ft), shows that repeated invasions of the sea occurred simply to account for the existing thickness, which must have required the evaporation of a phenomenal volume of water. Although these underground deposits are very important industrially, they obviously cannot affect surface relief.

Traced southwards along the outcrop through Yorkshire into Nottinghamshire the evaporites slowly thin and even the Magnesian Limestone disappears a few miles south of Nottingham. It seems to be replaced by pebbly sands of Bunter type: this at least is the opinion of some geologists. Others would say that the Bunter laps unconformably on to the eroded edge of the Permian. Assuming that the former opinion is correct, the equivalents of the Permian in the Midlands are the basal conglomerates and breccias, and possibly the succeeding aeolian dune sandstones usually reckoned to be Bunter in age. These deposits occur on the edge of the central highlands which ran east–west across England and whose northern edge was fractured into a series of horsts and structural troughs open to the north. In such an environment the combination of desert conglomerates and aeolian sandstones is a natural one.

This belt of dune sandstones may have extended a long way up the west side of Britain, through Lancashire where it is represented by the Lower Mottled Sandstone (usually reckoned to be Bunter), into the Vale of Eden, where its equivalent could be the 300 m (1000 ft) or so of Penrith Sandstone. The latter must be Permian because it is succeeded by a thin development of Magnesian Limestone. If these very tentative correlations are correct it may be necessary to revise the age of the dune sandstones. Alternatively the whole lot could be facies deposits.

In the north-west it is frankly impossible to separate the Permian from the Trias. At the base are the brockrams, which reach a thickness of 600 m (2000 ft) in the southern part of the Eden valley: they are breccias, usually of Carboniferous Limestone in a red matrix, derived from the limestone escarpment produced by the Pennine faults (Fig. 9.2) and probably representing a coalescent mass of scree and desert fan deposits. As might be expected they thin westwards and also occur at more than one horizon. These brockrams are replaced northwards by the aeolian Penrith Sandstone, i.e. in a similar way to that in which the pebbly deposits of the Midlands may change laterally into dune sandstones. Above, a thin Magnesian Limestone comes in, followed in turn by the St Bees Shale and the St Bees Sandstone and other red beds, all of which except for the St Bees Shale are usually recognised as Triassic.

This palaeogeographic picture of basal breccias and dune sandstones as the Midlands and west coast equivalents of the Permian of the north-east appeals to geographical probability but that does not make it necessarily true. However, the geomorphologist, mainly concerned with rock-relief problems, need not become too involved in these purely geological speculations about correlations.

The Permo-Trias of the Midlands has been conventionally divided into:

Keuper Marl
Keuper Sandstone
Upper Mottled Sandstone (Bunter)
Bunter Pebble Beds
Lower Mottled Sandstone (Bunter)
Basal breccias and conglomerates

The succession is very variable from place to place and the threefold division of the Bunter is only well-developed in Shropshire and Staffordshire.

The Permo-Trias has a very large outcrop, approximating to a thickened Y in shape, the two upper branches running parallel to the Pennines with the eastern branch much the more important. The Midlands occupy the thickened part where the outcrops join, and the tail runs south-south-west down the Severn estuary, where outcrops occur on both sides though mainly on the east, through east Devon down to the mouth of the Exe. As a whole the group of rocks is less resistant than either the Carboniferous below or the Jurassic above, so that it generally forms lowlands. Within the rocks there are some more resistant members, principally the Bunter Sand-stones and Pebble Beds, which form low rolling uplands such as Cannock Chase and Sherwood Forest. Locally, as at Nottingham, they are resistant enough to form bluffs and bare rock outcrops. These are highly permeable rocks on which dry soil creep is probably the dominant form of mass move-ment: they are, therefore, characterised by dry valleys and convex slope profiles, by light oak forest and heath vegetation. The Keuper Marls, which are mainly silty rocks and may be playa lake deposits, form the heaviest and lowest land. The Keuper Sandstones, which are finer grained than the Bunter, are occupied by drier and more fertile land. They often occur intercalated in marl sequences and may then give rise to local relief.

In the Keuper Marls massive salt beds occur, mainly in Cheshire, but also in the Midlands near Droitwich, in Somerset and in Yorkshire near Whitby. Gypsum is also present, for example in the Midlands between Nottingham and Leicester, where it is worked for the plaster industry in the Soar valley. Although such deposits often form part of evaporite sequences, they do not occur in the correct theoretical order in the Keuper: in Cheshire, there are no gypsum or anhydrite beds below the salt beds, for example. Unlike the Zechstein evaporites, which were produced by the complete evaporation of cut-off branches of the sea, the Keuper chemical deposits were probably formed in isolated and variable desert lakes. These deposits do not directly affect the landforms, but, at least in Cheshire, do so through the intervention of man. Salt is extracted in Cheshire not by mining but by pumping brine out and evaporating it. The result is a series of solution subsidence hollows occupied by meres. It is a wonder that a term has not been coined for such a landscape—for example, 'anthro-pohalokarst'!

In the tail of the Y-shaped Permo-Trias outcrop, except in east Devon where the variable Watcombe Clay (0–60 m: 0–200 ft) is found, the basal beds are conglomerates or breccias as they are in the Midlands and north-west of England. In the Vale of Glamorgan, the Mendips and the Bristol area it is the Dolomitic Conglomerate, which was largely derived from the local Carboniferous Limestone, while a former westward extension of the Trias may be evidenced by the gash breccias (collapsed cavern deposits) of south Pembrokeshire. Around the Quantocks and Exmoor the equivalent conglomerates contain an abundance of Devonian fragments. In Devon much of the debris was derived from the denudation of the Devonian lime-stones and volcanic rocks. Further away from Dartmoor sandstones of dune type are found, again as in many of the other Permo-Trias areas. In Devon, as in north-west England, the separation of the Permian from the Trias is impossible and the combined succession runs up into sandstones and marls. The most renowned bed is the Budleigh Salterton Pebble Bed, usually equated with the Bunter because of its lithological similarity. It forms a marked west-facing escarpment with a crest reaching 175 m (585 ft) running north-north-east for about 13 km (8 miles) from the coast between Exmouth and Budleigh Salterton. The presence of its distinctive pebbles has been used in the interpretation of various Tertiary and Quaternary deposits and modern beach gravels in southern England.

Permo-Trias also occurs in various parts of Scotland, for example in parts of the Southern Uplands, the Midland Valley, southern Arran, Skye and the adjacent islands and mainland, and on the Moray Firth around Lossiemouth.

The Rhaetic

The Rhaetic presents a geological problem. It is a thin series of beds, usually less than 15 m (50 ft) thick, that represents a marine invasion over the Keuper desert flats. As such, the British evidence would seem to attach it to the overlying Jurassic, which was deposited largely in epicontinental seas. But in the Alps of eastern Switzerland, ammonites found in the Rhaetic are more closely related to Triassic than to Jurassic forms. The beds in Britain are a thin series of limestones, shales and marls which may form a small feature where they overlook the Keuper outcrop. The limestones are in the Cotham Beds (these are compact calcite mudstones often with dendritic manganese dioxide markings, which give rise to a rock known as landscape marble) and the Langport Beds (also known as the White Lias) above: the latter do not extend north of the Midlands. The widespread nature of the Rhaetic is shown by the presence of marine Rhaetic as far north as Mull, though its absence from Skye and adjacent areas may indicate the limits of the transgression.

Exhumed Permo-Trias relief

Although the Permo-Triassic rocks might themselves be described as not distinguished in their contribution to the relief of Britain, the landscape in which the rocks were deposited may be revived in the relief of modern Britain as the soft Permo-Trias beds are eroded from it. There is, however, a certain ambiguity in many of the suggested examples.

The classic case is that of Charnwood Forest originally described by Watts. Here, in the area immediately west of Leicester, old Pre-Cambrian rocks are buried beneath Keuper Marls. Many of the present streams, though not all, for examples are known where they are incised into the Pre-Cambrian, are re-excavating desert valleys filled in with Triassic marl. Between them the Pre-Cambrian rocks emerge as small, rocky, tor-like features on the ridges. Are they exhumed desert inselbergs? Again the Mountsorrel granite to the east is buried beneath the Trias and is very fresh in appearance there as though attacked by mechanical weathering. Further, some of the exposed surfaces are grooved as though by wind blast— desert erosion of Triassic age? Some ambiguity exists here because freeze–thaw shatter in the Pleistocene could have produced the tor-like features on the ridges and Pleistocene wind action, or glacial action, could have grooved the Mountsorrel granite, at least above the level of the Trias.

Similar deductions might be derived from the various conglomerates and breccias found at the base of the Permo-Trias in many western parts of England. Some of the main relief features, e.g. the Pennine scarps, the Mendips and the southern scarped edge of the South Wales coalfield may be very similar to Permo-Trias relief features.

So too may many features on a much larger scale. O. T. Jones, using the evidence of the Permo-Trias conglomerate banked up against the south of the South Wales coalfield, and other features such as the reddening of rocks on the Welsh plateaus (which, however, could have occurred in later periods of tropical climate e.g. the early Tertiary), suggested that the Welsh plateau was essentially an exhumed sub-Triassic feature above which the coalfield syncline projected as a higher block. It is probably true that most geomorphologists do not believe in this hypothesis today, but there are other features in Britain which might have had this sort of origin, for example the downfaulted Eden valley between the Pennines and the Lake District.

Seismic work has revealed the existence of submarine Permo-Trias basins. One of these, described by Hill and King, lies in the western part of the English Channel between Cornwall and Brittany. The rocks are identified by comparing the velocities of sound through them with the known sound velocities of British rocks. The Permo-Trias trough does not coincide exactly with the western Channel and there is no question at all of the whole of the English Channel being an exhumed Permo-Trias feature. Nevertheless, in broad outline the western Channel occupies the site of a New Red Sandstone trough, although whether the erosion surfaces of

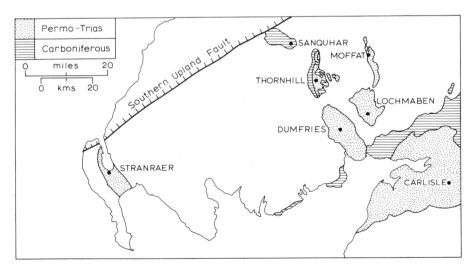

Fig. 10.1. Permo-Trias basins in the Southern Uplands

335

Devon and Cornwall are exhumed sub-Triassic pediments, as the authors suggest, is much more doubtful.

Features corresponding approximately with Permo-Trias features of an intermediate scale are present in southern Scotland. Some of the main valleys crossing the Southern Uplands contain thick, very little disturbed, Permo-Trias deposits and even Carboniferous deposits in the Thornhill and Sanquhar basins (Fig. 10.1). The Nith valley includes the Dumfries and Thornhill basins, and the small Sanquhar coalfield continues the same line. The Lochmaben and Moffat basins are drained by the Annan. The Stanraer basin forms an area of lowland, the rest of the trough being drowned in Luce Bay to the south and Loch Ryan to the north. Thus, it is again impossible to escape the general conclusion that in broad outline some of the present features of the Southern Uplands are very similar to their Permo-Trias predecessors.

Similar ideas can be applied very readily in Europe where a number of the Hercynian massifs, e.g. the Vosges and the Black Forest, have a crystalline core on to which laps a feather edge of Trias. Here, the crystalline region must represent in broad terms exhumed sub-Triassic relief. In general on the Continent geomorphologists are probably more conscious of the role of exhumed relief in the landscape than they are in Britain.

11

The Jurassic system

The Jurassic system illustrates admirably the absolute necessity for a thorough knowledge of the palaeogeography of the period for the comprehension of the lithology and hence of the relief produced on Jurassic rocks. The Jurassic was mainly a period of deposition in shallow continental shelf seas subject to varying degrees of uplift in different places at different times. From time to time parts of the sea were converted to land and at other times invaded by deltaic conditions. The result is an almost bewildering variety of rock types.

The three dominant elements in Jurassic deposition are clay, sandstone and limestone, often deposited in that order, whether on a short time scale as alternations within a given series of beds, or on a long time scale as in the succession, Kimeridge Clay, Portland Sand and Portland Stone. These alterations are too irregular and too variable in duration to be treated strictly as cyclothems. The total thickness of the clays within the system is much greater than that of either the sandstones or the limestones, but the last are the most important rocks in relief formation. The Jurassic limestones provide the variable series of scarps which constitutes the western half of the English scarplands.

The Jurassic succession in Britain, from the geomorphologists' point of view, is best described in the old British terminology, which does, in addition, conform approximately to the Continental divisions. The correlation between the two is shown in Fig. 11.1, on which the stages are shown as equal lengths of the column, though this must not be taken as implying that either the times involved or the maximum thicknesses of rocks deposited were equal. Because they are mostly non-marine, and therefore lacking in good zone fossils, there is doubt whether the topmost part of the Purbeck Beds should be included in the Jurassic or the Cretaceous.

The palaeogeography of the Jurassic is often spoken of as being dominated by axes of uplift (Fig. 11.2). There are usually reckoned to be three of these: a Mendip axis running along the line of those hills; a Moreton axis trending approximately north–south through Moreton in the Marsh; and an east–west Market Weighton axis. These three zones of unwarping dominated deposition in the Lower and Middle Jurassic. After that the

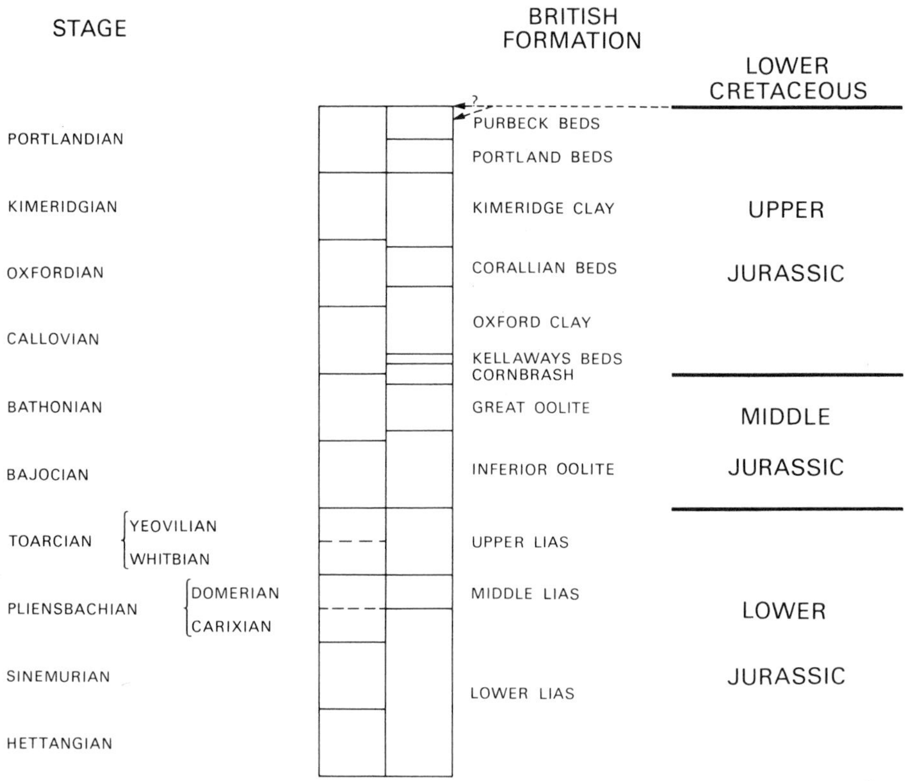

STAGE		BRITISH FORMATION	LOWER CRETACEOUS
PORTLANDIAN		PURBECK BEDS	
		PORTLAND BEDS	
KIMERIDGIAN		KIMERIDGE CLAY	UPPER
OXFORDIAN		CORALLIAN BEDS	JURASSIC
CALLOVIAN		OXFORD CLAY	
		KELLAWAYS BEDS	
		CORNBRASH	
BATHONIAN		GREAT OOLITE	MIDDLE
BAJOCIAN		INFERIOR OOLITE	JURASSIC
TOARCIAN	YEOVILIAN / WHITBIAN	UPPER LIAS	
PLIENSBACHIAN	DOMERIAN / CARIXIAN	MIDDLE LIAS	LOWER
SINEMURIAN		LOWER LIAS	JURASSIC
HETTANGIAN			

FIG. 11.1. Divisions of the Jurassic system

Market Weighton axis alone seems to have continued to function, while a general uplift zone came into existence in the Southern Midlands east of the Moreton axis. Rayner considers that the term, axis, creates a false impression of a linear feature, whereas the Jurassic upwarps were broad swells, which varied somewhat in location from time to time, and hence are better referred to either as shallows or swells. They probably sprang in part from movements associated with the landmass known as the London Platform. This applies especially to the Moreton swell.

An idealised diagram of zones of downwarping separated by zones of uplift, which did not emerge but only produced thinning of the deposits, is shown in Fig. 11.3. As long as it is realised that this is grossly oversimplified, the concept is useful in appreciating Jurassic lithological variation. It is true that there are zones where the Jurassic beds thin markedly especially over the Market Weighton swell, as shown in the diagram. It is true, and vitally important for understanding relief, that the lithology of the beds varies greatly from basin to basin because of their partial isolation. It might also be expected in an area of variable water depth that diachronous facies deposits would occur. On the other hand the diagram gives a false impres-

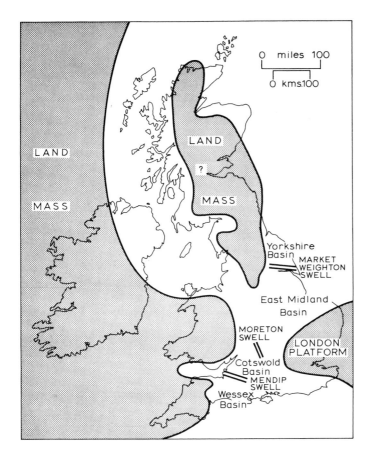

FIG. 11.2. Elements of Jurassic geography *(after Wills)*

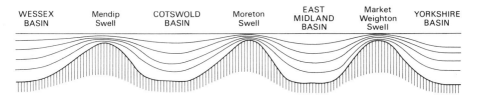

FIG. 11.3. Idealised section of Jurassic deposition in Britain

sion of the continuity of the deposits. Unconformities and non-sequences abound in the Jurassic. Furthermore, they are not confined to the main swells. This is where the concept of axes tends to create a false impression, for the unconformities in the basins imply that warping and shallowing must have occurred in them as well as on the swells. The separation into basins and swells is not, then, absolute but only a reflection of general tendencies.

339

From this it should be apparent that the variable lithology and relief to be described below, although more accurate than generalisations for the whole of the Jurassic, are themselves not valid for every part of the basin described. Because the dominant theme is the regional variation of Jurassic relief, the outcrop is treated in this fashion and not by considering relief changes along the outcrop bed by bed.

The Wessex basin

The succession of Jurassic rocks, especially in the far south in Dorset is probably the most complete in Britain, as it contains a full succession of Purbeck and Portland Beds which thin rapidly northwards. Yet much of the Middle Jurassic here contains far more clay and far less limestone than elsewhere in the outcrop. In fact the Portland Beds of the Upper Jurassic are the most important scarp former in the very south of the region.

The Lias, as is the case in most areas, has a very variable lithology. The Lower Lias, known as the Blue Lias, contains more limestone near its base than it does higher up. They are not massive limestones, however, but rapid alternations of limestones and shales, many of the former being marly and the latter bituminous. Such a succession is well-exposed in the cliffs at Lyme Regis. Although the individual limestone beds are resistant, being used in places as flagstones, the formation as a whole is not very resistant, though the limestones do form the low Polden Hills, trending north-west across the Somerset Levels (Fig. 11.5). Even in the Vale of Glamorgan, where the base of the Lower Lias is a littoral facies of pure limestone, the rocks form vertical cliffs (Plate 6) but their effect on relief inland is slight.

The Middle Lias is generally a much sandier formation with some limestone, but the top bed, the Marlstone, which is elsewhere a marked relief former, is not well-developed in this area. The Middle Lias as a whole, however, tends to stand out in relief above the lowland on the Lower Lias. It forms a low ridge encircling the anticlinal Vale of Marshwood in western Dorset, but this is dwarfed by the fringing scarp developed on the Bridport and Yeovil Sands above (Fig. 11.4). Outliers of Middle Lias on the Lower Lias form low hills, for example Glastonbury Tor, Brent Knoll, Pennard and Easton Hills, which run as a broken line parallel to and south of the Mendips (Fig. 11.5).

The Upper Lias contains a fine example of a diachronous formation. This is a group of generally soft sands known from north to south as the Cotswold, Midford, Yeovil and Bridport Sands. In the north they occur at the base of the Yeovilian and generally rise to higher stratigraphical levels southwards to straddle the Yeovilian-Bajocian boundary near the coast. North of Bridport they give rise to a tract of upland with convex slopes, steep-sided valleys and sunken lanes. At Bridport Harbour and Burton Bradstock they form fine vertical cliffs, with harder cemented layers

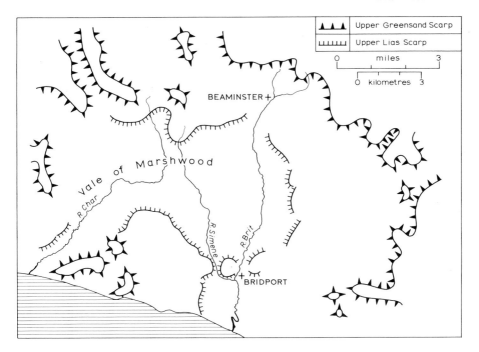

Upper Greensand Scarp

Upper Lias Scarp

miles

kilometres

BEAMINSTER +

Vale of Marshwood

R. Char

R. Simene

R. Brit

+ BRIDPORT

FIG. 11.4. Escarpments around anticlinal Vale of Marshwood

projecting in the face of the cliffs. Below these various sands the Upper Lias is mostly clay.

The Middle Jurassic, though far more argillaceous than farther north in the Cotswold basin, contains some limestone beds, most of which are not very thick. Through Somerset and Dorset they tend to form subdued scarp and vale features, the intervening clays forming the vales. The Inferior Oolite is very thin south of the Mendip swell and thins further southwards, a strange state of affairs for deposition in a basin and suggesting that at this stage the Wessex basin was choked up, possibly by the great thicknesses of Upper Lias sands deposited there. The Great Oolite consists of the Fullers' Earth below, an argillaceous formation with a middle harder bed known as the Fullers' Earth Rock, and the Forest Marble above, a group of limestones and shales.

Of these beds the Inferior Oolite tends to form low upland, especially in the north where it is thicker. It is separated from another low upland on the Forest Marble by a vale on the Fullers' Earth through which a median narrow ridge is developed on the Fullers' Earth Rock (Fig. 11.6). Within the Weymouth lowland the Forest Marble forms the innermost of the low ridges developed in the eroded Weymouth anticline (Fig. 11.7). But all these features are dwarfed by the Cretaceous Upper Greensand and Chalk escarpments east and north of them.

The lithology of the Upper Jurassic is much more constant along the

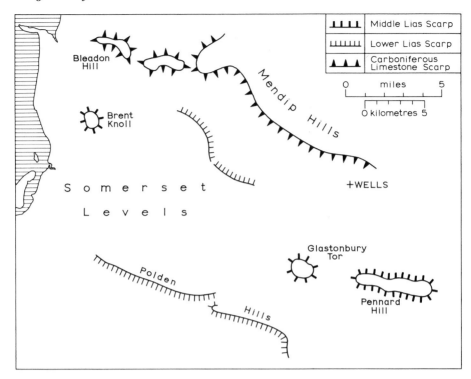

FIG. 11.5. Lias scarps in Somerset Levels

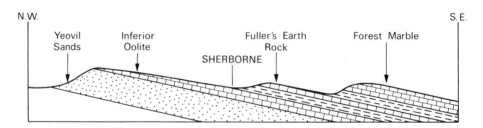

FIG. 11.6. Section across Middle Jurassic near Sherborne

outcrop than that of either the Middle or Lower Jurassic. This is because the swells were less pronounced and deposition was as a result more uniform. The Cornbrash and the Corallian, the former a rubbly limestone and the latter a unit of rhythmic sedimentation containing some limestone, form low uplands both in the Weymouth area and in the north–south outcrop through Dorset, Somerset and Wiltshire, except where it is overstepped by Cretaceous rocks. The Oxford and Kimeridge Clays are nonresistant beds forming lowlands. These alternations produce minor scarp and vale relief and, where they meet the coast as on the shores of the Fleet, an indented coast line (Fig. 11.7).

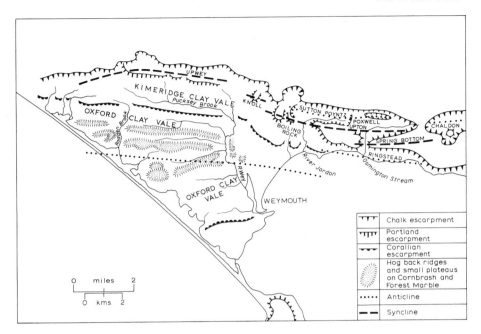

Fig. 11.7. Physical diagram of Weymouth lowland

The Portland and Purbeck Beds are only well exposed in south Dorset from Abbotsbury to Swanage. Farther north they occasionally appear from beneath the overstepping Cretaceous, but their effect on relief is less marked. The Portland Beds are divided into the Portland Sands below and the Portland Stone above: the latter consists of a Cherty Series below (the cherts are nodular and practically indistinguishable from the flints of the Chalk) and a Freestone Series above. The last, which is best developed in the Isle of Portland, provided the stone so extensively used in public buildings in southern England, but they also contain less valuable material, notably the Roach, a limestone full of the hollow casts of molluscs, which, however, seems to be coming into fashion as an ornamental building stone today. The Purbeck Beds are extremely variable and seem to have been deposited in a series of coastal environments, occasionally marine, but usually fresh or brackish water coastal marshes and lagoons. Their effect on relief is less marked than that of the Portland Beds.

The Portland Stone gives rise to the most impressive scarp in the Weymouth–Swanage district, although it is overshadowed by the fault-line Chalk escarpment on the northern side of the Weymouth anticline (Fig. 11.7). On the southern side of the fold the dip is less, the beds thicker, and here the Portlandian forms the tilted slab of the Isle of Portland with its crest reaching almost 150 m (500 ft) above the eroded Kimeridge Clay at its foot. Purbeck Beds occur on the dip-slope, but it is the Portlandian which is mainly responsible for the relief effects. On the northern side of the fold

343

the Portlandian forms a similar though somewhat lower escarpment (probably because of the increased dip and reduced thickness), the plan of which is complicated by the minor periclinal folds on the northern side of the Weymouth anticline. The relief of many of these folds, for example the Sutton Poyntz pericline (Fig. 11.7), is inverted, a rim of Portland Beds overlooking a lowland eroded on Kimeridge Clay. This erosion is facilitated by the spring-line, the most important in the district, that occurs at the junction of the permeable Portlandian and the impermeable Kimeridgian.

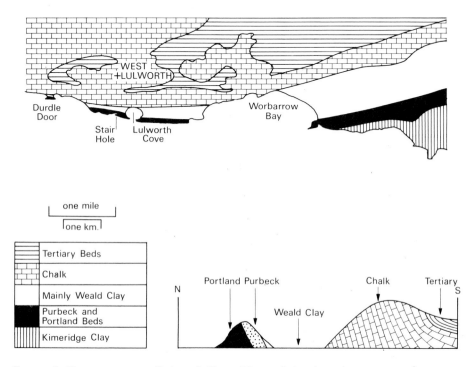

FIG. 11.8. Dorset coast near Lulworth Cove. Map and sketch section

Portland Beds, often with a capping of Purbeck Beds, dominate coastal relief to the east. Between Durdle Door and Worbarrow Bay they dip steeply inland and are succeeded by the non-resistant Cretaceous Weald Clay. Where the sea has breached the coastal ridge it has enlarged out Lulworth Cove on the Weald Clay (Fig. 11.8), while a little to the west the two small tunnels through the limestones at Stair Hole show a less advanced stage in the same process. Worbarrow Bay itself may be considered as a larger and more advanced stage of the process represented by the Stair Hole–Lulworth Cove series. East of St Alban's Head massive cliffs are formed on near-horizontal Portland Beds, again capped by the Purbeck.

344

The Cotswold basin

The dominant characteristic of the Cotswold basin is the great development of limestone in the two Middle Jurassic formations, the Inferior and Great Oolites. They give rise to the Cotswold Hills, which at their highest points exceed 300 m (1000 ft). Too often this is taken as the norm of the relief developed in the Jurassic belt, whereas it really forms less than a quarter of the outcrop. Nothing like it occurs in the Wessex basin, as we have already seen, while northwards other horizons and other lithologies are important. The Cotswold basin shares one feature with the Wessex basin, the absence of workable ironstones and so these two southern basins contrast with the two northern basins. Finally, the Cotswold basin, like those to the north and unlike the Wessex basin, has no real development of beds above the Kimeridgian.

The Lias in the Cotswold basin is very thick, about 500 m (1650 ft), and is mostly clay and shale. It forms for the most part the eastern margins of the vast vale that stretches from the foot of the Cotswolds across to hills such as the Malverns on the other side of the Severn. The exception is the Marlstone, a shelly ferruginous limestone, at the top of the Middle Lias. This is hard enough, although it is not very thick, to form a ledge in the main escarpment, for example near Cheltenham, and also flat-topped isolated hills, for example Dixton and Dumbleton Hills, as the dip is generally slight (Fig. 11.9).

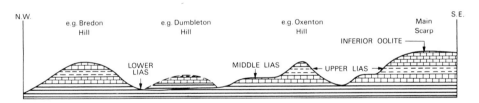

FIG. 11.9. Sketch section of Cotswold scarp (faults omitted)

The Inferior Oolite is almost wholly limestone of varying types. In the middle of the basin it is just over 100 m (350 ft) thick, which is not great compared with the Lias. The Inferior Oolite is, nevertheless, responsible for the main Cotswold escarpment. The full succession is nowhere present as the beds are full of non-sequences. The Great Oolite is much more calcareous than in the Wessex basin. Between the Fullers' Earth, which is again mostly argillaceous, and the Forest Marble, which at least in the north of the area is predominantly limestone, lie the 45 m (150 ft) or so of Great Oolite Limestones. The most interesting bed is the Stonesfield Slate, a thin, fissile, sandy oolite capable of being split into thickish slabs which provide the perfect roofing material for the limestone Cotswold buildings. The Great Oolite outcrops on the dip-slope of the Inferior Oolite, rather

345

like the Purbeck Beds on the Portland in the Wessex basin, and hence it is at the surface over a wide area.

One curious feature of the Cotswold escarpment is the comparatively large number of outliers capped by the Inferior Oolite. Bredon Hill is the most famous, but they are probably more common than on any other scarp in Britain with the possible exception of the Upper Greensand in Dorset. Although the dip is slight, an ideal condition for the formation of outliers, there are other escarpments with low angles of dip, e.g. many Chalk scarps, but with few outliers. Perhaps the greater hardness of the Inferior Oolite leads to the longer survival of outliers, whereas the Chalk is readily destroyed especially by freeze–thaw, but this does not seem a wholly satisfactory explanation.

The remainder of the Jurassic succession forms the relief that might be expected of it, with the Oxford and Kimeridge Clays as lowlands and the Cornbrash and Corallian as low uplands.

The East Midland basin

First of all the general characteristics which distinguish this basin of deposition from the others: the most prominent of these is that this is the only basin where the Lias provides the main relief-forming horizon, in this case the Marlstone of the Middle Lias. Second, workable ironstones occur at a number of levels and most of the British iron ore production at present comes from this region. There are ironstones in the Yorkshire basin, but the days of their exploitation are over. In the East Midland basin the Frodingham Ironstone of north Lincolnshire is Lower Lias, the Middle Lias Marlstone provides the ores of Banbury and north Oxfordshire, while the Northampton Sands, Inferior Oolite in age, are the source in the Corby district. Third, the Middle Jurassic, while still of importance in the relief, is less dominating than in either the Cotswold or Yorkshire basins. Fourth, the Upper Jurassic is unusually argillaceous and this, plus the low angles of dip, is responsible for the broad lowland between the Jurassic uplands and the Chalk escarpment.

The Lower Lias, 150 m (500 ft) thick in the south, forms the eastern edge of the Midland Plain next to the Trias. The angle of dip is low and the outcrop is wide. The beds thin northwards and the Frodingham Ironstone at the base, which nowhere exceeds 9 m (30 ft), is a condensed sequence representing 60 m (200 ft) of rock farther south. The Middle Lias is also thinner than in the Cotswold basin (45 m: 150 ft) and this, as do all the beds, thins northwards towards the Market Weighton swell. The Marlstone at the top is very important in the relief. South of Lincoln it forms a ledge at the foot of Lincoln Edge but increases in importance farther south (Fig. 11.10). South of Grantham the Marlstone, often capped by the Upper Lias, is increasingly dominant and forms a broad belt of uplands fringed by a scarp overlooking the Lower Lias and Trias lowland of the Midlands.

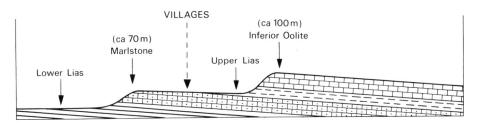

Fig. 11.10. Sketch section of Jurassic scarps north of Grantham

This is well-developed east of Leicester, but fine examples occur farther south, for example Edge Hill between Banbury and Stratford, where open-cast mining of ironstone is carried out on the dip-slope. The Upper Lias is mainly argillaceous and thins northwards.

The Middle Jurassic changes dramatically over the Moreton swell. The Inferior Oolite is reduced to 3 m (10 ft) and then thickens again in the East Midland basin but with a different facies. The Great Oolite thins less than the Inferior Oolite over the Moreton swell and is far less calcareous in the East Midland basin than in the Cotswold basin. The Middle Jurassic succession is:

Great Oolite Clay	
Great Oolite Limestone	Great Oolite
Upper Estuarine Series	
Lincolnshire Limestone	
Lower Estuarine Series	Inferior Oolite
Northampton Sand	

The Northampton Sand is usually less than 10 m (30 ft) thick, and its main importance is as a valuable iron ore, yielding 30–50 per cent iron, the main minerals being limonite, chamosite and siderite. Of the limestones the Lincolnshire Limestone (some 36 m: 120 ft thick) is more important than the Great Oolite limestone (usually less than 8 m: 25 ft thick). A number of limestone types are present in the former, including some important building stones (for example, Ketton, used in several Cambridge colleges where its pink and buff colours are distinctive, and Clipsham) and the Collyweston Slate at its base, a local bed very similar to the Stonesfield Slate of the Cotswold basin.

The Lincolnshire Limestone, together with the Marlstone, dominates the relief formation. It is most marked in Lincoln Edge, the linear narrow cuesta lying north and south of that city, but it also affects the relief elsewhere. It does not form karst, but solution holes are known in the Glen valley near Burton Coggles.

The two Estuarine series are misnamed. They represent coastal deltaic swamps smaller in scale but similar in type to those prevailing in the Upper Carboniferous Coal Measures. However, the succession rarely

347

reached the stage of coal formation and a black, carbonaceous shale and the marks of rootlets in the Lower Estuarine Series represent the nearest to it. Both Estuarine series are thin and non-resistant.

The search for iron in Northamptonshire has revealed a number of interesting minor structures in that region (Hollingworth, Taylor and Kellaway, 1944). The Inferior Oolite rests on incompetent Upper Lias clay and the streams have often eroded down to this. The weight of the Inferior Oolite on the ridges has been responsible for anticlinal valley bulges, as in the Millstone Grit of the southern Pennines (Chapter 9). In addition cambering of the harder beds of the interfluves into the valleys causes anomalous dips, while many small-scale, superficial slip structures, resembling step-faulting, are known.

The remainder of the Jurassic succession runs up to the Purbeck Beds, but the latter are thin and local, while the Portland Beds only occur near Aylesbury as very thin limestones. The Cornbrash follows the usual pattern, providing a belt of low uplands, good arable soils and village sites. The Corallian is a limestone formation in the south-west where it forms the hills, some 90 m (300 ft) high, that the Thames cuts through at Oxford: the stone itself was used largely in building Oxford colleges, but has proved inferior to many other Jurassic building stones in its resistance to weathering so that wholesale renovation was undertaken in the 1960s. A few miles to the north-east the Corallian undergoes a major facies change at Wheatley and becomes the Ampthill Clay.

As a result the Upper Jurassic through the East Midlands has three successive clay formations, the Oxford, Ampthill and Kimeridge Clays, in which other rock types play a very minor part: there is a limestone reef in the Corallian at Upware in Cambridgeshire, and at the base of the Ampthill Clay in western Cambridgeshire and adjacent districts a calcareous bed, the Elsworth Rock. Although the clay beds thin in Bedfordshire and adjacent counties, where a zone of shallows developed, the very low dip still ensures a very wide exposure. In southern Lincolnshire, Cambridgeshire and Huntingdonshire the ensuing lowland is some 50–60 km (35 miles) wide. It is in part occupied by the Fens, while the Wash represents a breach into it through the Chalk on its eastern margin. Northwards the lowland becomes pinched between the oolites of Lincoln Edge and the chalk of the Lincolnshire Wolds, as the beds thin towards the Market Weighton swell. Within this area the Oxford Clay is used in the brickfields of Peterborough and Bedford. Some of it is bituminous and this reduces the fuel requirements for brick manufacture, but probably the chief locating factors were the former Great Northern and Midland railway lines, which facilitated both the import of coal from the York, Derby and Notts field and the export of bricks to London in the days before road transport dominated rail.

The Yorkshire basin

The Jurassic of the Yorkshire basin differs very considerably from that of the other areas. The Lias and the Upper Jurassic, apart from the lack of anything above the Kimeridgian, probably differ less than does the Middle Jurassic, which here is predominantly a non-marine set of beds. The thinning of the Jurassic over the Market Weighton swell is most marked and the only outcrop over it is 30 m (100 ft) of Lower Lias, all the rest being either missing or overstepped by the Cretaceous.

Although the Lias is so thin over the swell it thickens northwards to almost 450 m (1500 ft). The whole series is predominantly shaly and non-resistant, although it becomes increasingly sandy towards the top. The Middle Lias contains the Cleveland iron ores, formerly worked as Britain's main sedimentary iron ore but now no longer economically worth while. The Upper Lias contains a number of unusual beds, but these are not significant in the relief. They include the Alum Shales, which were formerly worked, and the Jet Rock Series, which, although bituminous, is not a source of oil but of that hard, compact material derived probably from drifted wood and carved into elaborate, funereal ornaments by the Romans and the Victorians.

The Middle Jurassic is a thick series of deltaic beds with a number of marine incursions at different horizons. It reaches about 250 m (800 ft) thick at a maximum. The general environment was probably very much like that of the Estuarine series of the Midlands, which must not be thought of as an attenuated version of the Yorkshire beds, as they were probably derived from the London Platform whereas the source of the beds in Yorkshire is unknown.

The marine beds of the Middle Jurassic (probably about 60 m: 200 ft of the total) are themselves mostly calcareous and ferruginous sandstones, though an oolite at the horizon of the Lincolnshire Limestone occurs in the Hawardian Hills. The deltaic facies are predominantly clays and silts, though, just as in the Coal Measures, lenticular sandstones representing washouts caused by the distributary channels on the delta occur at different horizons. Even coal seams up to 0·5 m (2 ft) thick are known, but the quality is poor. Together these beds form the high ground of the North York Moors, which exceed 420 m (1400 ft) in elevation and are thus more impressive than the Cotswold Hills proper. But they are formed of sandstones and thus the moorland relief, diversified by occasional rock scars, is different in form and, more important, in vegetation, from the Cotswolds. The western border of these heather moorlands is the Hambledon Hills, where a terraced scarp formed on the sandstone–shale alternations overlooks the Vale of York. The same beds are again largely responsible for the Hawardian Hills at the western end of the Vale of Pickering.

The Cornbrash is unimportant in the Yorkshire basin and lithological changes affect the remainder of the Upper Jurassic. The Oxford Clay (including the Kellaways Beds) is about 75 m (250 ft) thick and only approximately the upper half is clay, there being a series of sandstones below that. It forms a belt of lower ground at the foot of the discontinuous Corallian escarpment, but nothing comparable with the broad clay vale of the East Midlands. The Corallian is exceedingly well developed with a total maximum thickness of the order of 200 m (650 ft), that is about double the thickness in the south of England. It consists of an alternation of cherty sandstones (Calcareous Grit so-called) and oolites, which include reef-like accumulations of corals and sponges. These beds, especially the Lower Calcareous Grit, form a marked but discontinuous escarpment overlooking the Oxford Clay. In its general form it is comparable with the dissected Aymestry Limestone escarpment developed on the dip-slope of the Wenlock Limestone escarpment in the Welsh Borderland. The Kimeridge Clay, 140 m (420 ft) or more of it, behaves in its usual non-resistant way to form the lowland of the Vale of Pickering.

Other Jurassic outcrops

Small Purbeck inliers are known in the Weald and Lias outliers in the Cheshire and Carlisle plains, but the largest outliers occur in the Inner Hebrides in and around Skye, the island of Raasay being especially important, and in south-east Sutherland on the opposite side of Scotland.

A considerable thickness of rocks is known in the Hebrides (900 m: 3000 ft) but it probably does not represent another basin. The facies is mainly deltaic and inshore marine, more like that of Yorkshire than the rest of the outcrop. Sandstone and shale are the main rock types. In the general setting of the resistant Pre-Cambrian and Tertiary igneous rocks of the area they cannot be expected to form the relief that they do in England. Nevertheless, in south-eastern Raasay 180 m (600 ft) of coarse sandstones of Inferior Oolite age form some very fine cliff scenery.

Around Brora on the opposite coast the lithology is similar, though the succession is less complete. The beds form the low ground between the Old Red Sandstone and the coast. As in Skye there is coal in the equivalent beds to the Great Oolite, but it is thicker (there is a 1 m: 3 ft seam at Brora) and worth working for local coal supplies. At one time this coalfield proudly held the record of the largest percentage increase in production since 1945 in Britain. The Kimeridge of this district consists of the submarine breccia of Old Red Sandstone blocks, probably formed when a submarine scarp was shattered by an earthquake and collapsed into the muds accumulating at its foot.

12

The Cretaceous system

At the end of Jurassic times deposition in Britain was confined to an area of southern and eastern England in which the Purbeck Beds, which are predominantly non-marine, were laid down. The Cretaceous opens with the continuation of the Purbeck delta-lagoon environments in the south-east separated by a land mass, the continuation of the Jurassic London Platform, from a wholly marine environment in Lincolnshire and Yorkshire. The oldest rocks in this northern area are in Lincolnshire but the

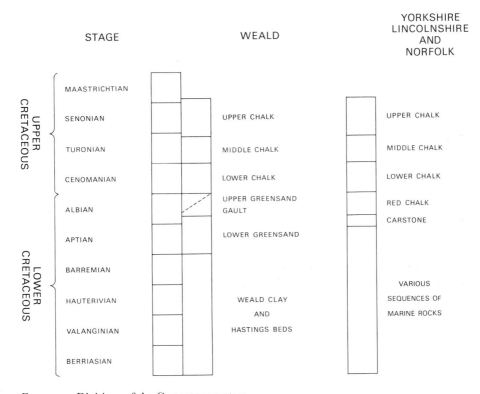

Fig. 12.1. Divisions of the Cretaceous system

sea spread shortly after into Yorkshire and Norfolk and later into Cambridgeshire and Bedfordshire. South of the London Platform marine environments came in roughly with the deposition of the Lower Greensand and thereafter continued, the link between the south-eastern and north-eastern areas of deposition being made finally in the later part of Lower Greensand times. This linking up represents the effective end of the London Platform as a major control on deposition: indeed, later on in the Tertiary era it was to become a basin and not a block. Another area of upwarp, the Market Weighton swell, ensured different deposition in the Yorkshire and Lincolnshire basins in the Lower Cretaceous, but finally ceased to affect sedimentation after the Lower Chalk.

This broad sketch of the palaeogeography should prepare one not only for considerable vertical variations in lithology, because of changing environments, but also for lateral ones. For example, the Lincolnshire and Yorkshire areas in the Lower Cretaceous have very different rocks and hence very different relief from that of the Weald. Again, to the west in Dorset and Devon, which were much closer to land, Cretaceous sediments and relief are different from those of south-east England. As the classic account of the relationship between Cretaceous rocks and relief is based on the Weald, it must be stressed that the account only really applies to the Weald and that even within that area lateral variations of lithology affect the relief to a large extent. The divisions of the Cretaceous are tabulated in Fig. 12.1, in which the vertical scale is conventional and corresponds neither to relative time nor to relative thickness.

The Weald

The Wealden rocks are an alternation of sands and clays, which are normally divided on a lithological basis as follows:

Weald Clay	Upper Weald Clay
	Horsham Stone
	Lower Weald Clay
Hastings Beds	Upper Tunbridge Wells Sands
	Grinstead Clay (impersistent)
	Lower Tunbridge Wells Sands
	Wadhurst Clay
	Ashdown Sands
	Fairlight Clay (impersistent)

All in all some 600 m (2000 ft) of strata are involved. The Wealden environment consisted of a freshwater lake to the south and a land mass, the London Platform, to the north. The beds are thicker in the centre of the Weald than farther south and it may well be that the Wealden lake was partly cut off from a similar freshwater lake in the Paris Basin by a submerged swell. Alternate deltaic advances, involving detritus from the

London Platform, and rises in level of the lake are responsible for the alternation of sands and clays. The clays, which are the lake beds, tend to be thicker in the south, as might be expected. Deltaic advances are much more strongly in evidence in the Hastings Beds than in the Weald Clay, which is more completely lacustrine except that brackish conditions, anticipating the Aptian marine transgression, came in at the top.

The Hastings Beds, alternating dominant sandstones and subordinate clays, are, like the rest of England roughly south of the North Downs line, folded and faulted fairly gently along east–west axes. This lithological and structural pattern has given rise to a series of ridges, mostly on the sands, and valleys, mostly on the silts and clays. Ashdown Forest, developed on the Ashdown Sands between East Grinstead and Uckfield, reaches almost 240 m (800 ft) near Crowborough. Elsewhere ridge elevations are usually between 120 and 180 m (400–600 ft). It is essentially a wooded and cultivated landscape, although considerable areas of heathland remain in Ashdown Forest. Only locally is the sandstone sufficiently well-cemented to form bare rock outcrops, for example near West Hoathly and Tunbridge Wells in the Tunbridge Wells Sands, where small tor-like features result from weathering along weak horizons and joints.

The Weald Clay is a much more homogeneous unit than the Hastings Beds. It forms a broad, irregular horseshoe of lowland between that stratum and the Lower Greensand escarpment. Within the lowland some relief variation is provided by freshwater sandstones and limestones, the latter being the well-known Paludina limestones, also known as Sussex Marble, used in ornamental work in the local churches. These beds may cause minor ridges and scarps within the Weald Clay, but not all the outcrops affect the local relief. The effects are probably at a maximum at Outwood some 8 km (5 miles) south-east of Redhill, where limestones and sandstones combine to form a south-facing scarp some 120 m (390 ft) in elevation. Yet in other places the beds hardly affect the relief. The structures which diversify the relief in the Hastings Beds probably continue in the Weald Clay, but they are very difficult to detect in a poorly-stratified homogeneous deposit. Structures are only reflected in relief when beds of varying resistance are involved.

The Lower Greensand is divisible into four units: Folkestone Beds, Sandgate Beds, Hythe Beds, and Atherfield Clay. Of these the Atherfield Clay is lithologically like the Weald Clay and needs no special consideration, but vertical and lateral variations in the other three units are responsible for the very great variety of Lower Greensand relief.

In east Kent the Hythe Beds have a 'rag and hassock' facies, consisting of alternations of thin beds of sandy limestone and loamy sand. This facies is of only moderate thickness and the ensuing cuesta rarely exceeds 150 m (500 ft) east of Maidstone. Because of its calcareous character the soil is fertile and orchards, arable and hop fields characterise the outcrop. Westwards the facies become sandier and stone bands, some of them cherty, appear in the Hythe Beds. The cuesta becomes higher (250 m: 800 ft) around Sevenoaks and Westerham and higher still in the western Weald, where Leith Hill, the highest point, reaches 290 m (965 ft). The exception to this lies between Reigate and Dorking where the beds lack stone bands and hence do not form high relief because of their lack of resistance. Throughout this section the soils are podsolic, acid heaths and conifer plantations characterising the outcrop. Owing to the incompetence of the underlying clays and the ease with which they become saturated, rotational slips are a feature of many parts of the escarpment. The plan of the escarpment is to some extent determined by folds, for example the big re-entrant in the far west where the Fernhurst anticline is denuded (Plate 51). On the southern flank of the Weald the escarpment remains prominent in West Sussex as far east as the river Arun, where an abrupt lithological change occurs: farther east not only the Hythe Beds but the whole of the Lower Greensand make only a minor feature.

The Sandgate Beds, if one can generalise satisfactorily about such variable strata, are more loamy or silty than either the Hythe Beds or the Folkestone Beds. Hence, on the whole they form low relief, the best example being in West Sussex where their outcrop is followed by the river Rother, the main subsequent stream of the area (Fig. 12.2). In west Surrey, however, they are much more cemented, usually with calcareous material but

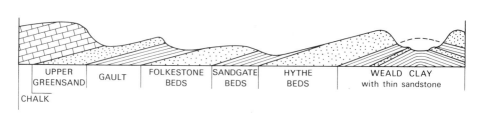

FIG. 12.2. Idealised section of Cretaceous relief in West Sussex

occasionally with siliceous. This facies is known as the Bargate Beds and forms plateaus in the Godalming area. Another variant is the development of fullers' earth between Redhill and Westerham at Nutfield.

The Folkestone Beds are more uniform than the other divisions and usually consist of coarse, uncemented, ferruginous sands. In places, such as in West Sussex, they form low heaths overlooking the strike valley on the Sandgate Beds. They are particularly infertile and the soils podsolic over

PLATE 51. Vale of Fernhurst, West Sussex. An anticlinal vale eroded in Weald Clay, overlooked by the Hythe Beds (Lower Greensand) escarpment in the background

much of their outcrop. Irregular ferruginous cementation occurs and in places is massive enough to affect the relief, for example in the low inselberg-like hills north of Hindhead known as the Devil's Jumps.

The two succeeding formations, the Upper Greensand and the Gault, must be taken together. Where both formations occur, for example in the western Weald and the Isle of Wight, the Upper Greensand lies above the Gault, but a wider view of the distribution of these two beds suggests strongly that they are facies. The Upper Greensand is missing in east Kent and the Gault is missing roughly to the west of central Dorset. The Gault represents a deeper water facies, the Upper Greensand a more terrigenous facies. This has important relief effects because the Gault is a non-resistant clay, while the Upper Greensand is a sandstone of varying type, sometimes glauconitic, sometimes cherty, but always sufficiently resistant to form a cuesta that increases in importance westwards and does not attain its maximum till well outside the Weald. Within the Weald it forms a low bench at the foot of the South Downs in West Sussex and a marked cuesta reaching about 150 m (500 ft) near Selborne in the western Weald, where the Chalk scarp rises above it to about 210 m (700 ft) (Fig. 12.2). In the eastern Weald there is no corresponding feature because virtually the whole deposit is Gault Clay.

It is sometimes alleged that the Gault forms a continuous strike vale at the foot of the Chalk escarpment and is drained by various subsequent streams. In fact, these longitudinal streams, which on small-scale maps look as though they should be subsequents on the Gault, often prove to be in anomalous positions. It has already been mentioned that the West Sussex Rother flows on the Sandgate Beds. Some of the other streams, such as the Len, the Darenth and the Tillingbourne are also not on the Gault and a perusal of the relevant 1-inch geological maps is a good corrective to hasty conclusions.

Finally, the Upper Cretaceous, which consists of the Chalk, is a much more homogeneous mass of rock than the Lower Cretaceous beds which were laid down in a variety of shallow water environments. The thickness of the Chalk is variable but it reaches 450 m (1500 ft) in places. Usually two-thirds or more of this are Upper Chalk, the remainder being equally divided between the Middle and the Lower. Various theories of origin have been put forward for the Chalk, including suggestions that it is a deep-sea ooze and that it is largely a foraminiferal deposit. Modern work indicates that much of the finer material consists of minute calcite plates derived from exceedingly small planktonic Algae known as coccoliths. The Chalk sea is usually thought now to have been shallow (of the order of 300 m: 1000 ft), and the generally small amounts of sand and clay fractions suggest low-lying or desert shores or both, so that little detritus was introduced

PLATE 52. Chalk escarpment of the South Downs west of the Arun gap, Sussex. In the foreground is the dip-slope east of the gap

by influent streams. However, this must not be regarded as the final word on the matter. One thing difficult to explain is the presence of occasional erratics: the greatest concentration of these is in the basal phosphate bed known as the Cambridge Greensand and confined to the neighbourhood of that town. They might have been introduced by floating ice, but fossils from regions adjacent to the main area of Chalk deposition suggest a tropical climate. They might have been derived from the roots of floating trees, a suggestion difficult to reconcile with that of desert shores. Or they might have been carried by masses of floating seaweed.

On the whole, the thick, generally homogeneous Chalk forms an impressive escarpment reaching to nearly 270 m (900 ft) in the South Downs. Where the Chalk is bare of superficial deposits the outcrop is characterised by dry valleys and a full development of upper convexities on the slopes, probably indicating the dominance of dry soil creep over rainwash, a result of the permeability of the Chalk (Plate 52).

Within the Chalk there are, however, certain lithological variations, which have been alleged to cause relief effects. These beds, which are usually conglomeratic, possibly intraformational conglomerates formed when the Chalk sea shallowed and its floor was affected by wave base, are found at three or four horizons. The Totternhoe Stone, which is more characteristic of the south-east Midlands than of the Weald, is in the middle of the Lower Chalk; in the Cambridge area it is called the Burwell Rock. The Melbourn Rock at the base of the Middle Chalk is fairly widespread. The Chalk Rock and Top Rock at the base and at the top of the lowermost zone of the Upper Chalk stretch as far north as Cambridge. The few field investigations that have been made of these beds have not substantiated any real relief effects due to superior resistance. However, the Totternhoe Stone is underlain by the Chalk Marl, best developed in Cambridgeshire and very argillaceous in character. The ensuing spring line controls the level of sapping so that the Totternhoe Stone often marks an important break of slope—but this is much more characteristic of the borders of East Anglia and the Midlands than it is of the Weald. Similarly, the Melbourn Rock is underlain by a few feet of marly chalk at the top of the Lower Chalk and so makes another but far less important spring line.

There are, however, important relief effects within the Upper Chalk, which seem to depend on lithological effects not easy to define. The lowest three zones of the Upper Chalk usually form the capping of the main escarpment (Fig. 12.3). This is typically flinty Chalk but it is unlikely that the flints contribute to the resistance of the Chalk, which is probably due to its massive nature and high permeability. Above this come two Chalk zones, which seem to be weaker, while the highest parts are again soft but so permeable as to form a second escarpment. This escarpment does not occur in the North Downs where the zones are not exposed but it is important in the South Downs as a dissected scarp half way down the dip-slope (Fig. 12.4). It is difficult to say why the eroded zones are weaker.

358

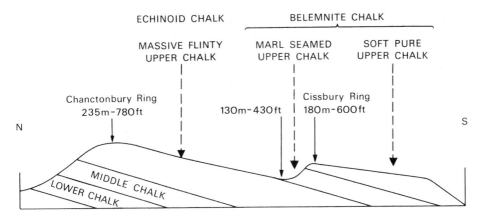

FIG. 12.3. The two Chalk escarpments of the South Downs

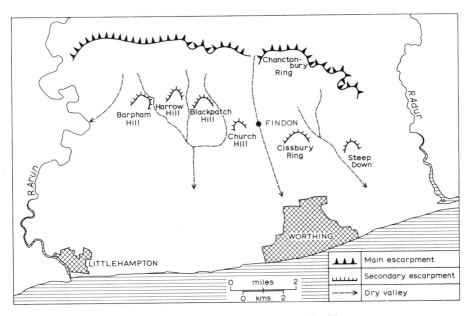

FIG. 12.4. Map of South Downs Chalk escarpments near Worthing

They contain more frequent marl seams and partings but these are exceedingly thin. They contain twice as much insoluble residue, but only 4 per cent against 2 per cent. When the water table was much higher these slight permeability differences may have caused differences in run-off and hence in erosion. Alternatively, they may have rendered the rock more liable to freeze–thaw action (see Chapter 14). The reasons are obscure, but the relief, an escarpment ranging up to 180 m (600 ft) with a relative relief of up to 60 m (200 ft), is obvious.

359

West of the Weald

Considerable changes in the Cretaceous formations take place to the west, so that the Wealden relief model is no longer applicable.

The Wealden appears in both the Isle of Wight and in Dorset. In neither place is it possible to distinguish between Hastings Beds and Weald Clay. In the Isle of Wight the whole of the formation is dominantly clays and divisible into the Wealden Marls below and the Wealden Shales above, the combined thickness being about 225 m (750 ft). They are lacking in resistance and form the centres of the southern lowland portions of the Isle of Wight. The formation becomes thicker and sandier in Dorset (690 m: 2300 ft near Swanage), but then thins to less than a tenth of this figure west of Lulworth. It becomes increasingly coarse westwards and includes some conglomerate. The source of the sediment obviously lay to the west, and on the basis of heavy minerals and rock types, it appears to have been in Dartmoor and adjacent parts of Devon and Cornwall. The Wealden is overstepped by the Albian in west Dorset. In Dorset it is conformable on the Purbeck Beds and forms an easily eroded zone, e.g. at Lulworth Cove (Fig. 11.8).

The Lower Greensand is present in the Isle of Wight, where it is at its thickest (240 m: 800 ft) in the south near Atherfield. On the whole it forms low undulating country, although there is a broken escarpment developed on it inland from Brixton Bay in the south-west. Only 60 m (200 ft) of Lower Greensand is left at Swanage and the formation disappears west of Lulworth. Indeed, the Aptian was a period of periclinal folding and faulting in the Weymouth area, thus creating an important unconformity between the Wealden and the Albian.

The Albian, i.e. the Upper Greensand and the Gault, occurs in the Isle of Wight and westwards almost as far as Dartmoor. The changes observed to the west in the Weald, namely the increase of Upper Greensand at the expense of Gault Clay, continue until the Gault is virtually eliminated in Devon. In the Isle of Wight the two facies are about equally represented, the Upper Greensand forming ridges and uplands depending on the degree of dip. However, it is in Dorset and Devon that the Upper Greensand really comes into its own as a relief former. The bed is not very thick, 60 m (200 ft) in east Devon and half this in the Haldon Hills to the west, but it contains irregular siliceous concretions, chert beds and in many places a massive nodular grit, the Calcareous Grit, occurs at the top. Practically all the main uplands of this region (Fig. 12.5) are capped by Upper Greensand, the westernmost being the Haldon Hills, where the beds are pebbly and contain shallow-water compound corals and thick-shelled Mollusca, both indicating the proximity of the contemporary shoreline.

The Chalk in the Isle of Wight, the Weymouth region and the Hampshire Basin is normal in character and forms the expected type of relief, variation being introduced by changes in the angle of dip, especially in the

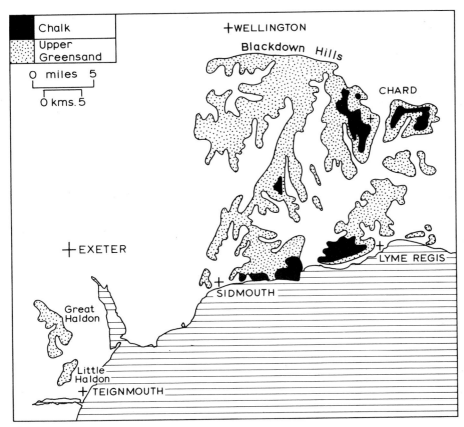

FIG. 12.5. Upper Greensand hills of Dorset and Devon

Isle of Wight and Purbeck, and by the dissection of the various periclinal folds that run through the outcrop. But farther west in Dorset and Devon the Chalk is much thinner and restricted to a number of outliers often capping the hills in which the Upper Greensand occurs. There are signs in its lithology of proximity to the shoreline: the Lower Chalk becomes glauconitic and gritty and is not unlike the Calcareous Grit below, while in the lower part of the Middle Chalk the Beer Freestone is a massive bed, up to 4 m (13 ft) thick, of coarse shell fragments, which hardens sufficiently on exposure to be used for ornamental interior stone work.

North of the Weald

North of the London Platform conditions throughout the Lower Cretaceous were marine and the area was separated from the predominantly deltaic Wealden region until Upper Aptian times. The two main basins of deposition in Lincolnshire and east Yorkshire were completely separated by the Market Weighton swell until about the same time. Thus it is in the

361

Lower Cretaceous that there is once again the greater variability in conditions of deposition.

The Wealden reappears in north Norfolk as the Sandringham Sands below and the Snettisham Clay above, the two probably being separated by a non-sequence. The total thickness is only 40 m (130 ft) and the beds on the whole are non-resistant, the sands tending to stand out as low heaths.

Across the Wash in Lincolnshire the corresponding beds are twice as thick at the southern end of the Lincolnshire Wolds and much more varied in lithology. They consist of a series of clays, limestones, sandstones and ironstones, none of them individually thick and not all of them persistent along the outcrop. The more indurated beds form ledges and hills in front of the Wolds Chalk escarpment, for example the line of hills reaching 135 m (450 ft) running some 16 km (10 miles) north-west of Spilsby. These beds all thin northwards and disappear before the Market Weighton swell.

Their equivalents reappear north of this swell at the foot of the Chalk escarpment of the Yorkshire Wolds on the coast at Speeton, where the Speeton Clay is some 90 m (300 ft) thick and forms a series of slumped cliffs.

The fossils in all these northern 'Wealden' beds have affinities with those found in the Lower Cretaceous of North Germany and Russia and not with the southern French province to which the Wealden area was marginal.

The Lower Greensand is very variable in thickness in these northern outcrops. It is thickest (65 m: 220 ft) in Bedfordshire where the Woburn Sands are probably the equivalent of the Folkestone Beds of the Weald. The outcrop extends from Leighton Buzzard to Sandy and locally forms an escarpment 150 m (500 ft) high, which declines north-eastwards, but still forms a marked features south of Bedford. The Lower Greensand thins to about 3 m (10 ft) in the area between Cambridge and Ely, where it can in places be detected as a small feature, near Cottenham for example. It thickens northwards to about 13 m (40 ft) and from Downham Market to Hunstanton occurs as the Carstone, a series of ferruginous sands and grits, well-exposed on the foreshore at Old Hunstanton, where the sea has etched out the rectangular joint pattern. It continues as a thin bed across Lincolnshire and Yorkshire.

In Bedfordshire there are some 70 m (230 ft) of Gault Clay. This thins steadily northwards, though its outcrop remains broad because of the very low dip. Its character changes at about the latitude of Kings Lynn and a few kilometres south of Hunstanton it has become the Red Chalk or Hunstanton Red Rock, a rough red or pink nodular limestone. It extends all through the outcrop in Lincolnshire and Yorkshire and is always thin, maximum 8 m (25 ft), but often less, for example at Hunstanton where there is about 1·5 m (5 ft) of it. Its red colour is thought to have been caused either by an admixture of fine lateritic mud derived from a tropically-weathered land surface or from the Trias on which the Red Chalk rests uncomfortably in places beneath the North Sea.

The Chalk throughout these areas is much more comparable with the southern outcrop, though some differences in relief are to be observed. Many of these are related to the deposits of glacial drift on the East Anglian Heights and Lincolnshire Wolds and are therefore not strictly lithological effects produced by the Chalk itself. In Cambridgeshire the lower half of the Lower Chalk is the Chalk Marl, a rock so impervious and non-resistant as to be best classed with the underlying Gault Clay, for the two together have been eroded into a strike vale by the river Cam. Locally at its base over a distance of 80 km (50 miles) from east Bedfordshire to Soham occurs a very thin glauconitic sand with phosphatic nodules, the Cambridge Greensand. This bed is usually less than 0·3 m (1 ft) thick and is not important in the relief, but it was an important source of phosphate fertiliser in the nineteenth century when labour was cheap and easily-worked rock phosphate deposits abroad not yet discovered.

The Chalk Marl has thinned and disappeared by the Norfolk coast at Hunstanton and is replaced by a few metres of hard, resistant chalk. It fails to reappear in Lincolnshire and Yorkshire. In Lincolnshire most of the Upper Chalk has been eroded from the Wolds, but it is very thick in Yorkshire, 300 m (1000 ft) or slightly more. On the whole it is harder and better-cemented than in southern England, contains flints only in its lower half, but probably has more thin marl seams than in the south. Many of its fossils show affinities with those of north-west Germany.

The Upper Cretaceous reappears in Northern Ireland and western Scotland, the outcrops in the latter area being minute and mostly of geological interest. The greatest thickness known in Ireland is 145 m (480 ft) in a boring west of Lough Neagh, but elsewhere much of the Chalk has been eroded. The beds are preserved beneath Tertiary basalt flows and consist essentially of glauconitic sandstones (greensands) overlain by White Chalk which is comparable in lithology with that of Yorkshire. The facies are diachronous and there is evidence of two advances of the sea separated by a withdrawal. The most extensive transgression, which deposited the White Chalk, appears to have taken place very late in Senonian (Upper Chalk) times. Here, as in Devon and Dorset, there is strong evidence of rapid lithological changes in the vicinity of the shoreline of the period.

The presence of these beds in Northern Ireland raises the question of the former extent of the Chalk sea. Many of the highlands of Britain, Wales, the Lake District and Scotland, have anomalous drainage patterns which could be superimposed—indeed, it is rare to find a well-adjusted, 'normal' drainage pattern. If one looks at the main Chalk outcrop, with its generally constant lithology, it is possible in imagination to carry it over the whole of highland Britain, warp it at will to account for the major lineaments of British drainage, and often to infer that a surface touching the highest hills would represent an exhumed sub-Chalk surface. Maybe this is the truth: in the absence of any evidence it is impossible to say. It is possible to hold a contrary view. In an area of tranquil deposition such as the Chalk sea marginal facies changes occur within short distances, as we have seen in Devon and Dorset and again in Antrim. None of the main Chalk outcrops is remotely near the main highlands and it is possible that the Chalk changed to greensands and petered out against the highlands rather than went over them in one grand sweep of unchanging lithology. Further, in Northern Ireland, the only area in the highland zone of Britain where the sub-Chalk surface can be approximately known, that surface is found to be intensely dislocated and not gently warped and tilted. The former extent of the Chalk cover remains and will remain a fruitful source of hypothesis and discussion.

13
The Tertiary system

The Tertiary used to be classed as an era like the Primary and the Secondary. With the Quaternary it constituted the Cainozoic, comparable with the Palaeozoic and the Mesozoic. But the time scale is really altogether shorter and the old systems into which the Tertiary era was divided are very much smaller than those into which the other eras are divided. An outline analysis of divisions and events is shown on Fig. 13.1. It should be noted that neither the ranking of the various divisions nor the inclusion of the Pleistocene in the Neogene is universally accepted.

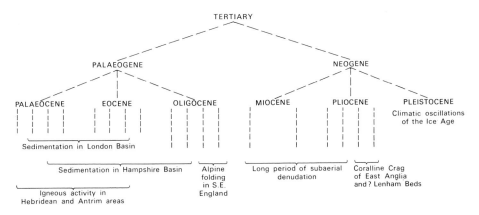

Fig. 13.1. Main divisions and events of the Tertiary system

Sedimentary cycles in the London and Hampshire basins

The essence of Palaeogene palaeogeography was a shallow gulf over southeast England and the Paris Basin, the Anglo-Parisian Cuvette, opening out northwards over the present site of the North Sea and occasionally connected south-westwards to warmer seas. The London Basin and the Hampshire Basin both lay on its western flank, and hence, as the gulf waxed and waned, they experienced alternating marine and continental conditions. There is some doubt whether they were separated by an incipient, updomed Weald. From this it is obvious that conditions in the east of these basins would have been more continuously marine than in the west where continental conditions prevailed more often; also, that because of its more westerly position, the sediments of the Hampshire Basin are more continental than those of the London Basin. The continental sediments are generally sands, which coarsen in the far west to gravel, while some of the marine sediments are clays, but there is no rigid division into continental sands and gravels and marine clays. The whole series is characterised by a great scarcity of limestone, a peculiar feature in view of the fact that the landscape being dissected was probably largely surfaced by Chalk. However, the calcium carbonate carried off in solution was probably precipitated mostly in the centre of the Anglo-Parisian Basin where it formed the important limestones of the central parts of the Paris Basin.

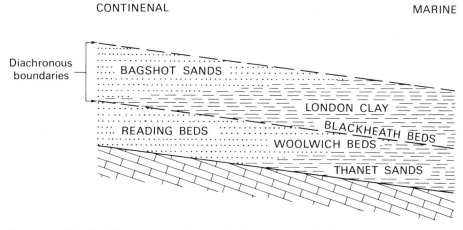

FIG. 13.2. Idealised Eocene deposition cycles in the London Basin

The elements of the sedimentation in the London Basin are shown in Fig. 13.2. The marine Thanet Sands are thickest, about 25 m (80 ft), in east Kent and thin to nothing half way between Leatherhead and Guildford in west Surrey. The succeeding Reading Beds are a continental facies with a thickness very much of the same magnitude as that of the Thanet Sands. They consist of two main rock types, sands with pebble beds and lenticular mottled clays. South of London they pass into an estuarine facies known as the Woolwich Beds, which change into marine sands in east Kent. Local irregular cementation in the Reading Beds sands and gravels produces sarsen stones, where patches of either sand or gravel have been cemented into hard masses with silica cement: the most famous examples have been formed from cemented pebble beds, for example the Hertfordshire Puddingstone, a superb example of natural concrete that few would guess to be other than artificial. Sarsens may also occur in other beds of similar lithology, for example the Bagshot Beds. These massive boulders often remain in the landscape after the rest of the incoherent beds containing them have been stripped away. They are usually found in residual deposits or glacial deposits derived from the Tertiary beds.

The succeeding cycle starts with the Blackheath Beds, a series of pebble beds reaching some 12 m (40 ft) in thickness and disappearing west of Croydon. They are marine and locally cemented into a puddingstone-like conglomerate. They are succeeded by the London Clay, occasionally pebbly or sandy at the base but generally a massive formation, 130 m (430 ft) thick at a maximum, and containing argillaceous limestone bands and nodules at various levels. The flora and fauna are tropical and give a good insight into climatic conditions and hence into probable weathering conditions early in the Tertiary. These have been used to suggest the presence of tropical denudation processes at that time and the attribution of certain features of the present landscape to such processes, for example the rotting of granite involved in the formation of tors (see Chapter 3). In Europe these ideas have been extended to include the formation of pediplains at this period on some of the Hercynian blocks, the Central Plateau of France for example.

The effect of London Clay on the landscape is shown by the spread of London, for large parts of the outcrop were not built on until comparatively late, largely in the post-1918 period. It also facilitated the construction of the deeper tube railways, which would have been much more difficult and expensive, if not impossible, had London been floored with granite.

Succeeding the London Clay are the largely-continental Bagshot Beds, a sandy formation some 35 m (120 ft) thick at a maximum. They are well-developed in the western part of the London Basin, where they form the heathlands largely taken over for military purposes. Nearer London they form local hills, for example Hampstead and Harrow and, farther east, the Rayleigh Hills in Essex. East of London they are mostly marine in origin though lithologically similar. There are patches of even younger beds in the London Basin, but they are of no significance in landforms.

Generally the British Palaeogene rocks are not cemented and hence their resistance to denudation is a function of their permeability and resistance to processes such as freeze–thaw. The local cementation into sarsens and puddingstones merely produces isolated resistant boulders, which may remain long in the landscape but which do not produce major landforms. In places where the Tertiary Beds are very permeable they may form an escarpment resting on the Chalk: the example in west Kent and east Surrey (Fig. 2.3) has already been quoted. Elsewhere they produce heathland, for example the Bagshot Beds in the western part of the London Basin and the Reading Beds at Puddletown Heath in the Hampshire Basin. The very sandy or gravelly facies of these beds are apt to give rise to podsol soils and it is possible for the development of an iron pan so to impede drainage that the permeability of the bed is much reduced and ponds and marshes result.

In the Hampshire Basin, as shown in Fig. 13.1, a much longer sequence of Palaeogene sediments is preserved on the mainland and in the northern half of the Isle of Wight. There are basic similarities with the London Basin. The sedimentation is more marine in West Sussex and the east of the Isle of Wight than in the west of that island. The cycles of sedimentation in practice are very much more irregular than those suggested in an ideal scheme and the marine invasions are of varying duration. The Eocene and Oligocene beds have usually local names, and correlations between various parts are difficult because the freshwater and shallow marine faunas are dominantly facies faunas and of no great use in correlation. The great majority of the rocks are sands and clays, often capped with Quaternary terrace gravels. They generally possess little resistance and form a large tract of terraced lowland in the southern part of the Hampshire Basin and the northern part of the Isle of Wight. Occasional beds of freshwater limestone, of which the 6 m (20 ft) thick Bembridge Limestone is probably the best example, may form local features inland and ledges and reefs on the coast.

Igneous activity in the Hebridean and Antrim areas

Concurrently with the sedimentation in the London and Hampshire Basins there occurred extensive igneous activity in north-western Britain. This has already been mentioned in connection with igneous rocks in Chapter 3. From a morphological point of view there were three phases of activity: plutonic intrusive complexes, minor intrusions (dykes and sills) and lava sheets. The first is exemplified by the gabbro of the Cuillin Hills of Skye and the Arran granite. Dyke swarms are characteristic of Mull and Arran, though their effects on relief are not great. Sills are common in Arran: they usually are much thicker than the dykes and, because of this and their inclination, have a much greater effect on relief, especially where their jointing is columnar. Lava sheets are found in Antrim and the north of Skye. Although two types of igneous activity are often found together, very rarely are all three in the same area. Skye, for example, has well-developed major intrusions and lava flows, while Arran has major and minor intrusions but no lava flows.

Other Tertiary events

Apart from the sediments and igneous rocks noted above there are some small problematic Tertiary deposits in Britain, but their problems are mainly geological ones of correlation. There is, for example, the well-known example of the Bovey Tracey basin in Devon with its infilling of lignites, sands and clays. It is usually thought to be Oligocene, but there is no positive correlation.

As shown in Fig. 13.1 much of the middle of the Tertiary system was taken up with folding, uplift and denudation. Indeed, block uplifts of many parts of western and northern Britain probably occurred at various phases from the beginning of the Tertiary onwards. The block mountains so formed, together with the gentle folds affecting the Cretaceous and Tertiary beds south of the North Downs line in the Weald and Hampshire Basins, were the last main earth movements to affect Britain. The Miocene and later periods were primarily phases of denudation of these structures. The importance of all these events in the evolution of British relief can hardly be exaggerated, for they formed the block from which the present landforms have been sculptured, but their discussion does not fall squarely within the scope of a book on rocks and relief. Similarly, the evaluation of the Pleistocene deposits, and of the Crag deposits of East Anglia which are now almost all included therein, is a fascinating problem of Quaternary research. But the effects of these unconsolidated deposits on relief, apart from those caused by permeability differences and the scenic differences produced by different vegetation on different lithologies, are small. At the present time knowledge about the Pleistocene, especially about the interpretation of the earlier parts, is in a state of flux, from which the elements of a coherent interpretation are beginning to emerge.

14
The future

The title is ambiguous. This chapter is not concerned with the stratigraphy of the future, but with the future of research into rock-relief relationships.

In this book the main concern has been with mainly coarse-scale and readily-observable relationships. Although many geomorphologists in the past were satisfied with a broad understanding of real landforms, there is an increasing tendency today to analyse in greater depth and to attempt detailed explanations of minor landform variations. Once one attempts to get beyond broad correlations the problems become very great and progress slow and tedious. Yet further real advance must depend on such work.

The essence of the problem lies in the multitude of variable factors involved. On the one hand there is the rock with its individual physical, chemical and mineralogical features. Complex questions arise: is the greater susceptibility of a rock to chemical weathering likely to offset its fewer physical weaknesses or will the reverse hold good? Will this relationship under one set of weathering conditions be the same as under another? The combinations of variables on the weathering side is as great as on the rock side, so that very few valid generalisations may be made from first principles; hence, the preface to all statements about rocks and relief, 'other things being equal, which they never are. . . .'

The usual arguments about rocks and relief are essentially circular. The rock is resistant to denudation because it forms high relief; it forms high relief because it is resistant to denudation. Somewhere one should be able to break in to test independently whether resistance to denudation always results in high relief or whether geomorphological accidents might not be involved.

There are really two ways of attempting greater understanding. The first is through laboratory experiments: the second lies in attempting detailed rock and relief correlations in the field. Both approaches, especially the latter, may need statistical treatment.

In the laboratory the problem lies in simulating natural conditions. Experiments on landform models are more complex than those on rocks because questions of scale as well as of time intervene, but even simple rock problems involve the time factor, the significance of which is very difficult to guess. The question cannot be shelved because the whole object of laboratory experiments is so to shorten the time element as to bring the process within the range of human experience.

Certain simple properties, such as compression strength, tensile strength and shear strength, which can be determined in the laboratory, have their uses, as has been shown in Chapter 2. But the reactions of a rock to denudation depend upon the whole complex of properties, among which rock strength probably plays usually a subordinate role.

A more useful method is to test whether a suggested explanation of disintegration can be duplicated in the laboratory. The practical problems of apparatus, selection of material and presentation of results are very great as will be apparent from the examples below.

The classic experiments of Griggs, already mentioned in Chapter 2, on heating and cooling a block of granite in the laboratory illustrate this. Following some earlier experiments by Blackwelder, Griggs set out to test the effect of repeated temperature changes on a polished block of granite, placed in front of an electric heater and subject to rapid temperature changes with a range of 110 degrees C. Assuming that his temperature fluctuations represented diurnal changes, Griggs weathered his block of granite for the equivalent of nearly 250 years, at the end of which he was unable to see any changes, either macroscopic or microscopic. However, when the cooling was effected by a spray of tap-water containing dissolved oxygen and carbon dioxide, $2\frac{1}{2}$ years of scaled weathering produced detectable alteration of the granite surface.

It is very easy either to believe implicitly that the results of such experiments are valid for natural examples, or to dismiss them vaguely as not being representative of nature. Neither attitude will do. One must attempt to assess critically the significance of the results.

First, it must be said that the main disruptive force tending to cause surface disintegration would be the shear stress caused by the differential expansion of the surface and immediately subjacent layers of the rock. This would depend on the temperature gradient in the rock which must have been as high in the experiment as in nature, because scaling down the time factor would not have allowed the heat to penetrate so far into the rock. Secondly, one might ask whether a cube of granite with free sides might not expand more freely than a natural granite surface, where expansion might result in a tendency for the outer layers to arch upwards and hence disintegrate. Thirdly, one might ask whether a completely dry experiment was natural.

The real differences might lie in the third objection. Griggs may have proved that completely dry heating and cooling had no effect, but this is an oversimplification of the natural problem. The moisture in deserts may not be equal to a spray of tap-water, but it is not completely lacking. Again temperature variations, by controlling evaporation and condensation, might affect the solution and crystallisation of salts, which are well known to be potent destroyers of natural and artificial rock. We are back to the viewpoint that natural problems involve many interacting factors, and it is insufficient to oversimplify the problem when setting up experiments.

Finally, we have ignored the effect of time. How far can one safely speed up the time scale? Small stresses applied over long periods may ultimately break something by the cumulative effect of that little-understood factor, fatigue. This is well known in the case of aircraft frames and vintage motor cars. One suspects that something similar may occur in rocks.

Some more recent experiments on the effect of freeze–thaw on chalk might also be cited as a splendid illustration of all the difficulties involved in experiments of this type. They were carried out by Dr R. B. G. Williams, now of the University of Sussex.

As was stated in Chapter 12 the English Chalk is peculiar in that certain quite pronounced relief features seem to have very obscure explanations. The classic one is the secondary escarpment found on the dip-slope in the Hampshire Basin and the South Downs. It was tentatively suggested twenty years ago (Sparks 1949) that slight differences in frequency of marl seams and in percentages of insoluble residue might have affected permeability and hence erosion when the water table was much higher than it is now. However, even the author seems not to have been very convinced by his own hypothesis.

Another possibility is that the potential destructive effects of freeze–thaw might vary from horizon to horizon. It was to test the susceptibility to freeze–thaw of different types of rocks that Williams set up his experiments. He used chalk, because it is highly susceptible to this process, is easily handled and cut, and the stratigraphy is well known owing to the great interest amateur and professional geologists have taken in the rock over so many years, so that accurately stratified samples can be fairly easily obtained. Let us look at the practical difficulties. These are admirably emphasised by Williams himself.

First, the apparatus presented a series of headaches. Commercial refrigerators need modifications to give the range of temperature required. They need subsidiary controls to give the oscillating temperature pattern necessary for freeze–thaw. Even when modifications have been made, lack of reliability can wreck experiments which depend on careful standardisation of conditions so that the results are comparable. Power failure, or even someone accidentally turning the wrong switch, can vitiate laborious experiments and necessitate their repetition. Finally, there are not unlimited funds available for work, the interest of which is entirely academic and which is unlikely to affect greatly how long we live and how much we earn. Williams's own comment on his apparatus was that it broke more frequently than did the rocks!

Having set the apparatus up, we must select our samples.

It is not only a question of looking up the relevant research papers so that samples from the right Chalk horizons are chosen. It is not even sufficient to cut the samples from the interiors of large blocks so that the weathered exteriors are avoided. How can one be sure that the blocks of chalk left unweathered at a particular horizon are not still whole simply because they are atypically resistant for that horizon? If one selects them, the experiment may be carried out on the exceptionally resistant and not on the average rock.

However, let us assume that average samples of chalk have been successfully selected. They must be prepared for freeze–thaw experiments. They have to be saturated in water. Should they be merely immersed in water to soak up what they will in a given period, or should some vacuum or boiling method be used to try to ensure total saturation? The answer here would seem to be natural soaking, as the ratio between the amounts of water required for natural and total soaking may be significant in freeze–thaw. The smaller this ratio, the greater the unsaturated pore space and, hence, the more space available for water to occupy when it expands on freezing. Theoretically the degree of freeze–thaw damage might be expected to decrease with the increase in unsaturated pore space.

How are the samples to be introduced into the refrigerator? If they are put in as they are, they tend to dry and the water appears as ice on the sides of the refrigerator. Accordingly they were immersed in water and put in flexible plastic containers so that the freezing of the surrounding water did not exert a disruptive effect on the rock samples, as the expansion would be absorbed by the container.

One must then decide the range and the period of the temperature fluctuation to which the samples will be subjected. These should be as representative of periglacial conditions as possible. Too rapid oscillations result in the freezing not penetrating to the heart of the rock specimens. Finally, after trial and error, a five-day cycle was chosen with ranges of $+15$ to -30 degrees C, and $+15$ to -10 degrees C. With the lower temperature two effects occur: first an expansion when the water changes into ice and then a thermal contraction at lower temperatures. Both might have a disruptive effect. With the higher temperature of -10 degrees C only a slight thermal contraction would occur. On the whole lower temperatures produce more shatter.

Obviously one must measure from time to time the amount of shatter. Williams started with 3 inch (8 cm) cubes of chalk, these being comparatively easy to cut. It is difficult to define an index of shatter. One might take the size of the largest fragment left unbroken. Thus, if a cube was simply split into two pieces it could be defined as 50 per cent shatter. But if 6·25 per cent of the cube were shattered from each corner, the size of the largest fragment left would also be 50 per cent of the cube. But are these two 50 per cent figures indicative of equal destruction? Ideally, the total surface area of all the fragments would probably provide the best measure of shatter, but it is almost impossible to obtain. Arbitrary measures based on the proportion of fragments of different sizes have to be used, as this can be fairly readily determined by sieving.

Increasing the number of freeze–thaw cycles produces fine debris at more than a linear rate. Contrary to views that have been held by others, slow freezing does more damage than rapid freezing. Curiously, this conclusion is in line with the experience of frozen food experts, who use rapid freezing because it does less damage than slow freezing. Arguing from first principles one might expect rapid freezing to do more damage because there would be less time for water to escape and so reduce the pressure caused by the ice: in fact, slow freezing probably encourages the growth of larger crystals which do more damage.

Williams's results generally confirmed that those parts of the Chalk with higher percentages of insoluble residue were shattered more easily by freeze–thaw than the rest. But there are all sorts of snags which might lead to unusual results for any given specimen. Each cycle of freeze–thaw probably produces weaknesses in the rock which greatly exceed the actual shatter at that stage. Thus, when the material is first selected its previous exposure to freeze–thaw might cause variations in the degree of experimental shatter produced during the first cycle. Again, the geomorphologist is mainly interested in the average resistance of fairly thick horizons, whereas there may be in fact considerable variations within those horizons. Thus, two horizons may differ significantly in resistance, but the range of resistance figures for samples from one horizon could well overlap that for another. Chalk is not an isotropic medium and should the specimen be taken across, for example, a slightly more marly horizon, then a major split might be provoked early in the experiment. Again, the presence of large shell fragments within the sample could well provide weaknesses, for the way in which rock tends to weather naturally away from fossils is well known. If fossils appear in section on the edge of the cut block the specimen can be discarded, but it is quite possible for them to be included within. Accordingly, it may be necessary to repeat the experiments over and over again until a number of results large enough for statistical analysis is available.

One can imagine setting up experiments on chemical weathering and meeting comparable difficulties. One might percolate acid solutions through rocks and measure the rate of solutional loss. But one would also have to assess the damage caused by the solution loss, and then the problem becomes very similar to that discussed by Williams.

In field studies the usual method is to try to arrive at a correlation, which may or may not be statistical, between an observed distribution of relief and one or more rock properties.

A paper often quoted in this connection is Schumm's work on the different effects of the Chadron and Brule formations in the badlands of South Dakota. The Brule overlies the Chadron and has relief of a significantly different type. The mean maximum slope angles are of the order of

44 degrees and the slopes are usually straight and meet in sharp interfluve crests. These properties are retained even in residuals on the Brule formation so that parallel slope retreat seems to hold good here. On the other hand, the underlying Chadron formation has mean maximum slope angles of 33 degrees and much more convex interfluves. The residuals on the Chadron formation generally show a decline in slope angle, so that it seems that here slope angles decline with age.

There can be no question about slope evolution under different climates for the formations lie one above the other, though the parallel retreat of the Brule slopes would seem to be typically sub-arid and the decline of the Chadron slopes normally humid temperate in type. Further, although the Brule produces sharper relief and steeper slopes it seems overall to be more susceptible to denudation than the Chadron, if one may judge by the way it has been stripped back from the latter.

Schumm related the differences in form to differences in dominant processes, the latter governed primarily by lithology. On the face of it the Brule looks more resistant than the Chadron. When dry it is hard and cracked compared with the loose mass of clay aggregates at the surface of the Chadron, which therefore looks much more erodible. However, run-off occurs at a much lower precipitation value on the Brule, so that surface erosion can start sooner. This was tested by spraying water on to an area of about 0·6 sq. m (6 sq. ft). On the Brule run-off occurred almost immediately and, in every case, before one gallon was sprayed on to the sample area. On the Chadron formation $4\frac{1}{2}$ gallons were needed to produce run-off. The water flowed in sub-surface channels until all the clay aggregates had become saturated to form a creeping mass which slid over the underlying impermeable fine-grained surface.

It was concluded that rainwash was the dominant slope process on the Brule formation, but creep on the Chadron formation. Thus arose the differences in slope forms and denudation rates, because denudation would always start at a lower precipitation level on the Brule. It is interesting to note that, with varying precipitation, the relative rates of denudation of two such rocks might alter appreciably. If rainfall never exceeded the threshold value for the Brule it could be that the Chadron would be more erodible by other factors: if rainfall always lay between the threshold run-off values for the two formations the Brule would seem to reach its maximum denudation rate compared with the Chadron: if the values of both thresholds was always exceeded by the rainfall intensity, then the differences between the two should decrease and might even reverse.

Another rock characteristic, which is very important in relief control and also susceptible to field testing, is joint frequency and inclination. Its importance has been discussed in Chapter 2, mainly in connection with inselbergs. However, in the field on real rocks the accumulation of data on joints is by no means easy. Rarely is the joint pattern as regular as represented in conventional diagrams. Joints often occur naturally at irregular intervals and with irregular inclinations so that field measurements involve subjective elements of generalisation and weighting, because some joints are obviously more important than others. Except where outcrops are bare it is rarely possible to observe joints over wide areas. Indeed, it is often difficult to distinguish between jointed blocks in situ and jointed blocks glacially transported and dumped, especially when they are partly vegetated. In some rocks it might be possible to substitute some other measure, for example scatter diagrams and other statistical measures of the weight of joint blocks. Such a technique would obviously suit a rock subject to an irregular and close jointing pattern, as at many levels of the Chalk, if one may quote a familiar example. One can imagine happy hours spent in chalk pits with a spring balance, but the imagination and the muscles boggle at the thought of attempting to apply such a technique to a granite intrusion, or a massive sedimentary rock, where the joint blocks might range up to a cubic metre in size.

An interesting attempt at a precise assessment of the effect of rock structures is the recent work on the form of corries by Haynes. Basically the author faced the question of the relative parts played by process (in this case glacial erosion) and rock structure in the formation of corries. The fact that it was possible to produce a general logarithmic curve which fitted well to the profiles of 81 per cent of the corries studied is strong evidence that process is the overall control, but a constant had to be introduced to cover the steeper forms of some corries. The effect of this constant on the profile is equivalent to a vertical exaggeration of scale. There tends to be a regional variation in the value of the constant which suggests some sort of structural or lithological control of corrie form, but this is not universal as some areas show a range of forms.

In detail, joints and bedding planes can exert a strict control on the form of a corrie. Where the rocks dip gently up the corries, then so do the corrie floors in almost every case. These relationships are well-exemplified by the Torridonian Sandstone of the Applecross area. The sandstones are massive, but the bedding planes are weak because of the presence of thin shale partings so that they are readily stripped by glaciers. The floors of the corries tend to coincide with the bedding planes, which dip into the corries on the average at 10 degrees and so give rise to rock basins holding small lakes. Where the rocks dip gently down the corries, reversed slopes are rare and hence corrie lakes do not occur. It seems that structural control dominates where the angles of dip are low. Theoretically, two sets of joints dipping in opposite directions at 45 degrees tend to exert minimum structural influence, but in practice, Haynes remarks, dips between the upper 20s and the lower 60s result usually in the effects of the two sets of joints cancelling out.

Another method of approach to the rock and relief problem is to measure various properties of the rock and to attempt an areal correlation with the relief pattern. A simple example of this type of analysis has been applied to the Lower Greensand of the south-east Midlands by Chorley. Virtually it consists of comparing a trend surface fitted to the relief with trend surfaces fitted to the areal distribution of the properties measured, so that the correlation between the two may be statistically analysed. Crude examples of trend surfaces are the generalised contours sometimes used by geomorphologists in trying to reconstruct the initial relief of an area: proper trend surfaces are mathematically determined best-fit surfaces but their discussion is beyond the scope of this chapter. They exclude subjective judgment, because of their mathematical nature, but at the same time, unless they are almost impossibly complex, they cannot easily reveal a complex natural surface, for example a stepped landscape or other forms of discontinuity.

Chorley divided relief controls into three groups:

(*a*) Structural controls, such as total bed thickness and average dip.
(*b*) Lithological controls, such as resistivity, joint frequency and permeability.
(*c*) Spatial controls, for example relative position in the drainage basin.

Not all of these are easily measured, and in the example quoted he correlated total thickness, distance from an arbitrary point in the upper Ouse drainage basin, and median grain size and proportion of silt and clay (this to give some measure of permeability) with the height of the Lower Greensand ridge. Not surprisingly bed thickness won hands down as the major control on relief.

In such an exercise as this difficulties of measurement are as apparent as in Williams's freeze–thaw experiments. Permeability could be controlled by the frequency and thickness of argillaceous seams, as Chorley recognised, rather than by the proportion of fines, so that if the wrong parameter was measured a wrong conclusion about the effects of permeability might emerge. Measurements of joint frequency, and chemical and physical resistance, as has been suggested above, are extremely difficult. It is also possible that the present relief may be largely related to different processes in the past, so that chemical and physical resistance to a variety of processes should ideally be measured.

One must beware, too, of extrapolating the results of such an analysis to other rocks and areas. Chorley showed that bed thickness dominated Lower Greensand relief in one part of the outcrop in England. The uninitiated must beware of applying this conclusion to all outcrops, a generalisation which would go far beyond Chorley's own conclusions. This can probably best be illustrated by a series of rash estimates. If I wanted another example to illustrate the dominance of bed thickness in overall relief, I would investigate a correlation in the height of the Cotswold escarpment with the thickness of limestone in the Inferior Oolite. I might also be tempted to apply a similar correlation between relief and the thickness of the Magnesian Limestone along the strike. On the other hand, should I wish to show another factor, in this case angle of dip, dominating bed thickness, I would investigate the North Downs Chalk between Dorking and Farnham or the Chalk ridge of the Isle of Wight. If I were seeking an example to illustrate the dominance of permeability, I might look at the glacial deposits of north Norfolk, where the highest ground is largely on coarse outwash sands and gravels. Finally, if I merely wished to put forward an awkward example, I might suggest that a correlation of controls and outcrop height should be applied to the Chalk from the Chilterns to Hunstanton. What the result would be I have not the faintest idea.

More seriously, the real value of this type of analysis will be apparent in cases such as the last, where no coarse differences allow the result to be guessed before the application of refined methods.

Whether one works in the laboratory or the field, further investigations of rock-relief relationships are going to require detailed work, which will certainly be very time-consuming and, in many cases one suspects, out of proportion to the results obtained. But there is no alternative.

References

Chapter 1

BAGNOLD, R. A. (1941) *The physics of blown sand and desert dunes*, London, Methuen (2nd edn. 1954).

BEARD, J. S. (1953) 'The savanna vegetation of northern tropical America', *Ecological Monographs*, **23**, 149–215.

BROWN, E. H. (1960) *The relief and drainage of Wales*. Cardiff. University of Wales Press.

BRYAN, KIRK (1940) 'The retreat of slopes', *Annals of the Association of American Geographers*, **30**, 254–67.

KING, L. C. (1950) 'The study of the world's plainlands', *Quarterly Journal of the Geological Society*, **106**, 101–31.

KING, L. C. (1953) 'Canons of landscape evolution', *Bulletin of the Geological Society of America*, **64**, 721–52.

LEWIS, W. V. (1944) 'Stream trough experiments and terrace formation', *Geological Magazine*, **81**, 241–53.

MARTIN, E. C. (1920) 'The glaciation of the South Downs', *South-Eastern Naturalist*, **25**, 13–30.

WOOLDRIDGE, S. W. and LINTON, D. L. (1955) *Structure, surface and drainage in south-east England*, 2nd edn. London, G. Philip.

Chapter 2

BAIN, G. W. (1931) 'Spontaneous rock expansion', *Journal of Geology*, **39**, 715–35.

BALK, R. (1939) 'Disintegration of glaciated cliffs', *Journal of Geomorphology*, **2**, 303–34.

BATTEY, M. H. (1960) 'Geological factors in the development of Veslgjuv-botn and Vesl-Skautbotn', *Royal Geographical Society Research Series No. 7* (ed. W. V. Lewis), 5–11.

BLACKWELDER, E. (1926) 'Fire as an agent in rock weathering', *Journal of Geology*, **35**, 134–40.

CHAPMAN, C. A. (1958) 'Control of jointing by topography', *Journal of Geology*, **66**, 552–8.

CHAPMAN, R. W. and GREENFIELD, M. A. (1949) 'Spheroidal weathering of igneous rocks', *American Journal of Science*, **247**, 407–29.

COTTON, C. A. (1944) *Volcanoes as landscape forms*. Christchurch, New Zealand. Whitcombe and Tombs.

DODGE, T. A. (1947) 'An example of exfoliation caused by chemical weathering', *Journal of Geology*, **55**, 38–42.

DURY, G. H. (1959) *The face of the earth*. London, Penguin (Pelican).

FARMIN, R. (1937) 'Hypogene exfoliation in rock masses', *Journal of Geology*, **45**, 625–35.

GEIKIE, J. (1908) *Structural and field geology*, 2nd edn. Edinburgh. Oliver and Boyd.

GILBERT, G. K. (1904) 'Domes and dome structures of the High Sierra', *Bulletin of the Geological Society of America*, **15**, 29–36.

GOLDICH, S. S. (1938) 'A study in rock-weathering', *Journal of Geology*, **46**, 17–58.

GRAHAM, E. R. (1941a) 'Acid clay, an agent in chemical weathering', *Journal of Geology*, **49**, 392–401.

GRAHAM, E. R. (1941b) 'Colloidal organic acids as factors in the weathering of anorthite', *Soil Science*, **52**, 291–5.

GRAWE, O. R. (1936) 'Ice as an agent of rock weathering: a discussion', *Journal of Geology*, **44**, 173–82.

GRIGGS, D. (1936) 'The factor of fatigue in rock exfoliation', *Journal of Geology*, **44**, 783–96.

HARLAND, W. B. (1957) 'Exfoliation joints and ice action', *Journal of Glaciology*, **3**, 8–10.

HATCH, F. H., WELLS, A. K. and WELLS, M. K. (1961) *The petrology of the igneous rocks*, 12th edn. London, Murby.

HILLS, E. S. (1953) *Outlines of structural geology*, 3rd edn. London, Methuen.

HOLMES, A. (1965) *Principles of physical geology*, 2nd edn. London, Nelson.

JAHNS, R. H. (1943) 'Sheet structure in granites: its origin and use as a measure of glacial erosion in New England', *Journal of Geology*, **51**, 71–98.

JEFFREYS, H. (1929) *The earth*, 2nd edn. Cambridge University Press.

KELLER, W. D. (1957) *The principles of chemical weathering*, rev. edn., Columbia, Missouri, Lucas Bros.

KING, L. C. (1948) 'A theory of bornhardts', *Geographical Journal*, **112**, 83–6.

LAKE, P. and RASTALL, R. H. (1927) *A text-book of geology*, 4th edn. Cambridge University Press.

LEWIS, W. V. (1954) 'Pressure release and glacial erosion', *Journal of Glaciology*, **2**, 417–22.

LINTON, D. L. (1955) 'The problem of tors', *Geographical Journal*, **121**, 470–87.

MCCALL, J. G. (1960) 'The flow characteristics of a cirque glacier and their effect on glacial structure and cirque formation', *Royal Geographical Society Research Series No. 4* (ed. W. V. Lewis), 39–62.

MATTHES, F. E. (1930) 'Geologic history of the Yosemite valley', *United States Department of the Interior, Geological Survey, Professional Paper No. 160*.

MATTHES, F. E. (1937) 'Exfoliation of massive granite in the Sierra Nevada of California', *Proceedings of the Geological Society of America for 1936*, 342–3.

MEAD, W. J. (1925) 'The geologic rôle of dilatancy', *Journal of Geology*, **33**, 685–98.

MERRILL, G. P. (1906) *A treatise on rocks, rock-weathering and soils*, 2nd edn. New York.

PARKER, J. M. (1942) 'Regional systematic jointing in slightly deformed sedimentary rocks', *Bulletin of the Geological Society of America*, **53**, 381–408.

READ, H. H. (1962) *Rutley's elements of mineralogy*, 25th edn. London, Murby.

REICHE, P. (1950) 'A survey of weathering processes and products', *University of New Mexico Publications in Geology, No. 3*. Albuquerque.

REID, C. (1899) 'The geology of the county around Dorchester', *Memoirs of the Geological Survey*.

RICH, J. L. (1911) 'Gravel as a resistant rock', *Journal of Geology*, **29**, 492–506.

SHAININ, V. E. (1950) 'Conjugate sets of en echelon tension fractures in the Athens Limestone at Riverton, Virginia', *Bulletin of the Geological Society of America*, **61**, 509–17.

SITTER, L. U. DE (1964) *Structural geology*, 2nd edn. New York, McGraw-Hill.

SMALLEY, I. J. (1966) 'Contraction crack networks in basalt flows', *Geological Magazine*, **103**, 110–14.

SMITH, L. L. (1941) 'Weather pits in granite of the southern Piedmont', *Journal of Geomorphology*, **4**, 117–27.

SPARKS, B. W. (1960) *Geomorphology*. London, Longman.

SPARKS, B. W. and WEST, R. G. (1964) 'The drift landforms around Holt, Norfolk', *Institute of British Geographers, Transactions and Papers*, **35**, 27–35.

TAMM, O. (1924) 'Experimental studies on chemical processes in the formation of glacial clay', *Sveriges Geologiska Undersökning*, Series C, No. 333.

THORNBURY, W. D. (1969) *Principles of geomorphology*, 2nd edn. New York, Wiley.

TYRRELL, G. W. (1930) *The principles of petrology*, 2nd edn. London, Methuen.

WHITE, W. A. (1945) 'Origin of granite domes in the south-eastern Piedmont', *Journal of Geology*, **53**, 276–82.

WILLIAMS, J. E. (1949) 'Chemical weathering at low temperatures', *Geographical Review*, **39**, 129–35.

WILSON, G. (1946) 'The relationship of slaty cleavage and kindred structures to tectonics', *Proceedings of the Geologists' Association*, **57**, 263–302.

WOOLDRIDGE, S. W. and MORGAN, R. S. (1959) *An outline of geomorphology*. 2nd edn. London, Longmans.

Chapter 3

COTTON, C. A. (1944) *Volcanoes as landscape forms*. Christchurch, New Zealand. Whitcombe and Tombs.

DANA, E. S. and FORD, W. E. (1932) *A text book of mineralogy*, 4th edn. New York, Wiley.

GILBERT, G. K. (1880) Report on the geology of the Henry Mountains, 2nd edn. *U.S. Department of the Interior*. Washington.

HARKER, A. (1909) *The natural history of igneous rocks*. London, Methuen.

HATCH, F. H., WELLS, A. K. and WELLS, M. K. (1961) *The Petrology of the igneous rocks*, 12th edn. London, Murby.

HUNT, C. B. (1953) 'Geology and geography of the Henry Mountains region, Utah', *U.S. Geological Survey Professional Paper*, **228**.

KING, L. C. (1949) 'A theory of bornhardts', *Geographical Journal*, **112**, 83–6.

LINTON, D. L. (1955) 'The problem of tors', *Geographical Journal*, **121**, 470–87.

MACGREGOR, M. and MACGREGOR, A. G. (1948) 'The Midland Valley of Scotland', *British Regional Geology, Memoirs of the Geological Survey*.

MARTONNE EMM. DE (1947) *Géographie universelle, VI, La France, Pt I, France physique*, 2nd edn. Paris, Colin.

PALMER, J. and NEILSON, R. A. (1962) 'The origin of granite tors on Dartmoor, Devonshire', *Proceedings of the Yorkshire Geological Society*, **33**, 315–40.

READ, H. H. and WATSON, J. (1962) *Introduction to geology*, Vol. I: *Principles*. London, Macmillian (2nd edn, 1968).

RICHEY, J. E. (1948) 'Scotland: the Tertiary volcanic districts', *British Regional Geology, Memoirs of the Geological Survey*.

SMITH, B. and GEORGE, T. N. (1961) 'North Wales', 3rd edn. *British Regional Geology, Memoirs of the Geological Survey*.

SPARKS, B. W. (1960) *Geomorphology*. London, Longman.

TURNER, F. J. and VERHOOGEN, J. (1960) *Igneous and metamorphic petrology*, 2nd edn. New York, McGraw-Hill.

TYRRELL, G. W. (1930) *The principles of petrology*, 2nd edn. London, Methuen.

WARD, W. T. (1951) 'The tors of central Otago', *New Zealand Journal of Science and Technology*, B, **33**, 191–200.

WEST, W. D. and CHOUBEY, V. D. (1964) 'The geomorphology of the country around Sagar and Katangi, M.P.', *Journal of the Geological Society of India*, **5**, 41–55.

WILHELMY, H. (1958) *Klimamorphologie der Massengesteine*. Braunschweig, Westermann.

Chapter 4

HARKER, A. (1950) *Metamorphism*, 3rd edn. London, Methuen.

RAMBERG, H. (1952) *The origin of metamorphic and metasomatic rocks*. Univ. of Chicago Press.

READ, H. H. (1962) *Rutley's elements of mineralogy*. 25th edn. London, Murby.

READ, H. H. and WATSON, J. (1962) *Introduction to geology*, Vol I: *Principles*. London, Macmillan (2nd edn, 1968).

TURNER, F. J. and VERHOOGEN, J. (1960) *Igneous and metamorphic petrology.* 2nd edn. New York, McGraw-Hill.

TYRRELL, G. W. (1930) *The principles of petrology*, 2nd edn. London, Methuen.

WILLIAMS, H., TURNER, F. J. and GILBERT, C. M. (1954) *Petrography.* San Francisco, Freeman.

Chapter 5
General

HATCH, F. H. and RASTALL, R. H. (1965) *Petrology of the sedimentary rocks*, 4th edn (revised by J. T. Greensmith). London, Allen & Unwin.

HODGSON, J. M., CATT, J. A. and WEIR, A. H. (1967) 'The origin and development of Clay-with-flints and associated soil horizons on the South Downs', *Journal of Soil Science*, **18**, 85–102.

JUKES-BROWNE, A. J. (1906) 'The clay-with-flints: its origin and distribution', *Quarterly Journal of the Geological Society*, **62**, 132–64.

KRINSLEY, D. H. and FUNNELL, B. M. (1965) 'Environmental history of sand grains from the Lower and Middle Pleistocene of Norfolk, England', *Quarterly Journal of the Geological Society*, **121**, 435–62.

PETTIJOHN, F. J. (1957) *Sedimentary rocks*, 2nd edn. New York, Harper & Row.

TYRRELL, G. W. (1930) *The principles of petrology*, 2nd edn. London, Methuen.

WILLIAMS, H., TURNER, F. J. and GILBERT, C. M. (1954) *Petrography.* San Francisco, Freeman.

Limestone relief

AUB, C. F. (1964) 'Karst problems in Jamaica, with particular reference to the cockpit problem', *20th International Geographical Congress, Karst Symposium* (duplicated).

BÖGLI, A. (1960) 'Kalklösung und Karrenbildung', *Zeitschrift für Geomorphologie, Supplementband* **2**, 4–21.

BÖGLI, A. (1964) 'Mischungskorrosion—Ein Beitrag zum Verkarstungsproblem', *Erdkunde*, **18**, 83–92.

COLEMAN, A. M. and BALCHIN, W. G. V. (1959) 'The origin and development of surface depressions in the Mendip Hills', *Proceedings of the Geologists' Association*, **70**, 291–309.

CORBEL, J. (1957) 'Les karsts du nord-ouest de l'Europe', *Revue de Géographie de Lyon* (special volume).

CORBEL, J. (1959) 'Erosion en terrain calcaire', *Annales de Géographie*, **68**, 97–120.

DRZAL, M. and SMYK, B. (1964) 'The influence of micro-organisms in the development of karst phenomena', *20th International Geographical Congress*, Abstracts of Papers, p. 105.

FAGG, C. C. (1958) 'Swallow holes in the Mole gap', *South-Eastern Naturalist and Antiquary*, **62**, 1–13.

FORD, D. C. (1964) 'Origin of closed depressions in the central Mendip

Hills', *20th International Geographical Congress*, Abstracts of papers, pp. 105–6.

FORD, D. C. and STANTON, W. I. (1968) 'The geomorphology of the south-central Mendip Hills', *Proceedings of the Geologists' Association*, **79**, 401–28.

GARRELS, R. M. and CHRIST, C. L. (1965) *Solutions, minerals and equilibria*. New York, Harper & Row.

GILEWSKA, S. (1964) 'Fossil karst in Poland', *Erdkunde*, **18**, 124–35.

GUILCHER, A. (1953) 'Essai sur la zonation et la distribution des formes littorales de dissolution de calcaire', *Annales de Géographie*, **62**, 161–79.

GUILCHER, A. (1958) *Coastal and submarine geomorphology*. London, Methuen.

HINDLEY, A. (1965) 'Sinkholes on the Lincolnshire Limestone between Grantham and Stamford', *East Midland Geographer*, **3**, 454–60.

JENNINGS, J. N. and BIK, M. J. (1962) 'Karst morphology in Australian New Guinea', *Nature, Lond.* **194**, 1036–8.

JUKES-BROWNE, A. J. and WHITE, H. J. O. (1908) 'The geology of the country around Henley-on-Thames and Wallingford', *Memoirs of the Geological Survey*.

KERN, R. and WEISBROD, A. (1967) *Thermodynamics for geologists*, San Francisco, Freeman.

MONROE, W. H. (1964) 'Lithologic control in the development of a tropical karst topography', *20th International Geographical Congress, Karst Symposium* (duplicated).

PITTY, A. F. (1968) 'The scale and significance of solutional loss from the limestone tract of the southern Pennines', *Proceedings of the Geologists' Association*, **79**, 153–78.

SAUNDERS, E. M. (1921) 'The cycle of erosion in a karst region (after Cvijić)', *Geographical Review*, **11**, 593–604.

SMET, R. E. DE and SOUCHEZ, R. (1964) 'Evolution comparée de deux massifs dolomitiques: Catinaccio et Sella', *Revue belge de Géographie*, **88**, 157–86.

SMYK, B. and DRZAL, M. (1964) 'Untersuchungen über den Einfluss von Mikroorganismen auf das Phänomen der Karstbildung', *Erdkunde*, **18**, 102–13.

SOUCHEZ, R. (1963) 'Le relief de la région de Couvin-Nismes en tant que paléokarst de climat chaud et humide', *Bulletin de la Société belge d'études géographiques*, **32**, 269–80.

SPARKS, B. W. (1960) *Geomorphology*. London, Longman.

SWEETING, M. M. (1950) 'Erosion cycles and limestone caverns in the Ingleborough district', *Geographical Journal*, **115**, 63–78.

SWEETING, M. M. (1958) 'The karstlands of Jamaica', *Geographical Journal*, **124**, 184–99.

SWEETING, M. M. (1966) 'The weathering of limestones, with particular reference to the Carboniferous limestones of northern England.

Essays in Geomorphology, ed. G. H. Dury. London, Heinemann. pp. 177–210.

SWEETING, M. M. (1968) 'Some variations in the types of limestones and their relation to cave formation', *Proceedings of the 4th International Congress of Speleology in Yugoslavia*, vol. iii, 227–32. Ljubljana.

SWEETING, M. M. and others (1965) 'Denudation in limestone regions: a symposium', *Geographical Journal*, **131**, 34–56.

THOMAS, T. M. (1954) 'Swallow holes in the South Wales coalfield', *Geographical Journal*, **120**, 468–75.

THORNBURY, W. D. (1969) *Principles of geomorphology*, 2nd edn. New York, Wiley.

THORNBURY, W. D. (1965) *Regional geomorphology of the United States*. New York, Wiley.

TROMBE, F. (1952) *Traité de spéléologie*. Paris, Payot.

WENTWORTH, C. K. (1938–9) 'Marine bench-forming processes', *Journal of Geomorphology*, **1**, 6–32 and **2**, 3–25.

WILLIAMS, P. W. (1963) 'An initial estimate of the speed of limestone solution in County Clare', *Irish Geography*, **4**, 432–41.

WILLIAMS, P. W. (1966) 'Limestone pavements with special reference to western Ireland', *Institute of British Geographers, Transactions and Papers*, **40**, 155–72.

WOOLDRIDGE, S. W. and KIRKALDY, J. F. (1937) 'The geology of the Mimms valley', *Proceedings of the Geologists' Association*, **48**, 307–15.

ZENKOVICH, V. P. (1967) *Processes of coastal development*, ed. J. A. Steers. Edinburgh, Oliver & Boyd.

Chapters 6–13

There are many books available on stratigraphy. The following selection may be useful in all these chapters.

DONOVAN, D. T. (1966) *Stratigraphy*. London, Murby.

GIGNOUX, M. (1950) *Géologie stratigraphique*, 4th edn. Paris.

HARLAND, W. B., SMITH, A. G. and WILCOCK, B., ed. (1964) 'The Phanerozoic time-scale', *Quarterly Journal of the Geological Society of London*, **120 S**.

RAYNER, D. H. (1967) *The stratigraphy of the British Isles*. Cambridge University Press.

WELLER, J. M. (1960) *Stratigraphic principles and practice*. New York, Harper & Row.

WELLS, A. K. and KIRKALDY, J. F. (1967) *Outline of historical geology*. 6th edn, London, Allen & Unwin.

WILLS, L. J. (1952) *A palaeogeographical atlas*. London, Blackie.

WOODFORD, A. O. (1965) *Historical geology*. San Francisco, Freeman.

For the regional stratigraphy of Great Britain, the reader is referred to the British Regional Geology series published by H.M.S.O. for the Institute

of Geological Sciences. These are by a variety of authors and revised editions appear from time to time. The titles are:
South-West England
The Hampshire Basin
The Wealden District
London and the Thames Valley
The Bristol and Gloucester District
The Central England District
East Anglia
East Yorkshire and Lincolnshire
The Pennines and adjacent areas
Northern England
The Welsh Borders
North Wales
South Wales
The South of Scotland
The Midland Valley of Scotland
The Grampian Highlands
Scotland: the Northern Highlands
Scotland: the Tertiary Volcanic Districts

For much greater detail there are memoirs covering the 1-inch Geological Survey maps. These include details of individual exposures and are indispensable in detailed local studies. In addition there are various special publications, such as reports on water and mineral resources.

A two sheet geological map on a scale of 1:625,000 is available for Great Britain and most of the country is covered by ¼-inch and 1-inch maps, but by no means all of these are in print. More detailed maps will probably not be required for general studies.

For Chapter 11 only

HOLLINGWORTH, S. E., TAYLOR, J. H. and KELLAWAY, G. A. (1944) 'Large-scale superficial structures in the Northampton ironstone field', *Quarterly Journal of the Geological Society*, **100**, 1–44.

For Chapter 13 only

WEST, R. G. (1968) *Pleistocene geology and biology*. London, Longman.

Chapter 14

CHORLEY, R. J. (1969) 'The elevation of the Lower Greensand ridge, south-east England', *Geological Magazine*, **106**, 231–48.
GRIGGS, D. T. (1936) 'The factor of fatigue in rock exfoliation', *Journal of Geology*, **44**, 781–96.
HAYNES, V. M. (1968) 'The influence of glacial erosion and rock structure on corries in Scotland', *Geografiska Annaler*, **50**, Ser. A, 221–34.

SCHUMM, S. A. (1956) 'The role of creep and rainwash in the retreat of badland slopes', *American Journal of Science*, **254**, 693–706.

SPARKS, B. W. (1949) 'The denudation chronology of the dip-slope of the South Downs', *Proceedings of the Geologists' Association*, **60**, 165–215.

WILLIAMS, R. B. G. (1969) 'Periglacial climate and its relation to landforms', unpublished Ph.D. thesis, University of Cambridge.

Index

395